The Tenement Revealed

Published by

Whittles Publishing Ltd,
Dunbeath,
Caithness, KW6 6EG,
Scotland, UK
www.whittlespublishing.com

ISBN 978-184995-570-6

Cover design & Typesetting by
KERRIE MONCUR
DESIGN & TYPESETTING SERVICES
kmdesigntypesetting@outlook.com

Printed and bound by CPI Group (UK) Ltd, Croydon, CR0 4YY

John Gilbert

The Tenement Revealed

History, Design & Construction

Whittles Publishing

Contents

Drawing of a tenement

Preface

I always wanted a book on tenements.

I wish I had had more understanding about how tenements were put together when I was a young architect. Tenement maintenance, or maintenance of any kind, was not taught at my architecture school in Belfast. Now I have had experience of working in Scotland, I thought it would be worthwhile setting down my understanding of how and why tenements were built, why they vary from place to place and how they changed over different periods, in the belief that this knowledge will help us better understand and maintain them. For younger architects I have also tried to describe the many changes, often badly done but with a positive intent, that we did in the 1970s and 1980s to try and repair and upgrade tenements.

Having been trained in Belfast, I left that illustrious city for Glasgow in 1976. I had no experience of tenement living. During my six years of architectural training, at no time was 'housing' ever on the curriculum and the closest we ever got to designing a house was 'A Holiday House' in year one. I did the project after a Norwegian fellow student helped me. It was fairly awful.

However, I did get involved in voluntary work with community groups, in Sandy Row, the Markets and the Divis area, mainly to fight the Belfast urban motorway which was planned to encircle Belfast. My eyes were opened to the conditions that people lived in and how they were dealt with by the local authority. The Divis Flats (Large Panel Concrete, deck access flats) were being built then by Laing, demolishing Victorian terraces which fractured local communities, in the vain hope that 'Comprehensive Redevelopment Areas' were the way forward.

The signboard on Laing's site was soon graffitied to read 'Laing Kesh'.

The flats were unpopular from the start and further Comprehensive Development Areas (CDAs) were planned for other areas like the Markets. Sandy Row was mostly threatened by the proposed urban motorway. At the time the area was a thriving protestant enclave with a long shopping street (lots of butchers' shops for some reason), and many densely packed Victorian terraced houses which needed to be upgraded. Only nobody ever considered such a possibility in 1972 (in 1976 the Northern Ireland Housing Executive did produce a slim report on 'Gradual Renewal'). But that is another story.

Arriving in Glasgow in the mid 1970s, the tenements I saw were largely blackened. Previous owners had applied paraffin wax[1] to the stone frontages, possibly thinking it would preserve the stone. The city smoke soon adhered to the stone, forming a black skin over the surface.

The biggest changes occurred in Glasgow during the decline of the traditional industries of the city and the consequent loss of jobs. Tenements needed to be maintained and the limit on rents[2] led to a decline of investment. Perhaps Glasgow was unlucky as wetter weather would also lead to extensive decay. Overcrowding was a clear problem and the approach taken was to demolish old tenements and to build new housing on the outskirts, in peripheral schemes.

I joined ASSIST Architects in 1976. At that time they were based in Govan and were formed as a Practice Unit of Strathclyde University's Architecture Department run by Jim Johnson. Peter Robinson had shown how bathrooms could be installed in a tenement and Raymond Young, another Strathclyde architecture student who lived in Govan, made it into his final year thesis. By making use of Home Improvement grants, a process of voluntary flat improvement got underway. Key to it was working with the tenants and owners (which was essential as toilet stacks passed through every flat). The Govan office of ASSIST gave support to local residents and eventually evolved into a Community Based Housing Association which was the start of the movement. Govanhill and Denniston soon followed Govan's example.

I started to work in Paisley in Ferguslie Park as a research assistant with ASSIST, who were appointed to carry out an estate-wide condition survey of the council houses in Ferguslie Park. The houses were a mix of pre-war and post-war tenements and cottage flats, of a reasonable size, but very poorly maintained. We worked with the Community Development Project on Ferguslie Park Avenue where I had my office.

I was staying with a good friend in Pollokshields at the time, but was clearly needing to find my own flat. Eventually I bought a top floor maisonette flat in Kenmure Street, continued working at ASSIST, and soon joined Pollokshields Housing Association.

There were repair grants available for roof replacement and repair, so eventually the owners in the close agreed to reroofing in concrete tiles, although I reslated the bay roofs and we restored the rear wash house which was reslated.

After Ferguslie Park I started to work in ASSIST's Govanhill office at 81 Inglefield Street. My first job was to handle a reroofing contract with Lafferty builders. Health and safety was not practised in those days. I remember that the builder Lafferty used a crane which lifted a skip high up to the roof which was then filled with all the old slates and chimneys which were chucked into the raised skip.

We did a lot of tenement repair and rehabilitation work to tenements and reformed many backcourts, always working closely with the residents. In hindsight, much of the work was not ideal, but it did start to change attitudes about what was possible in terms of saving communities and showing how tenements could be restored.

1 Often erroneously thought to be linseed oil, although boiled oil was also used.

2 From 1920 to the 1950s, rent increases were illegal. This caused a lack of investment in tenements; even though the 1954/6 Rent Act allowed realistic rents to be charged, it was conditional on repairs being carried out.

[above] John in his bed recess office at 91 Inglefield Street

[upper left] ASSIST staff in the 1980s: John Gilbert, Simon McIntosh, Judith Moores, Robert MacDonald. Josephine Mitchell, Lou Rosenburg

[left] John LeHarivel

[lower left] Judith Moores (now Judith LeHarivel)

In my free time I volunteered to help a number of tenants' associations, often trying to get evidence to take their council to court over dampness caused by condensation which was rampant then. I worked with Jim Friel to help with the Gorbals Anti Dampness Campaign. In 1983, with backing from the Scottish Consumer Council, Chris Orr, Stuart Hashigan and I produced *The Tenant's Handbook*. Because of tenant demand for the service, we formed the Technical Services Agency.

In 1992 I started my own architectural practice, John Gilbert Architects, and also with Annie Flint, we produced *The Tenement Handbook*[3] to help owners understand better how to manage tenement maintenance and typical common repairs. Eventually that led to forming 'underoneroof.scot',[4] a website which could be readily accessed by all property owners and interested parties.

In this book I try to cover both Glasgow and Edinburgh tenements as well as briefly featuring other towns and cities in Scotland. I admit to having limited experience of working outside the Glasgow area so I am sure readers will be able to correct me when I have missed an important feature regarding tenement construction in other areas.

Fundamentally, though, this book celebrates the Scottish tenement, not just those found in Edinburgh and Glasgow, but all tenements located across urban Scotland. It is in these other parts of Scotland that some really fascinating tenements are to be found. This is after all a distinct national building form, so should be celebrated for being iconic, just as the terraced house, the semi-detached villas and bungalows have long been in England.

The core objective is to show how tenements were constructed and why that knowledge will help owners and property managers to understand, treasure and maintain tenements. It is not, however, a guide to undertaking tenement repairs, though some of the knowledge imparted will certainly help with such an endeavour. In essence, it is a developmental history of this iconic Scottish building type. My ambition is to explain the details of the key technical developments that have impacted tenement construction over time and, as part of that, explore the practical, social and legal influences which helped bring about such changes.

Books of Scottish architecture tend to appraise the style of buildings and to value churches, stately homes, castles and public buildings, but often avoid discussing the tenement form that is all around us. We suspect this is because the tenement form is so common in Scotland that we fail to notice it, and, in the past, value it. But this is changing now and people are increasingly beginning to appreciate the importance of tenements and want to know more about them. Climate change is also now also part of this narrative, given the longevity of these structures, and the fact they were originally designed to cope with a quite different climatic context and use of energy.

This book is largely about how tenements were constructed. It begins with the context that we are now in. Climate change and the need to reduce carbon and energy use has never been more important, and tenements in Scotland are an asset which needs to be maintained and improved if they are to meet future needs (see Introduction). By describing the construction of the different elements, I hope owners will better understand what elements need our attention when repair is needed.

3 Published by ASSIST Architects.

4 www.underoneroof.scot

As stated above, the book does not show how tenements should be repaired; for that I must advise owners and landlords to make use of the website www.underoneroof.scot which has the advantage of being updated and offers many ways to best repair and manage the maintenance of a tenement, or to visit www.historicenvironment.scot which has more detailed technical help.

I have to thank the following for help in writing about tenements:

Mary Taylor

Mary Taylor is an academic who started as a housing practitioner, worked as an academic – teaching and researching – and finished her career leading the representation of housing associations at the SFHA. She managed and led a housing association in an inner-city tenement neighbourhood, later helping to initiate and develop some new associations in rural areas and peripheral estates. She also served on various boards and committees, supported students to gain work experience in associations and researched the activities and performance of associations. This chapter draws on written sources, personal knowledge and information from fellow practitioners in the sector.

Following maternity leave and a short career break, during which she helped to found or support new housing associations, Mary moved to Stirling having taken up a teaching post on the relatively new postgraduate housing studies programme at Stirling University. In recent years she has completed various housing renewal projects in a personal capacity, including her current home rescued from near dereliction.

Having various research interests, sabbaticals and secondments over her 20 years at the university kept her partly in the world of housing practice, and she left academic life in 2010 to lead the SFHA. Under her leadership, SFHA returned to be the credible representative voice of housing associations in Scotland. In retirement from full-time work, Mary undertakes occasional writing and research projects about housing and heritage, amongst a wide array of other interests.

Her chapter draws on written sources, personal knowledge and information from fellow practitioners in the sector.

Mary Taylor

Annie Flint

Annie is a housing professional, and has worked, amongst others, with John Gilbert. In 1979 she wrote several versions of *Common Repairs, Common Sense*, which was published by the Scottish Consumer Council and distributed to all the local authorities. The *Tenement Handbook* was published with John in 1993. But times change, so she got together with certain local authorities and raised enough money to get the Under One Roof website operational. The site is also available to all owners. This book is an offshoot from that idea.

Annie Flint

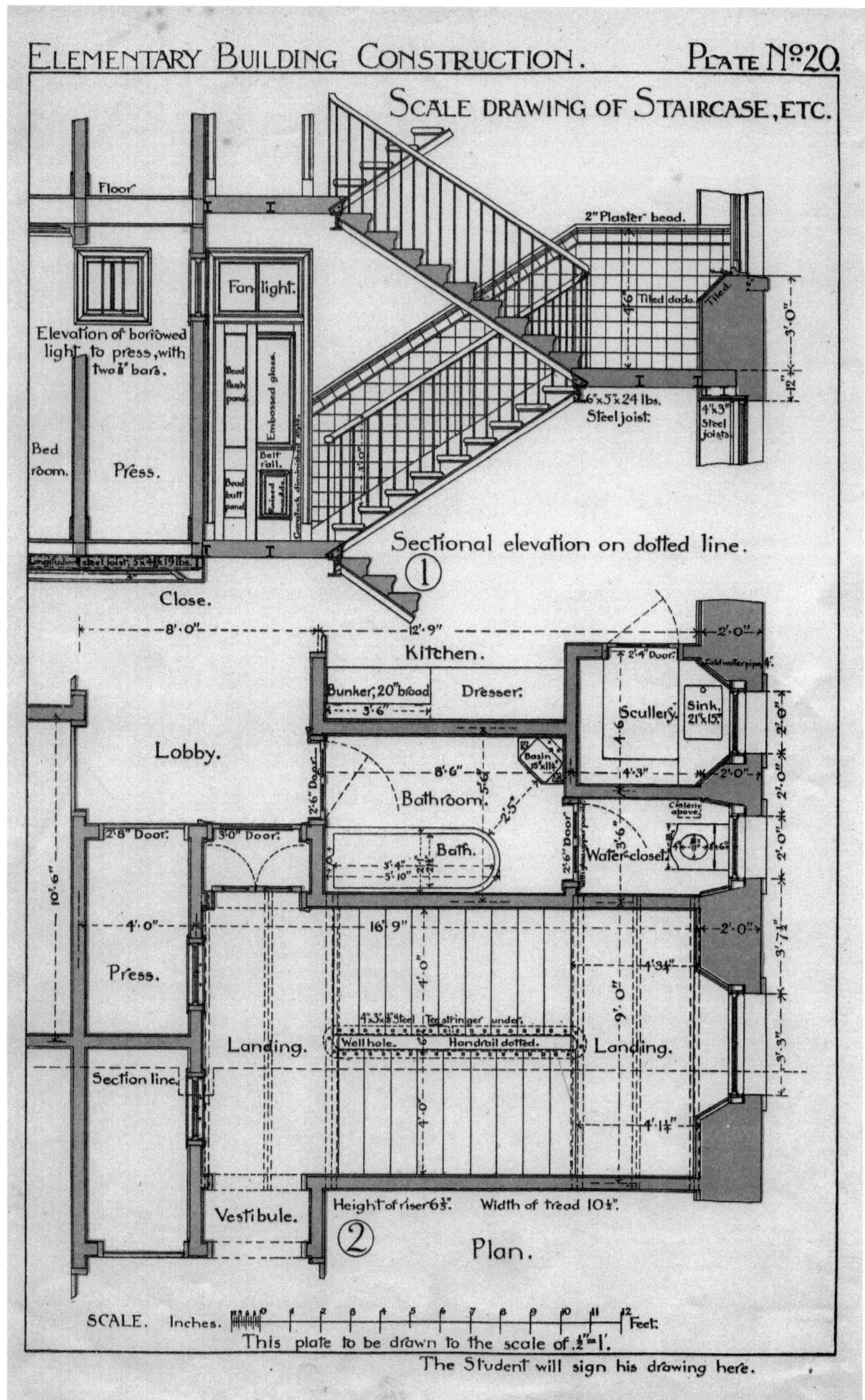
ELEMENTARY BUILDING CONSTRUCTION.
PLATE No. 20.
SCALE DRAWING OF STAIRCASE, ETC.
Floor
2" Plaster bead.
Fan light.
Elevation of borrowed light to press, with two ⅜" bars.
Tiled dado.
Tiled.
Embossed glass.
Bead flush panel
Belt rail.
Bead butt panel
Raised mould.
6"x5"x24 lbs. Steel joist.
4"x3" Steel joists.
Bed room.
Press.
Sectional elevation on dotted line.
1
Close.
Kitchen.
Bunker, 20" broad
Dresser.
Scullery.
Sink. 21"x15".
Lobby.
Basin
Bathroom.
Bath.
Water-closet.
2'8" Door.
3'0" Door.
2'6" Door.
2'4" Door.
Press.
Landing.
Well hole.
Handrail dotted.
Landing.
Section line.
Vestibule.
Height of riser 6¾". Width of tread 10½".
2
Plan.
SCALE. Inches.
Feet.
This plate to be drawn to the scale of ½"=1'.
The Student will sign his drawing here.

Other books on tenements

Frank Worsdall's *The Tenement: A Way of Life*

The one book that stands out as being important in understanding the Scottish tenement was written in 1903 by Charles Gourlay,[5] who taught at the Glasgow and West of Scotland Technical College. He produced *Elementary Building Construction and Drawing* with detailed drawings for students describing a typical Glasgow tenement. The book was not designed for the lay public, but it gives an essential insight into how late Victorian tenements were built in Glasgow.

Gourlay wrote that: *'To build tenements is undoubtably a cheap way of building houses, but the work in these tenements is often poor, both in materials and in construction. With simple but well designed mouldings, properly placed, a tenement need never be ugly. It may, indeed, have quite a pleasing exterior, with, in its way, even a measure of dignity.'*

The other essential book about Glasgow tenements is Frank Worsdall's *The Tenement: A Way of Life*, published in 1979. This book covers the legal and social background as well as the main features of tenements.

The Glasgow West Conservation Trust also produced a series of guides covering the repair and maintenance of Victorian and Edwardian tenements in the west end of Glasgow (now sadly out of print).

For Edinburgh, the book we must turn to is *The Care and Conservation of Georgian Houses*, by Andy Davey, Bob Heath, Desmond Hodges, Mandy Ketchin and Roy Milne, first published in 1978 (the 4th edition, 1995, is the edition I would recommend) and still very relevant. The bibliography at the end lists the full range of books which may be of interest to the reader.

5 *Elementary Building Construction and Drawing for Scottish Students.*

Acknowledgements

I have to thank Annie Flint, Mike Heffron and Mary Taylor for sticking with me in this difficult time.

To Tom Colburn and Stewart Parker, who awakened my understanding of how architecture was related to political choice, also for Tom, who introduced me to Rena McAllister, who I soon married and who supported me when I set up John Gilbert Architects, and her father, David McAllister who, as a quantity surveyor, advised people whose properties were compulsorily purchased to make way for the M8. He gave me his copy of Gourlay's *Elementary Building Construction* for students.

To ASSIST Architects for taking me on in 1976 and where I worked until 1992. Thank you for the subscription for the book. To Josephine Mitchell at ASSIST for being so level headed in making my letter writing make sense.

To Marianne Hughes who was there when I needed her and read the many drafts of this book.

To Chris Morgan who wrote the final piece about the happening at 17 Niddrie Road.

To people who read and commented on the text and provided many helpful suggestions, most of which I have tried to take on, to Douglas Robertson, John LeHarivel, Phil McCafferty, Darren McLean, Annie Flint, Mike Heffron, Jim Johnson, Lou Rosenburg, Lissie Hughes, Scott Abercrombie, Marian Jacobs, Mike Thornley, John Burns, Steve Wood, Moses Jenkins, Robert Storey, Roger Curtis, Barbara Lantscher and to Les Milne. In particular I would thank Jocelyn Gray-Marshall for spotting many of my errors and helpfully advising me of the wonders of Edinburgh tenements. Also thanks to Stephanie Pickering who assisted in editing my book.

The errors are entirely mine, but writing things down makes me realise the many things that I still don't know about tenements, so be warned – each one is different!

Illustration sources

All other drawings and photographs are by John Gilbert

xviii cover of Duneland: IPCC;

xxiii John Gilbert Architects;

Page 3, 5, 31, 33, 43, 56, 210, 250 all Glasgow Museums GL_BACS_2562;

Page 10, 43 Perth and Kinross Museums and Galleries,Perth;

Page 19 Blackfriars Wynd, Edinburgh Museum;

Page 21 Lower Fleshmarket Close, Robert Hurd & Partners;

Page 34 Nightsoil buildings in Paisley, Springburn Museum Trust;

Page 38, 54,all photos, Esther Galbraith, no further details available;

Page 40 Painting by John Atkinson, Grimshaw on the Clyde, 1881, National Galleries;

Page 45 John Pringle painting, National Galleries;

Drawings on pages 95, 207, 209, Cambridge University Press, *Architectural Building Construction* Vol 2;

Page 85 Historic Environment Scotland for radar image of flues in gable from Fire Safety Management in Traditional Buildings;

Page 101, 195, 196 Darren McLean;

Page 135 both photos of slates taken by Alice Clayton for Easdale Slate Museum;

Page 163 Close tiles, Tenement Tiles website;

Page 249 Aftermath of the Great Storm of 1968 *Glasgow Herald*;

Page 259 After the fire in Albert Cross, Pollokshields, *Glasgow Herald*;

Pages 312, 313, 314 Fire service plans, Glasgow Fire Service;

Page 315 Charles Goad's fire insurance maps, National Library of Scotland;

Page 321, 323 British Geological Society;

Page 322 Splitting stones in Caithness; Joan Walsh, the Easdale Folk Museum;

Pages 342, 345, 346 Hodder & Stoughton for images from *Flats, Urban Houses & Cottage Home*, edited by W. Shaw Sparrow.

Introduction

Climate change has affected the weather in Scotland. Since the 1960s, precipitation has increased by 27 per cent, and in north and west Scotland, winter rainfall has increased by 70 per cent. Last year, the average winter rainfall in Scotland was 676mm, while in England it was just 329mm. Clearly these are averages and the west coast of Scotland suffers a lot more rainfall than the east. Sea level rise has also increased in the last two decades and now exceeds 3–4mm a year around Scotland. This makes flooding of buildings more likely.

Increase in the overall rainfall may mean, for example, that gutters need to be enlarged. Typically, Glasgow receives around 1,245mm of rain per annum, while Edinburgh receives 704mm per annum. In a stark example, new calculations using a high emissions scenario with around 3°C of warming by 2070 show that in Glasgow the number of days where 30mm of rain or more is recorded in an hour could be 3.5 to 4.5 times more likely by the 2070s compared with the 1990s.

Driving rain has a significant impact on the ability of sandstone walls to dry out, as the periods to allow drying reduce and the amount of moisture increases along with the frequency of rainfall. In winter, Scotland gets less sunshine than England (an average of 119 hours as against 194 in England), so people in Scotland often lack vitamin D. Scotland's mean temperature in winter is about 2.5°C whereas England's winter average is 4.8°C.[1] A difference of 2°C is a lot of extra heating and underlines the need for good insulation to conserve heat. Scotland is a lot windier, hence all the wind farms, although wind-driven rain is an increasing problem.[2]

Wetter winters also lead to rising ground water levels which can lead to dampness in basements and solums, so it's important to make sure there is also sufficient ground and surface drainage. Any embedded iron (often found in bay lintels and close entrances) if exposed to dampness will eventually corrode. This is beginning to happen in Glasgow where the sandstone adjacent to cast iron is showing signs of cracks.

In the summer, as clay in the ground starts to dry out, it can affect foundations and lead to cracks in tenements.

1 Met Office statistics for winter/ February.

2 See HES Short Guide no 11, 'Climate Change Adaptation for Traditional Buildings'.

This climate has influenced how we build. Windows are set back behind a wall opening, whereas in England windows are often flush mounted. Window sills and string courses are designed to throw moisture away from the face of a building. Window hoods, sills and string courses are projected to shed the rain. Roofs are slated with thick Ballachulish slates and fully sarked to strengthen the roof structure (the 'penny gap' between sarking boards allows the timber to dry out). However, higher winds could loosen ridges and ridge tiles as well as damaging chimney-heads if they are poorly maintained. Stone walls are made thick, not just to allow them to be built high, but also to withstand wind and exposure. In extremely wet weather, the dampness in stone walls can migrate inside and conditions are set up to allow rot to develop in any embedded timbers (like joists, wall plates and safe lintels).

Tenements in Scotland developed due to how they were financed, the legal requirements of the time and the feu duties, as well as a demand for relatively cheap housing. However, the Scottish climate has continued to be a factor in their maintenance and the need for adaptation, particularly with climate change.

IPPC[3] Report[4]

The recently published IPCC report (sixth Assessment) provided a detailed view of possible futures, that current policies put the world on track for an updated estimate of an increase in temperature of around 3°C by 2100, although some climate uncertainties mean that warming of as low as 2.3°C or above 4°C cannot be fully ruled out.

Section C7 states:

'In modelled global scenarios, existing buildings, if retrofitted, and buildings yet to be built, are projected to approach net zero GHG[5] emissions in 2050 if policy packages, which combine ambitious sufficiency, efficiency, and renewable energy measures, are effectively implemented and barriers to decarbonisation are removed. How ambitious policies increase the risk of lock-in buildings in carbon for decades while well-designed and effectively implemented mitigation interventions, in both new buildings and existing ones if retrofitted, have significant potential to contribute to achieving SDGs[6] in all regions while adapting buildings to future climate.'

At present, we are clearly not doing enough, in the UK – or the rest of the world – to bring warming down to under

Front cover of IPPC report with Duneland housing in Findhorn shown as a good example of the type of housing we can supply, but which is rarely produced by housebuilders

3 IPCC – Intergovernmental Panel on Climate Change.

4 IPCC Report. The first IPCC Assessment was carried out in 1990, and for some time it played down the risks associated with climate change and only with the 6th Assessment had they caught up with the science. See Eugene Linden, *Fire and Flood*, Penguin, 2022.

5 GHG are greenhouse gases; they are compound gases that trap heat or longwave radiation in the atmosphere.

6 SDG refers to Sustainable Development Goals which were adopted by all United Nations Member States in 2015, to provide a shared blueprint for peace and prosperity for people and the planet. There are 17 goals.

2°C.[7] This will require a step change in behaviour, transport and all services, but particularly our buildings. The UK has some of the worst insulated housing in Europe, and Scotland, which is about 3°C colder than much of England, is particularly at risk.[8] As Dr Katharine Hayhoe, chief scientist for the Nature Conservancy and a professor at Texas Tech University, said, the world is heading for dangers unseen in the 10,000 years of human civilization. She continued: 'This is an unprecedented experiment with the climate … the reality is that we will not have anything left that we value, if we do not address the climate crisis.'[9]

We have managed many changes in my lifetime, from external toilets and coal fires to slightly better insulation and gas heating. But this is not enough. We need to reduce CO_2 in many different ways, with personal and government action, although the latter is still dragging its feet.

Reducing CO_2

If Westminster politicians were seriously considering policies to improve housing, we would not be encouraging new-build, one family houses on greenfield sites where every owner is expected to have a car (or two). Such policies increase CO_2 emissions by increasing the use of the car and make cities and towns unsustainable as well as unhealthy. It also means that everyone travels by car, rather than walking or cycling anywhere; people do not meet their neighbours and become more isolated. In the city, road traffic makes cycling dangerous and crossing a road problematic and dangerous. To say nothing of air quality.

Road traffic produces half of our nitrogen emissions, 75 per cent of our carbon monoxide and 25 per cent of our carbon emissions in the UK. The UK income from vehicle taxes and fuel taxes in 2019 was £27.789 billion, whereas the expenditure on roads, maintenance, congestion, casualties, non-productive drivetime costs and other environmental degradation costs is estimated to be £92.359 billion.[10] We need to be approving inner city housing with only limited parking for disabled users and deliveries and taxing large car parks with more than ten cars.[11] The recently revised National Planning Framework NPPF2 focuses on economic growth not sustainable development and does not state the need to reduce traffic to lower carbon emissions. It largely ignores maintaining buildings or energy conservation.

The one thing that could help is investing in our existing houses by repairing and renovating them, yet we continue to penalise this through VAT at 20 per cent. It is simply perverse.

Insulating and draughtproofing

Retrofitting insulation in older sandstone tenements is important, but what is just as important is ensuring we have tenements that are well maintained, managed and adapted for the future. At present, we do not have a system that is fit for purpose.

7 'What is the use of having developed science well enough to make predictions if, in the end, all we're willing to do is stand around and wait for them to come true?' Sherwood Rowland in his speech accepting the Nobel Prize for Chemistry in 1995.

8 According to the thermostat maker Tado, British homes lose an average of 3°C of indoor temperature over five hours, compared with 1°C in Germany. Lack of thermostatic controls is common in the UK.

9 *Guardian* report on 1 June 2022 on the Climate Change 2022; Mitigation of Climate Change.

10 From David Williams, *Civilizing Cities*, Arena Books, 2021.

11 First proposed by John Prescott as deputy prime minister but then quietly dropped.

Draughtproofing a tenement flat can save as much energy as insulation can, or the installation of new windows, yet most window installations are carried out without any serious attention given to draughtproofing which can usually save about £45 a year.[12] Note extensive draughtproofing can save more but would need to be associated with controlled ventilation using mechanical heat recovery systems; these are increasingly needed as we generate more moisture than 50 years ago and our structures are not as vapour permeable as before. Poor levels of insulation and poor building practice creating cold bridges are common in much new housing, yet are rarely accounted for, as new housing, or retrofitting, does not require independent post completion surveys which could flag up such defects.

There is a hotly debated argument between those who wish to insulate more and those who believe that too much insulation and draughtproofing can cause harm to a traditional tenement.

The conservationists believe that old stone walls which have embedded timbers in them need ventilation to ensure the timbers don't succumb to rot. Sealing flues can mean that chimneys don't dry out so easily. Also, by insulating on the inside, the external stonework becomes colder, possibly less able to dry out and so more at risk from frost attack.

On the other hand, those that believe insulation and draughtproofing can bring considerable benefits believe that draughtproofing alone can save a considerable amount of energy and, provided the walls remain vapour permeable, they should still be able to dry out.

There are many ways of insulating walls (described in Chapter 12): some will still allow water vapour to pass through a wall, whilst others require a *vapour barrier*. To reach Passivhaus standards, increasingly achievable in new-build housing, requires any cold bridges to be eliminated, which can be very difficult in a tenement. External insulation is an easier option as it is less disruptive and maintains the thermal mass of the wall; however, it covers over any stonework and can present future maintenance issues.

A pilot project developed at 107 Niddrie Road in Glasgow by Southside Housing Association is testing this approach to make a tenement exceptionally well insulated. Tenants have been rehoused temporarily and the tenement is an extensive refurbishment to EnerPHit[13] standards. The rear walls have been overclad externally with insulation and the front stone walls have been internally insulated by reducing all cold bridges caused by joists bedded into the outer walls, then applying a breathable insulant to the inside. New triple glazed windows are installed and all joints are draughtproofed. Ventilation is provided by an all-house heat recovery ventilation unit. On completion, the flats will be monitored and one trusts that we will learn from this approach. (A more detailed report on the project is contained in Chapter 12.)

Any solution for traditional stone tenements has to recognise that individual owners and tenants are involved, so the work is more likely to be done on a flat by flat approach. That may involve compromises no matter what draughtproofing and insulation system is adopted. Internal insulation is more difficult as it is disruptive to any occupier. Different options for upgrading stone-built tenements are described in Chapter 12. Gas combination boilers will

12 Includes draughtproofing chimney flues. Energy Savings Trust.

13 EnerPHit is the new Passivhaus refurbishment standard, which needs to be met in any project needing to achieve Passivhaus standard.

Preparation for internal insulation using a wood fibre

[top left] Existing embedded trimmer joists removed at bay to prevent cold bridging

[left] Existing stonework plastered, using mesh to reinforce it

[lower left] Bay tied back with metal fixings isolated from inside then wood fibre insulation applied

[above] Completed job

eventually be phased out and it is difficult to know what the best answer is for tenements, but at least the individual flats will be better insulated and ready for adaptation to work whatever heating system is appropriate.

Any strategy will need to emphasise the need for maintenance and thus require owners to form Stair Associations to coordinate the work. The project at 107 Niddrie Road can help us understand what is most effective, with additional pilot projects showing us the different options possible to upgrade tenement flats in stages.

Retrofitting our existing stock of housing will be critical in the next 20 years, and we trust there will be more investment in this neglected area.[14]

The standard of house building

There are limits to how much a traditional tenement can be improved through insulation and improved heating systems. Retrofitting stone tenements to meet passive housing standards is difficult and expensive, although much can be done by thoughtful approaches to insulation, draughtproofing and ventilation.

We can expect the internal layouts of tenements and flats to change with time, although this will be constrained by their structure.

Replacement windows which are correctly installed will help to reduce heat loss and, in some cases, external insulation to the rear of tenements may be a possibility. The website www.underoneroof.co.uk explores the options to improve energy efficiency. Historic Environment Scotland have also produced a *Guide to Energy Retrofit of Traditional Buildings*[15] and have completed a number of pilot projects on tenements which can be found on their website.

Although environmental improvements were often carried out in the 1980s, times change and back courts have become the only spaces that flat dwellers can use, so they are increasingly important, no matter how small some are.

During the COVID pandemic, it was clear that people needed to get out of their flats, or at least breathe fresh air and see the daylight. Modern flats should be required to be built with balconies, as in Denmark, for example. In Scotland, these should be made as glazed sunspaces, so they are protected from the weather but still allow solar gain and the facility of having an open-air balcony. Indeed some tenement refurbishments have actually managed to add balcony sunspaces to south facing walls.

New housing in urban areas can be expensive because of the need to decontaminate the land and relocate all the existing services, such as substations, water mains and drains. Access to build can also increase costs.

The Scottish government questions why Glasgow's housing is more expensive than other areas as they think that if Glasgow didn't adhere to the specification set by 'The Glasgow Housing Standard' they would be able to reduce costs and therefore build more houses. The Glasgow Housing Standard is quite rigorous in setting detailed requirements, to meet '*Housing for Varying Needs*' standards. It does this well.

Even then, we simply do not invest enough in new housing to maintain a quality. Energy

14 See *Guide to Energy Retrofit of Traditional Buildings* produced by Historic Environment Scotland.
15 This supersedes short Guide 1: 'Fabric Improvements for Energy Efficiency in Traditional Buildings'.

standards have improved under the building regulations, but a fabric first approach is often overridden by quick fixes that developers use such as applying 4m^2 of solar PV panels instead of a well insulated fabric. It's called value engineering.

Low carbon housing

In 2007, the UK government introduced its '*Code for Sustainable Homes*' as a voluntary standard. Before this, many architects had already been trying to improve insulation standards, particularly for housing associations whose tenants lived in hard to heat homes. In 2010, the European Commission mandated that all homes within the EU should be 'nearly zero' energy buildings by 2020. In 2018 this was updated to require 'cost-effective' renovation of buildings aimed at decarbonising by 2050. Now we are out of the EU, the target of net zero seems to have been indefinitely postponed.

As things stand presently, we only have an estimated seven years left in which to reduce our carbon emissions by 50 per cent by 2030.

One problem is the complexity of the regulations which require targets of carbon dioxide levels less than nominal levels and SAP[16] calculations are required which are notoriously difficult to obtain, given the different types of construction. This is why architects have started to use Passivhaus strategies. For new housing, the heat losses of the building are reduced so much that it hardly needs any heating at all. What heat is required can be provided by the supply air if the maximum heating load is less than 10 watts per square metre of living space. If such supplied air heating suffices as the only heat source, then the building is a Passivhaus. Such an air supply is usually supplied by an air source MVHR[17] unit.

EnerPHit is the new Passivhaus refurbishment standard from the Passivhaus Institute and works on similar calculations. Here the critical losses occur at cold bridges where existing structural elements often join one anther and air tightness can be very difficult to achieve.

Finally, there is little leadership in housing construction and maintenance in Scotland, other than the top-down approach from Scottish government. The asset management approach used by the SSHA (Scottish Special Housing Association) which morphed into 'Scottish Homes' did provide some leadership, at least in encouraging more sustainable housing, but we need an organisation that will research housing issues and drive the

Closeburn Passivhaus – houses owned and managed by Nith Valley Leaf Trust a community-based organisation

16 SAP Standard Assessment Procedure. The UK government's recommended method system for measuring the energy rating of residential dwellings. The SAP runs from 1 to 100+, with dwellings that have SAP rating greater than 100 being net exporters of energy.

17 Mechanical Ventilation with Heat Recovery (MVHR) provides fresh filtered air into a building whilst retaining most of the energy that has already been extracted though a system of mechanical ventilation.

change that is needed. Hopefully this will lead to a much greater variety of housing types, particularly housing which fosters community involvement such as co-housing.

Housing associations (community based or not) and individual owners and landlords, need support and leadership from government as well as funding.

Book structure

The remainder of the book is divided into three sections. The first examines the history and development of the Scottish tenement; the second describes the different elements of the Scottish tenement in some detail; and the final section looks at some possible ways to ensure the Scottish tenement can continue to flourish into a sustainable future.

The evolution of the Scottish tenement

Chapter 1
The development of the Scottish tenement

Most Scottish houses built before 1500 would have been single storey, or at the most two storeys, with walls built from boulders and bonded with clay to keep the wind out. The roofs would be thatched. Boulders and turf were used up to about 1629, the walls daubed in clay, which was eventually replaced with lime mortar and limewash. From 1420 walls become thicker as, in high value areas, flats would be built on upper floors (Canongate in Edinburgh, for example). Stone and lime mortar was only used for prestigious buildings and mansions. The main period of tenement building was during the Georgian and Victorian eras, and so this book largely confines itself to these periods. We have only a few examples of tenements from earlier periods, but it is worth touching on them as they influenced later developments. They were built by small companies of builders as owners of land invested in building tenements to meet a growing demand.

The modern period is not covered although there were plenty of good – and bad – tenements built in the inter- and post-war years. Modern tenements may have simple frontages, but they have many more facilities, such as kitchens and bathrooms, and can be a lot easier to heat.

In Scotland, there are around 895,000 tenement flats: one third built before 1919 (300,000); a third more by 1982; and a third more by today's date.[1]

These building periods are now discussed with more emphasis placed on the main tenement building periods: medieval 1200–1500; Renaissance 1500–1714; Georgian 1714–1837; and Victorian 1837–1901. The influence of colonialism on the wealth used to build tenements is explored between the section on the Renaissance period and the section on the Georgian period.

Medieval period 1200–1500

The following is a description of a city in 1532:

1 Scottish Parliamentary Working Group on Tenement Maintenance.

'Pigs rooted round the grandest of houses, along with geese, ducks and stray dogs; we know they were always there because the city tried again and again to ban them. New rules in 1582 still allowed the use of alleyways between houses for horses, oxen and other beasts; but if you were moving animals, you had to be told to go to the nearest field with the least damage possible. In 1557 there were special orders that nobody was to interfere with the dog catchers when they were out killing. The city paid for three different kinds of night worker … "tenants of the cesspit", who checked the city's pools and pits of sewage and emptied them when the levels were too high. Those buildings leaning into the street might fall, and the stench of shit was everywhere. In times of plague, which were common, the city tried to clean up streets that were "very filthy". People had to be told not to throw out skin and entrails from domestic animals, the barber surgeons were not to throw blood on the roadways, there was a polite suggestion that the gutters by houses could in fact be cleaned, that dung from the constant rush of horses and carts could in fact be cleared away. There was a hopeful rule banishing any animal with a "nauseating smell". The old city had streets almost too tight and bent to hold all the business of the new city as it grew. They were lined with doors down to the cellars and vaults which undermined the roads … They were overgrown with canopies, galleries and jetties that jutted out one above another, so people with wagons and horses could pass only with difficulty … Shops spilled onto the same streets, their goods stored in the dark inside but sold outside in the light, shown in cupboards on the walls. On shutters, on trestles; you have to be able to see the goods, after all, which was hard inside because glass was too thick and rough to let in enough daylight and the more usual glazing was oiled cloth.'[2]

This was not describing Glasgow or Edinburgh at that time, but Antwerp.

At a similar time, in 1586, in Edinburgh:

'Acts were passed to keep the public wells and water untainted; midden piles were removed from the highways and placed in backlands and, if in a riverside or coastal town, filth was deposited in the river or the sea … Cleanliness was, however, an uphill struggle. The slaughtering of animals on the main thoroughfare, the siting of middens – illegally – on the main streets, the intermingling of craft workshops and domestic properties all added to squalor. And even middens placed correctly in the backlands were often too close to wells where contamination between these wattle lined pits was inevitable … Small children and the unborn were particularly vulnerable, Many died before birth, as has been shown from foetal and perinatal burials … the vast majority dying in infancy.'[3]

Edinburgh's tenements were developed prior to 1500, and located mostly in the Royal Mile and Canongate. A few examples remain, such as 8 Advocate's Close, which has rubble

2 Michael Pie (2021), *Antwerp, The Glory Years*, page 17.

3 Patricia Dennison (2018), *The Evolution of Scotland's Towns: Daily Life in the Middle Ages*, page 61.

walls and ashlar dressings and a vaulted chamber at ground level used by shops. The principal floor, the first floor, was entered originally from its own door accessed from an external fore-stair off the close. The ends of floor beams supported a jettied gallery on an upper floor, probably an influence from France or the Netherlands. At 6 Advocate's Close, a turnpike stair, probably incorporating an earlier fore-stair, provided access to both 6 and 8 when they were linked, possibly in 1590.

All windows would be shuttered, with glass panes still being uncommon. Ground floor walls might be constructed of stone, which has always been plentiful throughout Scotland, while the upper floors were of timber, or timber framed then clad in timber laths and plastered, or infilled with wattles[4] and daubed with clay render.

Although some more prestigious tenements might have been built in stone, most were wood framed and clad in timber. Scottish burghs formed part of the feudal system.

A decaying house and example of the old fore-stairs

The large section joists at Provand's Lordship from 1471

4 Wattles, usually hazel rods and twigs that were interleaved to form a base to apply the 'daub', which was a clay and straw mix.

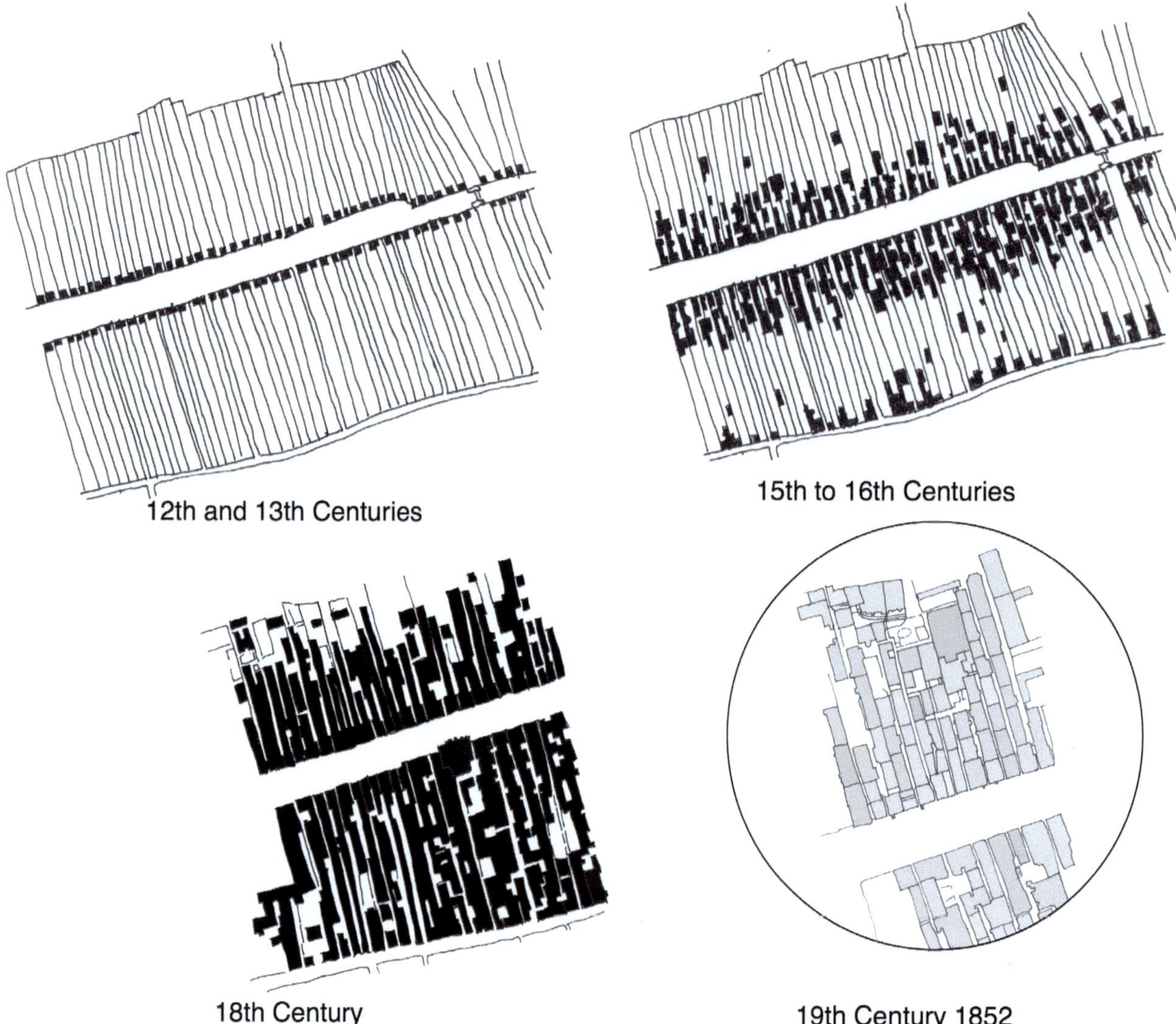

Typical plan development of riggs, with divisions still evident in 1852 (Edinburgh maps, 15th- and 16th-century conjectural)

How land ownership influenced the building of tenements

From the reign of David I (1124–53), Scotland's urban areas were known as burghs, with each burgh operating its own laws. Local regulation and decision-making came through landowners within the burgh who were known as burgers or burgesses. A burgess would own a plot of land, a toft or burgage plot. The average width of a plot along the main thoroughfare in Edinburgh was 25 feet (7.62 metres).[5] The 'tenement' was a Scottish legal term for a plot of land, a burgess plot, from whose ownership local tax obligations fell. These annual payments, in turn, bestowed local political power and influence through the voting rights which emanated from taxation obligations. The ownership of property was thus not just about habitation, but also about power and influence. Hence, such rights were long legally secured and protected within Scots law (King James VI, 1609). The burgh employed 'liners' to divide up building plots within the town, for which each burger would pay a rent. Across Scotland, the plots,

5 After the Reformation in 1559–60, plots lengthened to 8.4 metres.

known as burgage plots or tofts, tended to be long and narrow (known as 'riggs'). Within the burghs themselves, there is evidence of systematic layouts with each plot having a specific street width measured out to a common local unit, sometimes comprising two or more units depending on the wealth of the individual. Access to the rear of the plot tended to lie along the east edge of the plot though double plots with a central access point are not uncommon.

Burgh status often meant freedom from Crown taxes, but sometimes the burgh market was limited to certain trades or goods.

John Knox House on the Royal Mile.

Late medieval houses (say 1400–1500) tended to have the gable set against the street to economise on the use of space and because of the width of a plot. Tenements in country towns would often be built with fore-stairs directly from the marketplace. The fore-stairs would go to the first floor and other internal flights would be in timber (indeed the upper floors and walls were often also of timber construction).

The way land is owned in Scotland continues to influence tenement development into the present day.

Renaissance period 1500–1715

At this time (1500–1700) the majority of houses were wooden, some being survivals from the earlier period.[6] Around 1550 houses tended to have gables facing the street, often crow-stepped.[7] This was an influence from the many Flemish settlers in Scotland and trade with the Low Countries. The roofs of these gable fronted tenements might also have been finished with clay pantiles which were also imported from Middleburg and later from Rotterdam. To economise on space, fore-stairs and attics developed; windows would be shuttered or have an oiled cloth covering as glass was expensive. Ground floor walls might be stone and upper floors timber, or timber framed with timber lathes with wattle render.

Originally two storeys would have been common, with a *wynd, close or pend* through to the back. The ground floor was occupied by a workplace or housed animals, and the upper floor a house.

An Act of Parliament in 1621 required that new roofs and replacement roofs had to be slated or tiled, so thatch was banned although it may have remained a traditional roofing material for rural buildings in Scotland.

The techniques for quarrying stone had made the use of stone much more practical and cheaper. The gradual introduction of stone for external walls of houses was invigorated for three reasons:

- Quarrying techniques greatly improved with the introduction of gunpowder for blasting and the skills of sappers[8] honed by blasting castles.
- Fires in towns where all houses were timber were common, particularly where houses were built close together. Town legislation in 1800 forbade the use of fore-stairs in Glasgow, as they often caused obstacles to the use of the street.
- Timber became difficult to source in Scotland because of the high cost of transport from the Highlands as roads were poor, so it became easier to import timber by ship from Norway, Russia and the Baltic. 'Dundee had been importing structural timber from Norway and the Baltic for use through

6 'Edinburgh – a tenement city?', in Edwards and Jenkins (eds), *Edinburgh: The Making of a Capital City*, pages 106–8

7 Crowstepped gables were an influence from the Low Countries as were gabled facades.

8 *Sappers* were employed by the army to lay and clear mines. However, explosives could damage the stone, so it was actually little used in quarries where good-quality stones were needed. Quarrymen tended to use chisels and wedges to split the stone.

[right] Late Renaissance tenement (c. 1700) with wallhead dormers, cat slide garret windows and crowstepped gables

[below] Danish example of jettied timber construction that would have been used in medieval tenements (early influence on construction methods from housing in the Low Countries and Denmark)

Scotland since at least the 15th century.[9] Plague hit Scotland badly in 1640, but epidemics, such as typhus and cholera, were frequent, most likely caused by poor sanitary facilities and a lack of fresh water. In 1685, the throwing of refuse from windows was banned (apparently rather ineffectively) but from 1687 'streets wreaked and cleaned three times a week' and by 1692 the muck men 'were given the additional responsibility of patrolling the streets between 9 p.m. and midnight every Saturday to report on people pouring waste from their window'.[10]

Escaping from multi-storey tenements was problematic so fire was of huge concern, especially as many tenements were constructed from timber and roofed in thatch, and thus vulnerable to fire and the spread of fire. In 1652 in Glasgow, fire destroyed 80 closes and 1,000 people were made homeless (amounting to one third of Glasgow's population). As a precaution, candle makers moved to Candle-riggs, but after a fire in 1677 the Council decided that in future all building material front, back and gables should be stone rather than timber.[11]

Later, possibly as a response to a fire when the external timbers of tenements added to the spread of fires, the Council in Edinburgh began to encourage the enclosure of galleries, ideally with stone, although wattle and daub walls were still much in use. When the 17th century approached, stone buildings in some burghs took over from timber construction, at least on the ground floors of tenements. The upper floors would mostly be built in timber, based on solid timber sections (*balks*) forming a platform and raised in panels which would be either timber boarded or treated with 'wattle and daub'. These platform floor constructions were often jettied[12] out over the stone base to project into the street (see below also). The style was probably imported from Denmark and Germany where such timber frame construction can still be found. It is thought that the system allowed the structure to be raised and clad without external scaffolding, [13] something we will see repeated in many 19th-century tenements. Simplicity and speed of erection in raising these platform-like arrangements would simply mean the drive to higher height, particularly in Edinburgh.

In 1674 Edinburgh Town Council was attempting to encourage replacing timber fronted buildings by stone by granting a 17-year exemption from taxation, although this was not always successful. At an earlier stage they had actually encouraged citizens to extend the front of their houses by seven feet to use up timber they had felled.[14] As tenements rose

9 *Lost Dundee: Dundee's Lost Architectural Heritage*, Charles McKean and Patricia Wheatley.

10 Murray Pittock (2019), *Enlightenment in a Smart City*, Edinburgh University Press.

11 A fire machine was ordered in 1785. By 1740 all vents and chimneys had to be cleaned three times a year.

12 Jettied floors from old French 'jette'. This medieval technique for allowing timber framed buildings to project over the floor below was common in medieval French towns.

13 'Some works revealed that the platform frame also allowed the structure to be raised and clad without the need for extensive scaffolding since the cladding could be held and fixed from inside the building', Bruce Walker 'Getting your Hands Dirty', *Architectural Heritage*, 17, page 15 and *Architectural Heritage*, 22, pages 32–4.

14 '... the said Council, in order to feu out the Common Moor, cut down trees thereon, which amounting to a very large Quantity of Wood, they could not readily dispose of it: but to encourage the Citizens to purchase the same, were, by the said Act of Council, empowered to extend the Fronts of their Houses seven feet into the street on each side thereof; whereby, they soon disposed of the Wood at a good Price, at the Expense of the Street', William Maitland (1753), *The History of Edinburgh*.

higher, turnpike stairs developed, often at the front of the building but eventually moving to the rear. It became common to have stone tenements formed with the rear, party and gable walls in stone and with timber frontages, as found in Perth at Kinnoull[15] lodging on the Watergate.

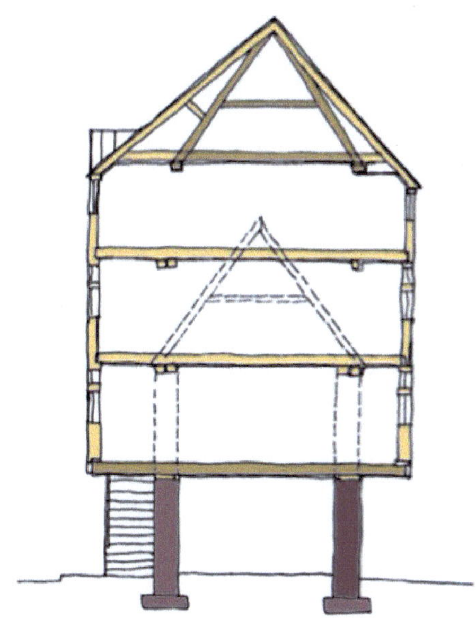

Based on a sketch by William Bruce 1883, it shows the way that the roof was used to integrate it into the four-storey tenement

Tenements developed in different cities at a different pace. In the 16th and 17th century, Edinburgh was still enclosed within the 140 acres of its 'ancient royalty' by the defensive walls[16], built mainly in the 16th century as protection against possible English invasion. Due to the restricted land area available for development, houses increased in height to accommodate a growing population. Small two-storey properties were rebuilt as stone tenements with a rear access via a turnpike stair. In Edinburgh, tenement heights increased; buildings of 11 storeys were not uncommon[17] and some rose as much as 14 levels.

They developed along the rigg system, at the rear of each tenement, so the Canongate and Royal Mile would have tenements and commercial buildings fronting the street and close entrances would go through to narrow wynds where densely packed tenements were built to make as much rental as possible from the feuing system. These tall tenements housed a cross-section of the entire society, nobles and judges rubbing shoulders with builders and shop workers on the common stair, as they had done in the medieval period.

By the end of the 17th century tall tenements that projected into the street and wynds were common in Glasgow and Edinburgh, restricting light as well as adding to fire risk, so the Scottish parliament passed two Acts restricting building height to five storeys above the carriageway in Glasgow and Edinburgh. This proved difficult in Edinburgh because tenements were built into steep hillsides, so some levels were well below the main carriageway. Stone was mostly reserved for important buildings such as the tollbooth, jails and even some shops,[18] but gradually it came to replace timber, which was increasingly difficult to obtain due to deforestation, but also sometimes because of legislation to help restrict the spread of fires.

In the Renaissance period, many houses were still of timber construction, or wood joists spanned between mutual party walls. Timber constructions were banned around 1674 in Edinburgh due to fire risk (thatched roofs were banned in 1624 by the Scottish parliament).

15 Demolished in 1968.

16 Defensive walls also existed in Stirling, St Andrews and Perth.

17 According to contemporary travellers' accounts, some tenements were even taller, as high as 14 or even 15 storeys.

18 By 1591 there was a four-storey tenement in Canongate; Dysart 1575; Musselburgh 1590 and Peebles 1572. Inverkeithing had a stone vaulted house in the 15th century (cleared away for war memorial) and two turnpike fronted houses on Spittal Street are still there from the 16th century. Darnley House in Stirling is a good example of three storeys with vaulted ground floors from this period.

[left] Drawing showing how the gable and rear walls and stairs were built in stone with the front structure built as timber platforms, projecting at each floor

[above] A picture of an early Perthshire tenement, now sadly demolished

Fore-stairs largely disappeared in the Georgian period after Burgh Acts which saw them banned and existing stairs taken down to ease access for goods, etc. However, some still exist in country towns, but mostly at the rear.

There appeared to be an aesthetic preference for timber at the front of buildings but due to the fire risks described, or to a shortage of wood, timber eventually gave way to stone. It was also common to have 'jettied' timber floors projecting into the front street and sometimes this was even encouraged. Successive upper floors would be jettied outwards with each floor level supported by a main bressummer beam. These projecting galleries were divided off by a partition lining up to the bressummer beam (a bit like modern sunspaces). The timber frame would be covered initially with wattle (hurdles made from interleaved hazel twigs) and clay or lime daub, and in later periods with vertical timber boards, although, as timber boards became scarce, stone and harling took their place.

Early roofs might have been formed with coupled rafters and *collars*, sometimes with a scissor-bracing. Sometimes an earlier two-storey house was added to with additional storeys, with each floor extended out in timber and the roof extended to ensure the whole frontage was covered.

By the 1700s, tenements were usually constructed out of rubble stone and harled with a lime mortar mix to protect the stone from the weather. The walls were solid, about 550–600 mm thick. Sometimes timber projections were added to enlarge upstairs rooms, but timber (and thatch) had been outlawed in most burghs by 1680.[19] The idea of using corbels continued when built with stone.

However, when 60 ft-high tenements were built on narrow 30 ft-wide streets, daylighting in some floors became so bad that gas lighting (or candle light) was needed for reading. Poor Law inspectors sometimes had to light a match or carry a torch to see the names on flat doors.

Early tenements were built to the street edge, with the ground floor usually formed into commercial premises and the owner and any extended family living above. (See plans of tenements showing their development up to 1915 in Appendix 4).

These 17th- and 18th-century tenements often had gables facing the high street, usually extended from the tenements behind. This was perhaps because of a Dutch influence. Crow-stepped gables with simple flat topped 'crow steps[20]' were likely influenced by Dutch or Swedish links. The roofs would then have been steeply pitched and sometimes attic spaces would be developed with *cat slide* dormer roofs.

[right] Thomson's Lodging, 2 Bank Street, Inverkeithing, 1617, with corbelled cap-house stair tower

[below] Tall tenements in Libberton's Wynd, Edinburgh

19 Edinburgh banned wood from façades in 1674 and in 1677 Glasgow banned wood front and rear.

20 Crow steps were square shaped stones on steep gables that allowed a roofer to climb up the edge to lay planks between crow steps to repair thatch. They were also easier to form than stones cut at an angle. The style mostly died out between 1750 and 1840 but returned with *Scots Baronial* style.

Earlsferry fore-stair

As a more classical style developed, the roofs ran parallel to the front facade. Stone margins (or scuntions)[21] framed the window openings and formed a check for the harling to finish flush with margins and sills. Once harled, the walls would be lime washed along with the stone margins, sills and lintels.

As demand for housing grew, tenements rose higher, usually with stone construction at ground floor level and timber framed upper floors. Access would then be from a *turnpike* stair, initially sited at the front, which led into a rather narrow passage to serve each house. The plan form might have had several rooms, although no toilets, as water supply was via a carrier and drainage was rudimentary, sometimes draining to the street or placed in a dung-stead. When tenements were timber, turnpike stairs were also made from wood, although none of these remain today. (See Chapter 6 for a fuller description of turnpike stairs.)

The Georgian and Regency periods followed (1714–1837). At this time mercantile capitalism was undergoing its transformation into industrial capitalism, greatly assisted by the influx of vast sums of capital, generated directly from the Atlantic slave trade, which then funded significant housing and infrastructure developments, especially in relation to transportation.

Colonialism and the urban form

The Atlantic slave trade was critical to Scotland's industrialisation, given that in 1808, 40 per cent of import trade and 48 per cent of export trade were accounted for solely by Scotland's trade links with the West Indies. Accessing colonial expropriation and exploitation had also been a critical driver in bringing about Scotland's union with England. This access to Empire provided a subsequent stimulus to property development, both in terms of its variety, scale and funding.

21 Scuntions are commonly termed window reveals.

[above] Steep roofs, cat slide dormers, harled and lime washed tenement in Inverkeithing, 1666

[upper left] Kirriemuir

[middle left] Brechin

[lower left] Forfar, gable fronts to street

It is in the Georgian period where we see both terraced town houses and tenements adopt that geometrical balance in form, with their imposing highly dressed stone facades studded with finely proportioned, six over six sash windows. Part of this development story also links to trade in timber, with expanded trade links ensuring access to longer and stronger lengths of timber, so expanding what it was possible to build, with tenements being a key beneficiary. These tenements are still very much part of the contemporary fabric of urban Scotland. While undoubtedly the best example of this imperial built form is Edinburgh's extensive New Town, similar significant pockets of Georgian housing can be found within most of Scotland's towns and cities (such as Perth, Stirling, Greenock, New Lanark). Power and money are always critical dynamics within any property development process, so their impact and influence are threaded throughout this book. The tenement, and its ever-changing design and physical form reflects these many changes.

During the 17th and 18th centuries, the Scottish Clearances led to large numbers of people migrating to the cities to find work. Increasing trade and industrialisation meant that Scotland changed from a rural economy to an urban economy. Much of this was the result of the Atlantic slave trade, first in tobacco (Glasgow imported more than London in 1760), then in cotton. After the American Wars of Independence, trade shifted to the Caribbean to deal in sugar, then to Cuba and Brazil. Scotland in 1630 was already an imperial power and had a colony in Nova Scotia at that time.[22] Scots were both traders and colonists, plantation owners and plantation managers, who actively participated in the transatlantic slave trade.[23] By 1763 on the Caribbean island of Antigua, 62 per cent of slave plantations were owned by Scots, mostly involved in the slave trade since the 1630s, as ship-building and trade had begun to help international transport of goods and slaves. By the 1670s the Atlantic slave trade had become an essential part of trade. After the Acts of Union in 1707, Scotland joined England under the same crown, allowing Scotland more trade opportunities. As David Alston notes in his book *Slaves and Highlanders*: '*... between 1750 and 1800 something in the region of 17,000 Scots left Scotland for the West Indies*.'[24]

Between 1755 and 1827, Scottish imports grew eightfold – from £465,411 to £3,948,233. Approximately 72 per cent of Scotland's West Indian trades went through Clyde ports.[25]

Scotland and the Caribbean 1775–1838[26]

It must be noted that much of Edinburgh's New Town was able to be built not only on the back of profits from slavery, but because, even after the abolition of slavery, Scottish slave holders received substantial compensation payments[27] which allowed them to build so well.

The trade in tobacco made the Glasgow Tobacco Lords very rich, and they tended to move west and out of the city. But trade in cotton, textiles, timber and sugar, which used slavery to procure the material, brought industry to Scotland and resulted in a significant rise in the size

22 Scotland lost this colony in 1632 because of supporting a war between England and France (on France's side).

23 'To terrify slaves, they needed people willing to inflict terror. These people were ordinary white men acting as overseers and bookkeepers on slave plantations.' Michael Fry, quoted in David Alston (2021), *Slaves and Highlanders*, page 18.

24 David Alston (2021), *Slaves and Highlanders*, page 266.

25 Stephen Mullen (2022), *Glasgow Sugar Aristocracy 1775–1838*, page 238.

26 *Mullen*, page 238.

27 Slavery compensation payments were only paid off by 2015.

of the urban population. One Glasgow firm, James Finlay and Co., traded in cotton bringing it back to be processed in Scotland. They then started to trade in tea, thanks to the colonial trade with India and were known to be still underpaying tea workers on tea estates in 2020.

The Leith merchant Sir John Gladstone,[28] whose son William Ewart Gladstone became prime minister, owned more slaves than any other Briton and so received the highest compensation of all slave owners after the abolition of slavery.

Not forgetting Henry Dundas, who was highly influential in ensuring a 15-year delay to the abolition of the slave trade.

Bute House,[29] in Charlotte Square in Edinburgh, the official residence of the first minister of Scotland, was built with money from Jamaican slave plantations.

After 1707, Scots took up posts in the East India Trading Company and by 1771 nearly half the company's writers were Scots. By 1790, jute was imported from India to Dundee, leading to its nickname 'Juteopolis'. Like other companies, the East India Company in the 17th and 18th centuries relied on slave labour. Although slavery is often associated with America, Scotland and England were also hugely involved.

It is also worthwhile considering the effect of the ideas that came through the *Scottish Enlightenment*.

The Medicis of Maryhill Cartoon by Marcus Patton. The Medicis and the investment that came from abroad

Plaque to Gladstone in Leith and Finlay's Tea AGM, 1975

28 His plaque in Leith was erected by the Leith Liberal Club in 1909. Recorded in 'Britain's Forgotten Slave Owners', www.bbc.co.uk as being Britain's largest slaveholder. (Retrieved 26 January 2021.)

29 Bute House, 1791, was originally the home of John Innes Crawford who had inherited his father's estate when he was five years old. The estate included a slave plantation producing sugar which gave him an income of £3,000 a year.

David Hume, in his essay 'Of National Characters',[30] wrote a footnote about inferiority and race. His contemporaries challenged him on this idea, but he didn't back down. The Enlightenment was viewed as 'rational', but essentially it wished to impose order on nature and underline the 'natural order' of things, essentially an Anglo-Saxon ethos which seemed to value white males over everyone else.[31]

Georgian period 1714–1837

Writing in 1689, Thomas More, an English visitor, described the housing in Edinburgh's Old Town thus:

> *'Their old houses are cased with boards and have oval windows (without casements of glass) which they open or shut as it stands with their convenience. Their new houses are made of stone, with good windows modishly framed and glazed, and so lofty, that five or six storeys is an ordinary height, and one row of buildings that is near the Parliament Close with no less than fourteen (storeys). The reason is their scantness of room, which not allowing 'em large foundations they are forced to make up in superstructure, to entertain comers, who are very desirous to be in, or as near as they can to the city … Most of the houses are parted into divers tenements, so they have as many landlords as storeys; and therefore have no dependence on one another …'*[32]

Tenement houses were populated by a range of classes. One commentator noted that the houses were shared by the rich and poor alike, although '*while the gentry and better sort of people dwelt in fifth and sixth storeys*'[33], rents varied within the same tenement. This approach was even set to rhyme in the 1830s:

> *'You may call on a friend of note, and discover him*
> *With a shoemaker under, a stay maker over him,*
> *My dwelling begins with a periwig maker:*
> *I'm under a corn cutter, over a baker;*
> *Above the chiropodist; cooker too;*
> *O'er that is a Laundress – e'er is a Jew:*
> *A painter and a tailor divide the eight flat,*
> *And a dancing academy thrives over that!'*[34]

The maintenance of building standards was the responsibility of the Dean of Guild

30 David Hume, *An Enquiry Concerning Human Understanding*, note 7.

31 Indeed, we are still living with people who use misogyny as a political tool.

32 Peter Robinson (2005), 'Edinburgh – a tenement city?', in Edwards and Jenkins (eds), *Edinburgh: The Making of a Capital City*, pages 106–8.

33 From Henry Graham, *The Social Life of Scotland in the Eighteenth Century*, quoted in 'Edinburgh – a tenement city?' by Peter Robinson, page 113.

34 *Philadelphia American Courier* quoted in *Castles of Edinburgh* by John Heiton, 1861.

Courts. An Edinburgh case brought to the Dean of Guild Court in 1775 concerned access to clean and maintain a chimneyhead:[35]

> *'The petitioner owns the third story of a tenement built by Alexander Fleming, wright upon an area marked H in the plan of the new Town. His complaint is against William McConochie, owner of the fourth and garret floors, who refuses to allow him access to heighten his vents or have them swept. The Court ordered the defender to allow access or to take down the additional superstructure he had built. The process also contains depositions from tradesmen, owners and chimney sweeps viz.: John Hamilton, chimney sweep who keeps eight boys to sweep and that his boys can sweep a chimney 8 inches square. He is also employed by Messrs. Butter, Cowan and Brown, wrights to clear out any loose lime in new chimneys. In a deposition by the other owners it is suggested that the system of using boys is safer than the use of a brush, pulled up and down by two men. There is a measure of the height from the pavement to the upper bed of the cornice by Alexander Ponton, measurer who refers to the property, Flemings land, on the corner with Queen Street.'*[36]

By the 17th century land had become subdivided into many smaller plots.[37] Scotland had three- to four-storey tenements (flats sharing a common entrance), whereas England, Wales and Ireland built, in the main, two-storey terraced housing. Legal, technical and geological influences explain the differences in early housing construction preferences between England and Scotland and also illustrate a stronger European tradition that influences Scotland's built heritage.

The influence of classical architecture was profound in the Georgian era and found itself expressed in features such as stone architraves around windows, pediments and parapets, window spacing and proportions, the use of ashlar on frontages, plinths and base courses which were often rusticated or banded.

The period was largely a reflection of new-found wealth, requiring the owner's status to be shown publicly in the classical form with doorways and entrances, closes, lobbies and stairs as well as internal layouts and finishes such as decorative cornices and roses, window design and shutters, skirtings and door architraves. Once the Georgian period arrived there was more of tendency for the richer tenants to relocate to more salubrious areas, possibly to single-family Georgian terraces or ideally to the more upmarket tenements.

From the late 1760s onwards, the professional and business classes gradually deserted the Old Town in favour of the more desirable 'one-family' residences of the New Town in Edin-

35 Dean of Guild Courts received many petitions, this one from 1775:
'WARRANT GRANTED: 19 July 1775.
PETITION LODGED: 3 July 1775.
PETITIONER: James Ranken, wright in Edinburgh.
SITE: Princes Street.
PROPOSAL: Complaint about access to vents.'

36 In Edinburgh Dean of Guild Court Part January to December 1775. BOX: 1775/30 warrant granted 19 July 1775.

37 John G. Harrison, 'Archival evidence and urban vernacular houses', in *Vernacular Building* 42, page 7.

burgh, with separate attic or basement accommodation for domestic servants.[38] The same migration occurred in Glasgow where the rich Tobacco Lords migrated to terraces and villas in Blythswood and the West End. The same pattern of development occurred in Stirling, Dundee, Perth, Aberdeen and smaller towns. Building increased as a result of the increased income to investors in the slave trade, who then invested in housing which attracted a steady rental income.

By the middle of the 17th century in Edinburgh, buildings were built higher within the city walls, and the tenement, by then almost universally built of stone with slate roofs, had become the standard form of housing. There are also examples of a new pattern of flatted courtyard dwelling, the earliest of which is Mylnes Court (1690s) which still stands and which involved the demolition of old multi-storey structures and buildings laterally across the narrow feus to make room for three massive tenements facing a forecourt. The narrow feus also meant that as fore-stairs were removed, they could only be replaced by a stair set just behind the front wall, as can still be seen in some High Street tenements in Edinburgh. The rents for fifth and sixth storeys in this period were often higher than flats on lower levels, and rented to 'better sort of people'.[39]

Around 1700, the stone would be quite roughly hewn, sometimes with margins around windows where stones would be cut and dressed to form *inbands* and *outbands*[40] to create a uniform opening for the windows. The rough rubble stone would then be lime harled and lime washed. Tenements could be quite high.

Glasgow also had some high tenements. Due to the risk of fire and collapse, the height was reduced in 1698 to six storeys, and eventually in 1862, to four storeys in Glasgow, although by then Glasgow had a few six-storey tenements. Six-storey tenements could still be built after 1862, because the tenement height was constrained by the width of the street it was on, so if the tenement was sited on an open corner, it could still rise to six storeys.

Early tenements generally got their stone from local quarries as transport was expensive. The Georgians adopted a classical style and sought to impress people by using finished ashlar to the frontages, as in the New Town of Edinburgh. This created an increasing demand for good-quality stone. By the 1750s, they would still have used rubble walls which were *lime harled*, but also ashlar stone margins, sills, lintels, quoins, pediments, cornices, string courses and plinths were all in greater demand for more expensive properties. By around 1730, lime harling fell out of fashion and ashlar faced tenements became more common although rubble stone was still used on gables and the rear of tenements. Mylnes Court in Edinburgh, built in 1690, had an ashlar front.

From 1830 onwards, there was an influx of people from Ireland escaping the famine, and from the Highlands, who were all seeking work in towns. Tenements in the central area which

38 This migration changed the social character of Edinburgh, which Robert Chambers, writing in the 1820s, described as: 'a kind of double city – first, an ancient and picturesque hill-built one, occupied chiefly by the humbler classes; and second, an elegant modern one, of much regularity of aspect, and possessed almost as exclusively by the more refined portion of society.' Traditions of Edinburgh (1825), page 8.

39 Around 1830 the very highest flats could be up to 12 to 14 storeys and lowest flats were 'possessed by artificers, while the gentry and better class of people dwelt in fifth and sixth storeys'. Steep slopes were a factor here, as normally the tenement was built higher, and below the ground also had tenants. *The Social Life of Scotland in the Eighteenth Century* (1833), by Henry Grey Graham.

40 Inbands were formed at window and door openings which went deep into the wall and alternated with shallower outbands; their purpose was to tie the wall together (see Chapter 7 windows section).

[left] Mylne's Court Building was improved in 1960; unfortunately they decided to close in underground rooms

[right] Blackfriars Wynd, Cowgate, Edinburgh, 1856. Photograph taken by Thomas Keith who was a gynaecologist surgeon working in Edinburgh

used to be lived in by the rich, had to be made down[41] to provide accommodation for poorer tenants; this was the main reason slums developed and were associated with tenement living. (Large tenements in the Gorbals in Glasgow and Dumbiedykes in Edinburgh were 'made down' to let them to more people.)

In Georgian times dressed ashlar stone[42] became fashionable in Edinburgh, rather than rubble stone and harling, or as in England, brick and stucco. From the 1800s, and for prestigious buildings, a stone *ashlar* finish was preferred. More emphasis was placed on creating a sense of place (or maybe social status), as front facades became faces to the street, windows were arranged symmetrically in a classical manner and base courses were often *rusticated*. The front stonework was formed from evenly cut and smoothly dressed '*ashlar*' stones, sourced from the best quarries.[43] The rear and gables of tenements were built using

41 A term to describe the conversion of family houses into a number of single rooms which would then be let to a family.

42 Ashlar stone is stone which has been hand dressed to a rectangular format and tightly bedded with thin mortar joints. Usually about 300mm high by 400–600mm broad.

43 In Edinburgh this was Craigleith Quarry, or Hailes Quarry.

16 Buccleuch Street, built in 1872 with rubble walls and ashlar sills and lintels

coursed and random rubble which was roughly hewn, but not dressed smoothly. Sills and lintels were simply formed. What tends to be different is that the Georgian tenements we see today were largely planned to take richer people out of the densely packed inner-city areas that existed in the 1600s and early 1700s, but poorer Georgian tenements also existed.

Edinburgh New Town was designed to mimic large London mansions, and some ground floor flats certainly were 'mansions' with a basement and first floor level (some containing 12 rooms).

However, when the burghs started to expand rapidly from the 1800s, despite such regulations, town layouts failed to maintain the health of the population. In 1801, Glasgow, Edinburgh, Dundee and Aberdeen contained 11 per cent of the country's total population. By 1901, this figure had grown to 35 per cent of a much larger population. The population of Glasgow alone grew by 350 per cent between 1801 and 1841 – partly due to the Highland Clearances.

Wider feus were introduced to burghs after 1760 which allowed four- and five-storey tenements with central staircases to be lit from a cupola above. These formed the typical layout for Edinburgh New Town and other developments further north. Stairs that were placed in the centre of a plan had to be lit from above.

There was a great mix of people living in such tenements, but as more salubrious houses were built in new areas, the richer tenants moved out and, as previously mentioned, the larger flats were *'made down'* to accommodate a greater number of people.

[left] Lower Fleshmarket Close. Drawing by Marcus Patton.

[below] Section through typical Edinburgh New Town tenement

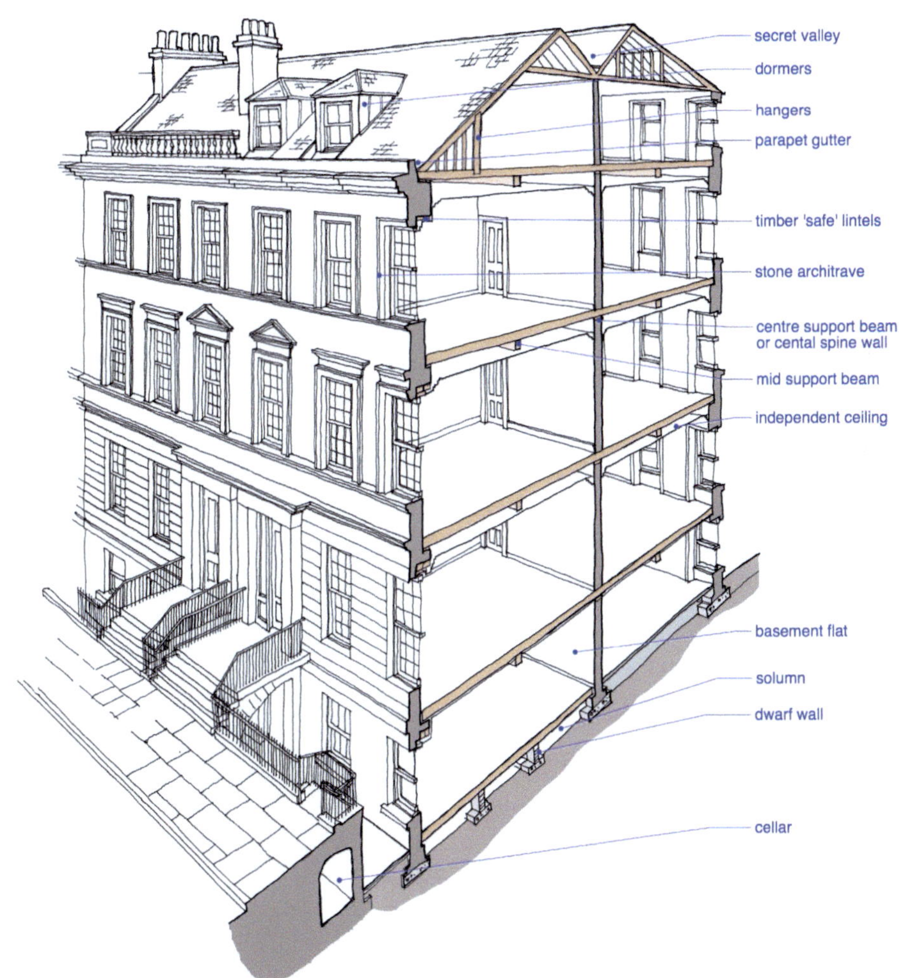

Around 1810, most people were living in one room, although it might have been part of a larger flat. However, houses of different sizes to cater for different markets were being built, so some tenements might have one and two bedrooms, and other tenements might have six rooms and even a bathroom by around 1840.

Staircases in Edinburgh's New Town were usually at the front or in the centre and top lit. The plan form would be mostly curved or D-shaped, resulting in incredibly elegant hanging stairs where each stair tread stone was set into the close wall and cast-iron balusters pinned the outer edge together. Some stairs were 60 ft high from the basement to the top landing.

Roofs which were slated were less steep[44] so the ridge level was lower, so we often get parapet gutters and cornices which hide the slated roof which in Edinburgh was usually quite low. Perhaps because the length of the timbers used was limited (ships of that period could only carry limited lengths of timber), the layout of Georgian tenements tended to be more cellular and roof timbers shorter in length; this encouraged roofs to be made in two parts with a central valley.[45] A large timber beam would support the joists and rafters at midpoint (usually hidden under the plaster ceiling which was supported on hangers from the joists, chapter 3).

'Sash and case' windows[46] were employed with sashes containing six or nine panes in Georgian times. The glass would have been crown glass which was formed by blowing a circle of glass and forming small sheets which had a distinctive radial pattern. In 1845, tax duty on glass was reduced and this lowered the price of glass by 75 per cent. This increased demand, and a cylinder sheet process[47] was introduced in 1834. Much of this cylinder glass was in use by 1903 when machine drawn cylinder sheet glass was used. In Victorian times this allowed larger sheets of glass to be used.

As the Georgian period ended, there was more demand for housing and speculators would build a variety of types, to cater for people with different incomes and class groups. Some tenements contained main door flats with only upper floor flats accessed by the close. In Edinburgh and Glasgow it was common to have three flats per floor. However, cheaper tenements for working class housing in Edinburgh would often be built four flats to a landing with each flat facing to either front or rear.

The classical Georgian window pattern gave way to more emphasis on the principal room, which in Georgian Edinburgh might have two or three windows, and could be to the front or rear. Parlours were lesser rooms, but other rooms in these New Town flats might be accessed from a private stair. In Edinburgh New Town and in places like Queen's Crescent in Glasgow, this front space was often formed by a retaining wall to allow for access to a basement storey, usually needed because the site being built on was hilly, although not always; sometimes it was simply reserved for servants' quarters. In general, the Georgians did not build terraced houses; they simply built larger, more comfortable tenements. Ground floor, main door flats

---

44 Downplayed, for classical reasons.

45 See Roof Construction. Chapter 5.

46 Sash and case windows appeared in Scotland at the end of the 17th century. The main innovation was to apply pulleys and weights to timber sashes. The design became popular very quickly, so people had them installed all over Scotland.

47 Cylinder sheet glass was in use from 1834 to 1900.

Edinburgh New Town frontage

in Castle Street in Edinburgh had about twelve rooms and the design of the facades was as important as the surrounding area. In New Town designs, the stair was placed at the front, with stair landings crossing the front windows, although in corner tenements it was more common to have the stair lit from a skylight or cupola above. Similar developments occurred in Glasgow, Greenock, Dundee, Perth and Aberdeen, but to a lesser extent.

String courses were a horizontal projecting moulding originally of bonding stones which bonded the inner and outer leaves of stone together to form a compound wall. They also acted to form a projection which helped direct water away from the stone, and hoods, pediments, cornices and sills also performed this function. In Georgian times they are mostly a simple shared edge projection, although in Victorian times the stone was often shaped to provide additional interest, and sometimes was even formed with a drip to ensure water was thrown off the face of the building. Parapets with cornices and pediments enhanced the classical facade.

Although the tenement form developed initially during the 16th and 17th centuries, Glasgow developed the construction over the period from 1750 to 1919. Glasgow in 1760 was transformed by tobacco barons, and the riches of the East continued to flow, exceeding expectations. By 1830, landlords were urbanising, taking advantage of the massive developments in coal and iron exploitation, together with general industrialisation and transport links. To house the workforces required, there would need to be tenements.

The planform that developed – starting with fore-stairs at the front (and the back), then turnpike stairs, then moving the turnpike stair to the rear to satisfy classical sensibilities, and gradually bringing the staircase inside the tenement, often at the rear, but sometimes the front and centre – is explained in 'How tenement plans developed' later in this chapter.

With growth of the cities (in Glasgow westwards and southwards, in Edinburgh, the development of the New Town), it became acceptable for the middle classes to live in a tenement, although initially usually a three- or four-storey development with basement and attic, with a main door access, and often little to distinguish it between three-storey town houses from an earlier period (now often converted into flats).

These early Georgian tenements were usually plain, with the tallest windows on the first floor (the 'principal floor'), the next biggest window on the ground floor and the smallest on the top floor, with the eaves formed by a projecting cornice.

When four-storey tenements were built for the working classes, the first and second floor windows were similar, with more emphasis on the first floor with stone pediments and architraves (often with a blank stone riser panel to increase the overall window height), the second floor with simpler pediments and the, usually smaller, top floor windows the simplest. A cornice or string course might separate the ground floor windows which were simpler and often set in banded or rusticated stone bases.

Victorian period 1837–1901

In Glasgow, the main era of tenement building was between 1850 and 1905. From about 1860 to 1912 there was a huge demand for housing, because of industrial growth and urbanisation. Houses were predominantly rented and often on a yearly lease. From 1801 to 1901 Glasgow's population grew tenfold, from 77,385 to 784,496. Tenements were the physical

Tenement in Royal Cresent, Glasgow

form of Glasgow's working-class housing, dominated by large street blocks. This housing was privately built and rented, by speculative builders and landlords.

The 1845 famine in Ireland brought further migration into Scotland. An increase in housing costs caused more overcrowding. By the 1890s, the cost of land in Glasgow was 25 per cent higher than in Liverpool. The cost of higher building standards in Glasgow also resulted in higher construction costs.

Larger landowners within the burgh and in adjacent burghs started to capitalise on urban growth by setting out feuing plans. Sometimes the standards sought to attract more affluent persons, and higher rents. The feu was granted initially to a builder who then carried out the construction and sold off or rented out the properties. Such 'feu-farming' helped to finance building in Scottish cities. James Gowans (architect and builder) reported that in the 1860s and 1870s: '*A builder looks forward to the town increasing, and he takes up a lot of land from the superiors at £50 an acre, and then by re-feuing or building himself he works it up to £200 an acre. That has been done with this city and large fortunes have been made out of it.*'[48]

Feu duties are succinctly described by Richard Rodger: '*As Feu-Duties were a first charge on a bankrupt's estate they were a coverable and secure basis on which to raise capital. The effect was to encourage builders to offer for sale the right to exact or "farm" inflated feu-duties in return for the lump sum capital advances which they then used to build the prop ties from which the duties would be obtained. Re-feuing in this way certainly added veto to the cost of the land, so much so that the land charges added 10–14 per cent to the gross rental of the tenement property.*'[49]

These feu superiors, as they became, not only set out the physical plans for development but also derived an ongoing income stream from the feu duties paid by each owner (see

48 Richard Rodger (1986), in Doughty, M. (ed), *Building the Industrial City*, page 173.
49 R Rodger, Building the Industrial City, page 173.

Appendix 6). And the feu superior also had powers to enforce the house and building rules they had set out for their developments. These powers could be exercised benignly – the feu superior granting land to the local burgh to build a school but with 'burdens' built in to ensure the land would be returned when no longer used for education. In time, however, the enforcement powers of the feu superior came to be used in increasingly lucrative ways.

However, while this system might have worked for the development of new middle-class housing, the pressures of urban population growth led to a huge health burden on the poorer sections of society. High housing costs resulted in properties being subdivided, and when richer people moved into Georgian terraces and villas and into the New Town of Edinburgh, their older and larger tenemental floor plans were then 'made down'. Tenants also took in lodgers to supplement their incomes. The back lands of the long medieval burgage plots or tofts were often developed into housing too. The high population density created conditions suitable for the spread of infectious diseases. Eventually, the government responded by the passing of the Burgh Police Acts of 1892 onwards. These applied to all housing in the almost 300 burghs across Scotland.

Building legislation in Victorian times was applied through the Dean of Guild Court,[50] applying power through various Glasgow Police Acts (see Appendix 7). They sought to control the maximum number of houses accessed off a close stair and restricted the height of tenements to no more than the width of the street the tenement was sited on. As the streets were about fifty-five feet wide, it meant that tenements had to be no more than four storeys high. However, where tenements were on corners, or adjacent to open space, five storeys were allowed.

Tenements were often laid out in hollow squares, sometimes with smaller tenements in the backlands, so not only was daylight restricted, but ventilation was a problem, particularly as smoke from coal fires was concentrated in these areas and toilets were not part of tenement living, so dungsteads would be formed at the rear. Removal of waste and cleansing were often limited in poorer areas.

Up until 1850 people in cities lived in houses[51] of one or two rooms and were often overcrowded, with external shared toilets (usually a midden in the back court). Internal toilets only became a requirement in 1862[52] and then they were often shared and inserted onto the half landings of close stairs. Later it became compulsory to have separate male and female toilets.

As towns developed in Scotland, the tenement form remained, with closes to the rear and turnpike stairs accessing floors above (and below), and densities increased by building higher. *Turnpike stairs* would provide access to each level. This tenement tradition of living remained, even when the densities became so high that some middle classes moved to suburban areas. But larger middle-class tenements were built around 1890 in Glasgow and earlier in Edinburgh. The principal room, often called a parlour, would have two main windows set close together, with a central mullion, and sometimes one large window with a smaller window on either side (tripartite windows). Two public rooms were sometimes

50 Precursor to the Building Regulations.

51 Flats; in Scotland, a flat is often referred to as a house.

52 The burghs of Scotland and The General Police and Improvement (Scotland) Act 1862 (The Lindsay Act).

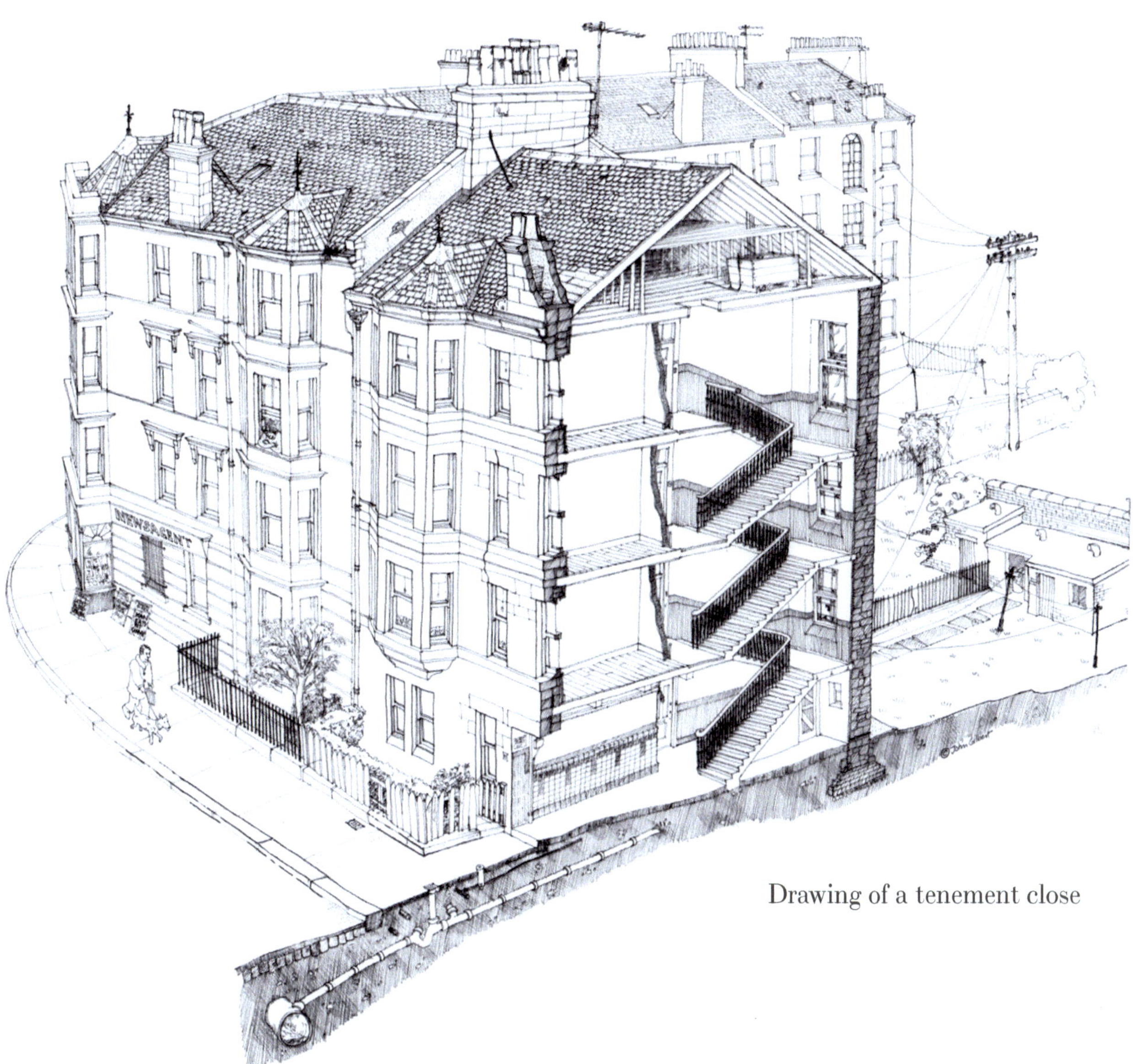

Drawing of a tenement close

provided and with it grew the importance of having two or three windows to each public room. Bay windows developed in this period provided there were also small front gardens or basement areas. Slightly downmarket tenements for artisans were built to the street edge and here oriels could be used to corbel out at first floor level. Rents were low and most closes housed a wide variety of people.

Flats such as Caledonian Mansions in Glasgow with a rear access deck and two escape stairs, built as an annex to Central Station; Hope Street Mansions, also in Glasgow built by Glasgow Corporation, also with rear deck access and a U-shaped layout plan; and Camphill Gate, with flat roofs as drying areas and concrete floors – all were typical of the period after 1900.

Tenements in Glasgow in Victorian times were predominantly four storeys but there were also often five storeys on corners and two and three storeys in some earlier developments. Most had a single pitched roof, mostly with cast-iron gutters, but a limited number of earlier Georgian tenements had dual pitched roofs. Close stairs were mostly at the rear with windows on half landings. Closes were originally without front close doors; tiled closes appeared later. Bed recesses became common in original construction.

Caledonian Mansions

Camphill Gate

Bay windows were common in early tenements, but oriels developed for tenements built to the pavement line without gardens. Also, in nearby towns, there were some two-storey properties with stairs at rear and additional attic flats; these are now mostly demolished, although some still remain in places like Wishaw and Coatbridge.

The height of rooms was typically 12 to 14 feet (higher than in English houses), and the height of the windows let a lot of light into the rooms. The problem in the design of these early facades was that although they were classical, the repetition of the window openings on long stretches of tenements led to a feeling of horizontal monotony. This sort of fenestration was common in the 1830s and 1840s. To avoid this repetitive monotony of windows, the Glasgow approach was to keep the windows the same size but distribute them differently; for example, grouping the windows 1+3+3+1, 1+3+3+1 etc. where the single window was the front close stair window. Eventually windows were coupled with a central stone mullion and a shared lintel, or triple windows with two mullions, a main central window with two smaller windows either side, although this pattern only took hold after 1840 as Victorian ideas overcame the classical influence.

The bow window was used in Hanover Street in Edinburgh by John Young as early as 1780[53] and although the canted bay window was used around 1840 it only became popular around 1880, perhaps due to the increasing demand for tenements to cater for the middle classes and artisans. Bow windows and bay windows were common (mostly appearing in tenements in Marchmont, Bruntsfield, Dalkeith Road and Comely Bank, in Edinburgh), and by the 1890s oriels (bay windows projected from the first floor) were often found, as described above.

However, given that builders were the main drivers, tenements had to be built quickly, so common systems of tenement construction were developed to ensure materials and labourers were used effectively.

Platties and gallery decks

Platties, also known as gallery access flats (and locally called 'pletties'), were built in several cities after 1893, becoming a traditional sight in Dundee, perhaps because Dundonians had been building the same form of deck access tenements since the 1850s. Many platties were built to house the influx into Dundee of jute and engineering workers.

They could be two to four storeys high with a close through to the rear and were accessed via turnpike stairs to individual flats from doors that were always open to the air from the deck. They did not have toilets originally, although these were added later as shared communal toilets accessed off the plattie decks.

After 1890, more gallery access tenements were built in Glasgow at 3–11 Bain Street; at Cathedral Court, Rottenrow; at Hope Street; and at Caledonian Mansions. Edinburgh also had some, such as 6–10 Ramsay Place in Portobello, which are still with us. There were objections at the time because of the loss of sunlight and although balcony decks made with glass blocks were suggested, they were too expensive.[54]

53 John Young (d.1801) learnt his trade as a wright and with his partners became a successful builder in the New Town of Edinburgh.

54 Matthew Withey (2003), 'The Glasgow City Improvement Trust: An Analysis of its Legacy. 1866–1910', unpublished PhD thesis, University of St Andrews.

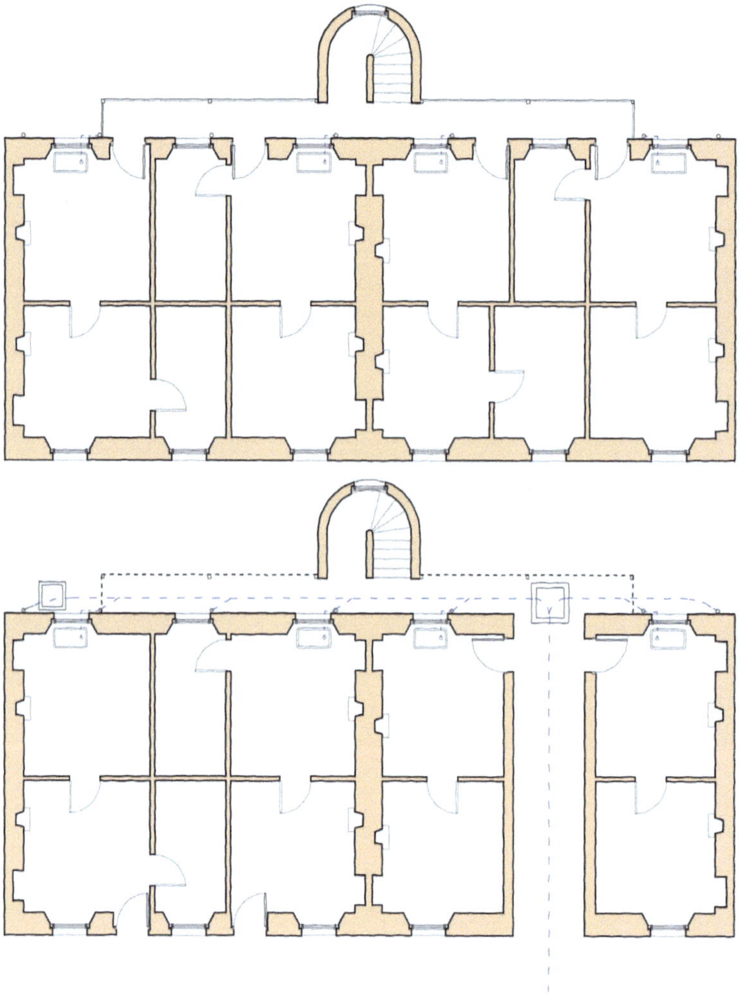

[left] Typical plattie plan at ground and first floors

[right] Dundee plattie

Ramsay Place in Portobello

The Kirk Flats in Glasgow

How health issues affected tenements in the 19th century

Frederick Engels wrote in 1844 that he had: '*seen wretchedness in some of its worse phases both here [in Britain] and upon the Continent, but until I visited the wynds of Glasgow I did not believe that so much crime, misery and disease could exist in any civilised country*' and that '*the poor in Scotland, especially in Edinburgh and Glasgow, are worse off than in any other region of the three kingdoms, and the poorest are not Irish, but Scotch.*'[55]

The form of tenement construction exacerbated bad health. Rows of 60 ft high four-storey tenements built on narrow streets no more than 30 ft wide, denied light and air to residents. So dark were the houses that tenants had to burn gas to light rooms. Also, tenements were often set out in hollow squares with only very narrow openings at the corners, thus preventing light and air from penetrating the houses. Development of the backlands made matters worse as this was where the night soil[56] was kept and collected using carts wheeled through the narrow alleys.

In contrast to housing in England, Scottish tenements, with their shared close and stair, communal wash house and drying greens, shared toilet[57] and refuse arrangements, made it difficult for a Scottish tenant to have any privacy or introspection, compared to a tenant in an English terraced house. However, the main cause of the slums was the vast number of poor people having to move to urban areas to find work and feed themselves. The cities provided such employment, albeit at terribly low wages, so the demand for housing was from families on very low incomes.

Hence squalor and disease were common in areas of high density and serious overcrowding was common. In November 1861 a tenement at Chalmers Close in Edinburgh collapsed killing 35 people, many of whom were living in windowless basement rooms. This collapse did focus

55 Frederick Engels (1887), *The Condition of the Working Class in England*. Written 1844–5, German edition published 1845; English translation 1887.

56 Human faeces or sewage.

57 See Chapters 5 & 9 on drainage and water supply.

the city's minds on the need for public health as well as the structural fabric, but it did not change the way of tenement building except to improve sanitation. Tenement flats were still regarded as the 'Scottish system of building'.

Dr Henry Littlejohn, who recorded the deaths, eventually became the city's first Medical Officer of Health (MOH) and his research and publication in 1865 led to a *Report on the Sanitary Conditions of the City of Edinburgh.* [58] Several thousand copies of Littlejohn's report were published, which led to a number of municipal interventions, particularly to improve drainage and water supply.

After 1860, as previously stated, the influx of immigrants from Ireland and the depopulation of the Highlands led to what were middle-class areas being 'made down' to provide smaller flats which became overcrowded – and eventually slums. Even a large single room might be *made down* to form two rooms, each room housing a family. These made down tenements existed from 1860 to 1910, providing poor-quality accommodation which eventually led to ticketing[59] controls in Glasgow.

Between 1871 and 1911 the population of Scotland's towns and cities grew exponentially, mostly attracting people from rural and highland areas where work was scarce and difficult, and from Ireland, where years of starvation and famine in the middle of the century (known as the Great Hunger, or Potato Famine) led to huge numbers of deaths and mass emigration. Clydebank grew by 275 per cent, Govan by 366 per cent, Hamilton by 236 per cent, Falkirk by 251 per cent and Rutherglen by 163 per cent. In contrast, Edinburgh only grew by 63 per cent as there was no work in industry. Given the rise of population, it is no wonder that in some areas, people (with different ages, sexes and different households) had to share rooms. Internal bathrooms and toilets only became common around 1885 although tenements were still being built with no bathrooms up until 1910 in some areas.

By the end of the 1860s, the medical officer for heath was trying to ensure that there was at least a piped water supply up to the second floor with at least one water tap for a minimum of ten families.

The focus was firmly on working class areas and the Victorians with their Improvement Acts made some inroads into reducing tenement density, but at a cost.[60] Police Acts were made to control public use of the streets and only later did Public Health Acts try to raise living standards and prevent disease.

Sanitary inspectors in Paisley had, since 1890, pursued changes to transform the privy midden system of conveniences into the water carriage or water-closet system. Some small towns in 1905 were described as dry closet towns and places; for example, in Galashiels in the Borders, people used a pail system. There was clearly a link formed that enteric fever[61] occurred when sewage was stored on the premises. Medical officers of health knew of the problem and steps were taken, with some difficulty, to improve water supply, drainage and the provision of WCs. This despite the Act of 1892 which required residents to have access

58 Dr Henry Littlejohn (1865) *Report on the Sanitary Conditions of the City of Edinburgh* page 5.

59 Ticketing was introduced in 1868. A metal plate was fixed to each flat door, along with the number of occupants allowed by law (300 cubic feet for every person over eight years old).

60 See D. J. Johnston-Smith (2019), 'Dislocation and Domicide in Edinburgh, 1950–1975', unpublished PhD thesis, University of Edinburgh: 'We never tried to push people out, unless it was for their own good.' This traces Edinburgh's clearance activities between 1950 and 1973, similar to earlier slum clearance projects.

61 Enteric fever is a type of typhoid fever.

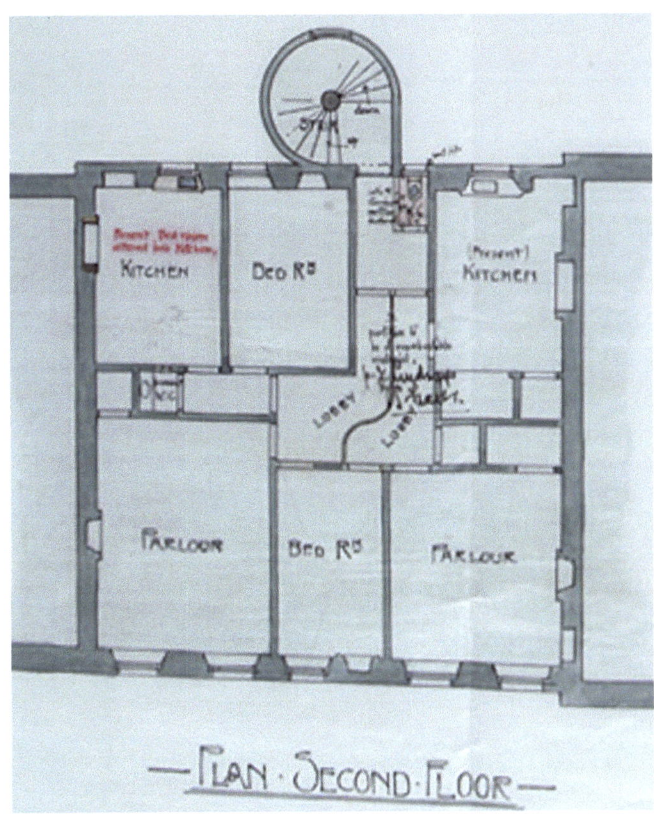

[left] Developers proposal to 'make down' a Gorbals tenement

[above] Ticketing tokens that would be fixed to each entrance

to an internal toilet (albeit shared on a landing), but some areas were still not connected to a drainage system or did not even have a piped internal supply of water.

Investment to build tenements, often came from middle-class people as well as wealthy land speculators. The church also invested in housing as the returns were profitable. After 1890, weak demand for workers' housing meant that building speculatively for the poor was unprofitable. To some degree this was replaced by a demand for more middle-class housing, leaving low-cost housing to be built by cooperatives, trusts and then by councils. Richard Rodger[62] believes that Scottish Building Regulations which required walls at least 2 ft thick prevented early adoption of improvements which would have helped improve public health. It was more difficult to supply water and waste systems to upper floors in a tenement than to a terraced house. However, there may be other reasons for Scotland's more robust construction, not least traditional details built up from years of experience of working in the west of Scotland's harsh climate as well as the materials to hand.

Wages were low in Scotland around 1900 so rents were all low, which led to lack of investment in maintenance and new building. As Richard Rodger explains: "*In a sense … the market economy had not failed, it was a success*';[63] but it led to appalling conditions suffered by the poor. Builders and landlords accordingly reduced their involvement in housing, and only cooperative ventures succeeded in building working class houses in this period. The condition of tenements declined after 1900 as investment switched to more lucrative schemes which commanded higher rentals.

62 Richard Rodger (1986), in Doughty, M. (ed.), *Building the Industrial City*, page 157, from a quote taken from 'Royal Commission on the Housing of the Industrial Population of Scotland, Rural and Urban', 1917, para 481.

63 Richard Rodger (1986), 'Victorian building industry and the housing of the working class', in Doughty, M. (ed.) *Building the Industrial City*, page 197.

Nightsoil buildings in Paisley

In 1902 the function of the Municipal Housing Commission allowed investment in the City Improvement Trusts which placed land in the council's ownership as a modest start to council housing. This was seen as a threat to landholders, factors and house agents who formed a Citizens Union to campaign against increased municipalisation. After 1904 there was a rapid rise in prices which affected labour costs and changes in Building Regulations affecting the cost of materials compounded by an increase in interest rates.

The growing discontent of tenants towards yearly lets, rent levels and sequestration of goods eventually led to the Glasgow Rent Strike of 1915. After rent controls were introduced in 1915 the private rental market was dead and was eventually replaced after the First World War with housing provided by publicly funded schemes.

Whilst we need to understand how tenement slums developed, we also need to recognise that many good tenements were built.

After the First World War

Inter-war years 1919–39

Building between 1920 and 1924 supplied a limited number of homes based on the Garden Cities movement which were known as 'Homes for Heroes'. From 1930 to 1939, homes were funded by governments but quality was reduced; three-storey tenements and side ends were common. The coming of the motor car and the consequent suburban sprawl with massive urban upheaval and a lack of investment in tenement stock led to decay.[64]

Post Second World War 1945–57, but most building from 1950 to 1957

Post-war reconstruction, development of peripheral estates with limited facilities.

64 Councils tended to blame tenants for the decay, as they were often rehoused from tenements which had to be cleared due to the slum clearance programme.

Later post-war 1957–76

System building and the start of high-rise and deck access flats.

In the 1950–70 period, tenements still had a bad name and extensive plans were made for wholesale clearance with the development of peripheral estates. Comprehensive Development Areas, combined with massive road building projects, also resulted in a massive demolition of tenemental areas. Few tenements escaped. Those that did were confined to urban pockets, where tenements were either in *grey areas*, which mortgage lenders avoided, or in middle-class areas.

Modern 1976–2023

The modern era including Low Carbon Buildings, multi-storey and deck access flats.

After it was shown that tenements without a bath could be improved, a movement amongst inner city communities grew which eventually led to the formation of community-based housing associations, firstly in Glasgow, but eventually growing in Glasgow and throughout the rest of Scotland.

Local communities formed to make use of environmental Improvement Grants and backcourts were improved, then tenement blocks were stone cleaned.[65] People began to see the tenement in a new light.

New flats to modernise our streets

Most of the tenement properties were densely built. Converting these to new flats would make getting to office or even working from home less of a strain; densely built housing means that people are closer to shops and culture. The whole of suburban development has been the dependency of a car. It's time to rule out the car, or have its place underground.

65 Some stone cleaning, particularly with sand blasting and acid cleaning, was very poorly executed and damaged the stone. It is best avoided wherever possible.

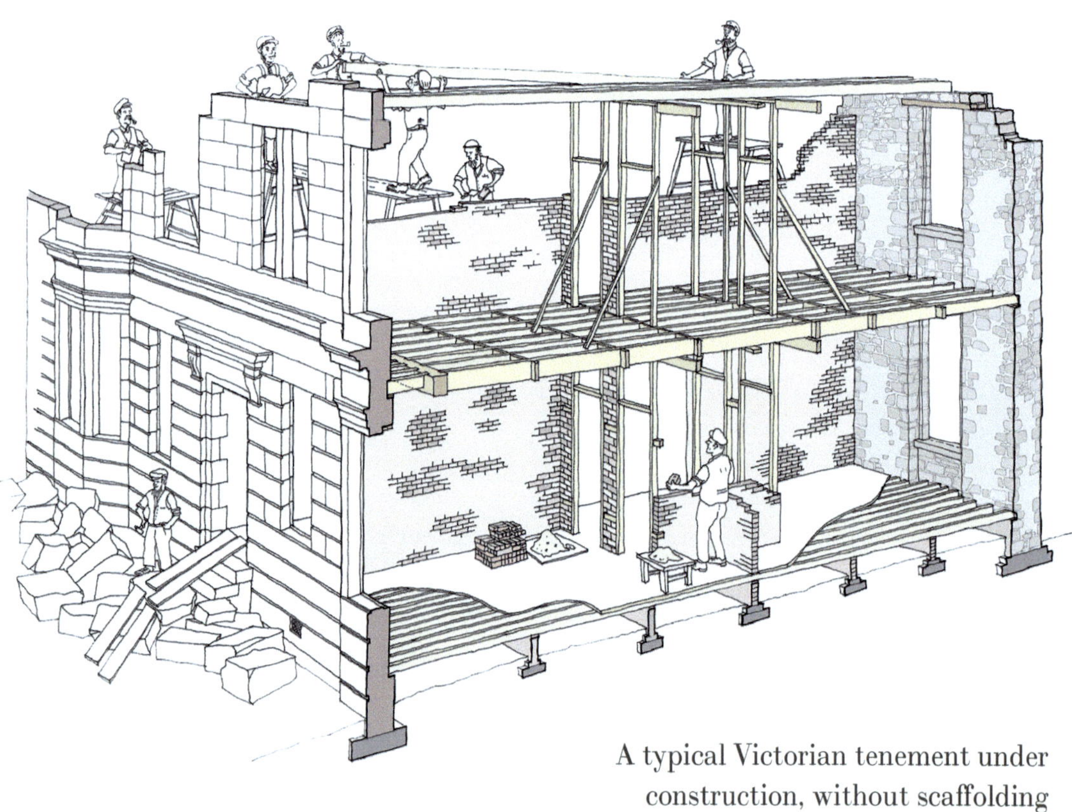

A typical Victorian tenement under construction, without scaffolding

Chapter 2
Building and construction of tenements, access ways and plans

In the Victorian period, tenement builders were typically small operators; over half of the house building in most burghs was undertaken as a one-off project. Tenement building teams sometimes consisted of only ten to twelve men. There was a large element of standardisation and tenements were built with speed[1] and efficiency. Stone mullions, lintels and sills would largely be cut in masons' yards and then transported to the site by horse and cart.

Most Victorian tenement builders appeared to build the external walls using each floor plate as a working platform. They would still employ cranes and jibs to move the stonework, but the final stone pointing would have to be done overhand. Localised scaffolding[2] to finish the gutters at eaves level and corner turrets might have been erected off the top floor and cantilevered out. Scaffolding would be used in some places, mainly on frontages where ashlar stones required some precision in their placement, but in lower status tenements scaffolding was rarely used.

Typical stonemasons and wrights

Between 1800 and 1850, most tenement work was done by stonemasons or wrights working together. Eventually such construction was carried out by builders, with masons employed to cut and dress stone and wrights doing the timber work. In Glasgow, the largest firm in the1820s was owned by Alexander Green who employed 187 masons, although the business collapsed in 1825. Some builders, such as Thomas Binnie and Robert Aitken, had their own quarries.

Around 1860, brick was more extensively used in tenement construction and builders who were also brickmakers expanded into tenement building.

After the famine in Ireland in 1845, a lot of Irish men came to work in Scotland as labourers. The influx of immigrants placed an increasing demand on housing and tenements, in particular.[3]

1 Thomas Binnie completed a four-storey tenement in High Street, Bridgeton, Glasgow, in 1819; he started building it mid-March and tenants moved in on the 28 May. In other words, it took about 11 weeks to complete. (*Memoirs of Thomas Binnie. Builder of Glasgow 1792–1867.*)

2 'Bracket' scaffolding projects from the top floor windows.

3 '[T]he ashlar is produced in square stones, 6 or 7 inches high and two or three feet long … Two Irishmen trembling under its weight, and at risk of their lives, carry it up the scaffold and deposit it beside the builders.' Alexander Greek Thomson, *The Light of Truth and Beauty*, page 45.

A team of tenement builders taking a break

Virtually all the traditional tenement housing was privately built and rented out by speculative builders and landlords. Tenements of three, four and, in some cases, five or more storeys were common in Scotland and blocks of tenements were to be found 'from Dumfries in the South to Lerwick in the far North'.[4] Two-storey properties with a workplace at ground level and one or two dwellings above, entered via a fore-stair or rear stair, were already common in Scotland and likely the precursor of the tenement style of living. Builders and masons often drew up tenement plans, and some even prepared working drawings. The builder Thomas Binnie drew up his tenement plans 'which were frequently adopted by other builders'.[5]

Journeymen[6] often formed the majority of the workforce. They were not freemen who were incorporated into the burgh, but hired hands who could travel and were not confined to the local area, and usually paid by the day. There was a 'Society of Journeymen' in 1700 in Edinburgh but wages were low and the working day lasted for twelve hours, six days a week.

One reason for the absence of records around 1790–1830 is the 'Combination Act' which made it dangerous to keep any records of local trade union clubs in the building industry. Edinburgh masons formed a 'combination' in 1764 and struck for higher wages. The magistrates and council responded saying their conduct was 'illegal, tumultuous and unwarrantable' and ordered them to return to work at such wages as 'the said masters shall think reasonable'.[7]

4 The Royal Commission on the Housing of the Industrial Population in Scotland (1917). Query 2232.

5 *Memoirs of Thomas Binnie. Builder of Glasgow 1792–1867.*

6 'Journeymen' possibly derived from *journée*, meaning 'whole day' in French.

7 *Scottish Journal of Topography, Antiquities, Traditions*, vols 1–2, 1848.

The first Combination Act[8] was introduced by Pitt in 1799, leading to an uproar from employers, so an amendment to the Act was then passed in 1800 and this remained in force until 1824 when it was repealed. The law was draconian; it sentenced any working man who combined with another to gain an increase in wages to two or three months of hard labour in gaol. The sentence was not given by a jury but by two magistrates who would belong to the employing class. Most tradesmen held meetings as Friendly Societies in pubs, but some took particular steps to ensure any minutes were secretly recorded.[9] George Wilde wrote: 'I find that the Society of Ironfounders which began in 1810 used to meet on dark nights on the peak moors and wastes on the highlands of the Midland Counties, and the books etc., of the Society were buried in the ground.'[10]

It was not until the 18th century that we find anything more than ephemeral combinations of journeymen, and indeed there are very limited traces of them before 1800. It was after this date that the capitalist system, as we call it, spread all over England and Scotland.

The repeal of the Act was down to Francis Place, a London tailor, who provided the MP Mr Hume with information and arguments to support his case.[11] Unions were formed again after 1825, but mostly on particular trades and although they had oaths and subscriptions, few were successful in improving working conditions.

The United Operative Masons' Association of Scotland was started in 1831 and Glasgow lodges were involved in a strike in 1833, but the Association soon ran out of money and had to appeal to England for money. Societies existed for bricklayers, carpenters, plumbers, glaziers and slaters, but administration of these societies was poor. There was a 'nine hours a day' movement which resulted in a strike and lock out in 1859, but the strike resulted in many societies folding, after which a more conciliatory approach was adopted.

It was a busy period; in the UK in 1831 imports of products amounted to £97,623,332 and by 1860 this had risen by 283 per cent to £373,491,000.

By 1830 Aberdeen masons had prevented stone being worked in the quarries and had even got a short Saturday, but, after a dispute with employers, they went bankrupt in 1870. Masons were still the main builders in Scotland because of the use of stone, although they did bricklaying as well and eventually turned to being 'builders'. At the time of a building boom in 1877, the masons had 13,759 members in 116 lodges. By1894 the Scottish operative masons had 8,224 members and £6,500 in the bank, by 1898 they had 12,000 members. However, they were poorly managed, and the secretary, James Craig, was old and sick.[12] The Edinburgh Lodge of masons created a sensation by striking in 1897 for an eight-hour day. However, they found that as other trades could not follow them, they did not start later or knock off earlier, thus losing an hour's pay under the new arrangement.

---

8 When two or more tradesmen combined to raise wages or to withdraw their labour, they were said to be 'combining'.

9 The Falkirk Society excluded any 'drunkard, swearer, or sabbath breaker'.

10 R. W. Postgate (1923), *The Builders' History*, page 17.

11 Postgate has a very full chapter on the repeal of the Combination Acts, as Francis Place explains: 'Taxes, machinery, laws against combinations, the will of the masters, the conduct of the magistrates, these were fundamental causes of all their sorrows and frustrations', pages 35–54.

12 His small salary was essential to his household, so his stepdaughter, Elizabeth Henderson, did all the correspondence, interpreted the rules and formed the policies. Craig was simply the figurehead. In 1895 she was taken from work to give birth to her child. Lest the deceit be discovered, she rose from bed three days after giving birth to prepare Craig's papers for a meeting and died, due to the strain on her.

A painting by John Atkinson Grimshaw on the Clyde, 1881

Just after the boom, the City of Glasgow Bank failed in 1878 with losses of £12 million and many builders and trades were ruined, as work ceased for seven years.[13] The only traders who were little affected were the slaters as they mostly did small jobbing work repairing roofs. Thomas Binnie (senior) lost an investment of £1,000 in the crash but was able to recover and by 1882 he had leased the brig Devron which imported timber from North America.

Having a particular trade at these times might affect your life expectancy. The average age at death of a plumber in 1890 was only 37 years, perhaps because they worked with lead or had to sort out drains which were associated with typhoid and cholera.

Masons, when they reached 40, were generally troubled with a cough, caused by stone dust. There is anecdotal evidence that it led to an epidemic of silicosis or tuberculosis among the stonemasons during the development of the New Town. The Royal College of Surgeons of Edinburgh studied the preserved lung of a contemporary stonemason who had worked on Edinburgh New Town, and confirmed the presence of silico-tuberculosis in it. The evidence showed that a major epidemic did occur, caused by a combination of factors. The size of the project attracted many stonemasons to work in Edinburgh over a period of almost 100 years, intensively cutting and dressing stone. The principal stone worked was Craigleith sandstone, a very high-quartz sandstone having properties that gave a high standard of finish to the ashlar. However, although stonemasons appeared to be aware of the risks of their trade, little was known about preventive measures and protective wear and good ventilation were often absent.

Changes in mechanisation and materials also led to a loss of craft trades, although some innovations simply reduced material costs. Machine-made nails were available after 1830. Brickmaking from 1830 used machines to grind clay, brick pressing, extruding clay and wire cutting, with the tax on bricks ceasing by 1850. Stone cutting and polishing, then concrete, affected the masons. Lead being replaced with asphalt for roofs and the introduction of electric light affected the plumbers (they used to install the copper feeds for the gas lights). Wrights were affected by the introduction of cast iron from 1830. Plastering skills were less required when fibrous plaster was introduced after 1856. Steel construction after 1850 affected most building trades.

Competitive building contracts were in use in Edinburgh by 1814 and commonly used in Scotland by 1830, although if a builder was also an investor in the tenement project, this rarely applied.

13 After the bank crash in 1878, builders were often laid off and tradesmen would sometime drop tools and go to the tenement across the street, for the tiniest of pay increases. There is some evidence that around this time corners were cut, such as limiting the amount of dwangs installed between floor joists.

Risk of fire and preventive measures

Because of numerous fires, timber framed walls (and *thatch*) were banned[14] and tenements had to be built in stone at least 3 ft thick. After a fire in Edinburgh in 1824 changes were introduced, banning the use of thatch in urban areas. To prevent fires travelling from one tenement to the adjoining one, via timber joists, floor joists were only allowed to span from front to rear walls,[15] although they could still be supported on internal bed recess walls and additional mid-support beams. From about 1850, long floor joists[16] would span between front and back walls and be fully bedded into the outer walls at about 15 inch (380mm) centres. These long joists would be temporarily supported on *H-frames* (the internal door frames), and intermediate timber beams across bed recess openings, to support the joists whilst the tenement was being built. The internal walls would then be built up in brickwork between the H-frames to strengthen the structure and support the bed recess beams. (See 'the humble door frame', Chapter 7.)

By the 1860s[17] fire precautions required that the party walls should rise through the roof to be about a foot[18] above any slating. This was also to prevent fire spread between tenements. These are the skews which have a stone coping and at the base have a *club skew* to prevent the skew copings from sliding down.

Victorian Glasgow faced significant fire risks, yet rather than compelling builders and architects to incorporate fireproofing into new buildings, municipal authorities tended to intervene in the built environment through public health legislation, establishing, in Glasgow, the city's Improvement Trust to undertake comprehensive slum clearing from 1866. The municipality preferred to invest in its fire extinction capabilities, including providing water on the high-pressure gravitation system, which, combined with an extensive fire hydrant network, meant that firemen could connect

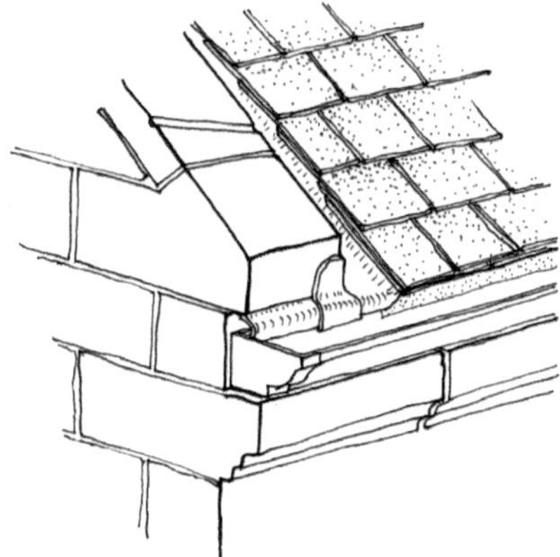

Club skew stops

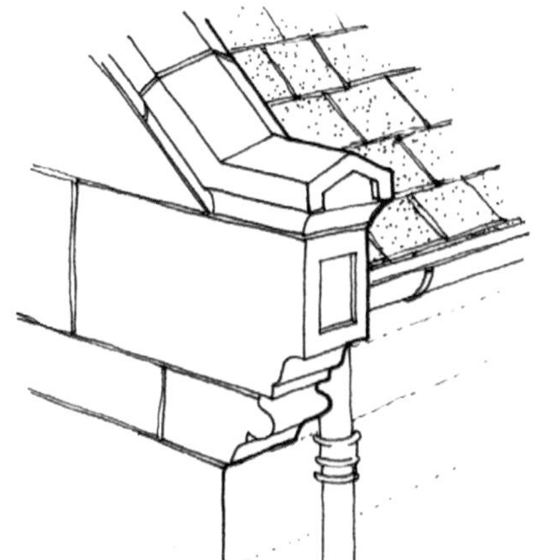

Projecting club skew

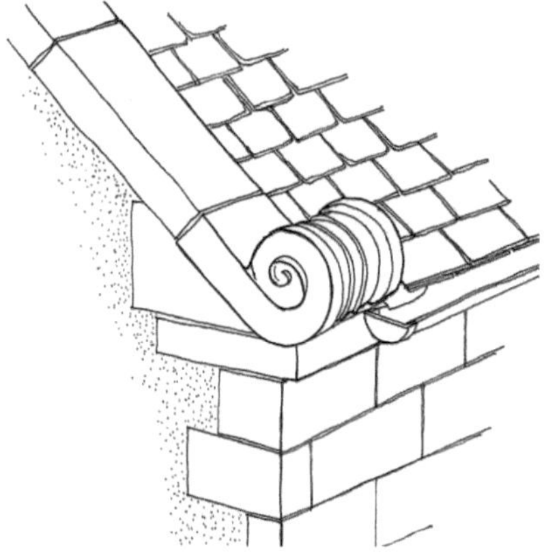

Scrolled skewputt

14 The Towns Improvement Clauses Act, 1847, stipulated the carrying of party walls through and above roofs to reduce the threat of fire spread, as well as the construction of walls and roofs of incombustible materials, while the Burgh Police (Scotland) Acts, 1850 and 1862, contained similar measures.

15 It was found that joists embedded in party walls led to fire spread to adjoining tenements, hence joists were only to span to outer walls.

16 Most of the timber was imported, often from Latvia, Norway or Russia, and so longer lengths of timber became possible.

17 Towns Improvement Clauses Act, 1847 Para 109.

18 In Edinburgh, the 1879 Police Act required the party wall to rise two feet above the roof timbers (clause 164).

their hose directly to a hydrant without having to use an engine in most outbreaks.[19] Of course, this non-intervention in building regulations could backfire, as it did when a strike of Glasgow's plasterers in 1876 caused practically all the buildings erected in the city during the tenement-building boom of 1873–7 to be lined with wood throughout, which significantly intensified the risk of combustion, a large proportion of them burning down shortly thereafter. Moreover, the feudal land system and propensity of small-scale building firms, combined with the working-class preference for cheap rented housing, meant that minimum regulations were difficult to enforce as they drove up construction and rental costs. The absence of a thoroughly coordinated insurance presence before the 1870s weakened any attempt to influence tenement building regulations in Scotland.

Given that houses were increasingly heated with coal fires in the 1860s and 1870s there were numerous fires caused by the adjacent floor timbers which supported the fireplace hearths. When the stone hearths heated up, the adjacent timber joists caught fire. A proposal in 1879 to change the Glasgow Building Regulations was defeated by landlords and architects who were concerned at the increased cost implications. It was not until the 1892 Glasgow Building Regulations were enacted that timber joists and bridles had to be kept clear of stone hearths by building brick supporting arches. (See Chapter 3 on chimneys and fireplaces.)

The same 1892 Act, further strengthened in 1900 with the prescribed installation of fire-resisting divisions in large mixed-use buildings, certainly recognised that fire was a critical threat to the city's rampant unregulated expansion, yet it also targeted sanitary improvements to working-class tenements. By the turn of the 20th century, Glasgow enjoyed some of the most demanding building regulations in Britain, insisting upon firm foundations, thick walls, stiff roofing materials and a high load-bearing capacity in tenements and warehouses alike. These regulations had the effect of reducing the destructive capacity of fires, which were more likely to be contained within a single room, rather than spread to other rooms or to adjoining properties. However, a number of experts such as Thomas Binnie (son of Thomas Binnie) gave evidence[201]to the municipality's commission on the housing of the poor in 1904 and argued that '*the strict enforcement of building regulations intended to make buildings more substantial and healthier*',[21] but raising building standards and improving sanitation had inevitably increased the construction and rental costs of working-class housing.

Builders were largely the main designers of tenements, often copying one another's plans. Architects in the early period rarely designed tenements, possibly because four-storey tenements were initially built for the working classes, although some builders became proficient and encouraged their sons to become architects and design tenements (like Charles Wilson). But most architects did some tenement work and some practices specialised in tenements, some raising the standard of tenement design to new heights such as John Burnet, James Salmon, 'Greek' Thomson and John Rhind.

19 C. Steven (1975), *Proud Record: The Story of the Glasgow Fire Service*, Glasgow: Glasgow Fire Services, page 37.

20 Thomas Binnie (senior) built a four-storey tenement in Main Street, Bridgeton in 1819. It was started mid-March and occupied by the end of May that year, two and a half months later. He did the mason work himself using his father's quarry and sub-contracted all other trades. His motto was 'Thorough'. From Thomas Binnie Memoirs of Thomas Binnie, 1882.

21 Report of the Royal Commission on the Housing of the Town and Rural areas of Scotland, 1917, pages 16–40.

Builders at work – note the exclusion of main scaffold

The 1892 Police Act introduced many improvements, not least to fire protection, and it also introduced gallery access flats and the requirement for internal sanitation.

The other big change in society was the introduction and spread of the railway system, which allowed much greater movement of people and materials. Red sandstone could be brought easily from Ayrshire and Dumfriesshire to sites in Glasgow and Edinburgh. The railways also turbocharged the export and import of materials.

The form of tenement plans developed differently in Glasgow and Edinburgh. Although Glasgow did have internal top-lit close stairs[22] and often single ends (small single aspect flats facing the front), the depth of the tenement tended to be limited to the roof span. In Edinburgh, tenements were slightly deeper, perhaps because of the tendency to extend to the rear of tenements, or because, being developed in the Georgian period, they liked to have a classical frontage and to keep the roof pitches low, resulting in the M-type roof. However, it may also have been because Edinburgh tenements developed in an earlier period when it was more difficult to source longer lengths of timber from the Baltic because of a levy of 275 per cent in 1807 on Baltic timber imports which lasted until 1820, although it's as likely that traditional construction from earlier periods tended to use shorter spans and mid supports which made Edinburgh tenements slightly deeper on plan.

22 Pollokshields, also Holyrood Crescent in Hyndland, have centrally placed close stairs and both have single pitched roofs although Glasgow does have some M-type pitched roofs, but they often tend to be Georgian terraces which have been converted into tenement flats.

Typical Edinburgh tenements developed, each floor having two flats front facing and two flats rear facing (sometimes six to a landing) as standard. The roof[23] would be formed in two pitches with a central valley, whereas in Glasgow, one pitched roof would be more common. However, tenements in both cities had roughly similar depths in the region of 40 ft (11.5–12.5 metres), though some Edinburgh tenements tended to be deeper, around 46 ft (14 metres).

Bay windows became a feature from around 1880 and oriel windows from the 1890s. Tenement flats were larger, each flat with two rooms and no single ends. Innovation was in the air after 1892 as Cathedral Court in Glasgow, a gallery access tenement, was built with cavity brick walls which were rendered externally and plastered on the hard inside.

Development of fore-stairs and turnpike stairs 1750–1919

Tenements were not always four storeys high. The ground floor rooms became used for workplaces, eventually shops, and the upstairs rooms then were developed to have separate accesses up external stairs. In the 15th century these might have been via 'fore-stairs' at the front, or through a pend at the back and then up external stairs.

In two-storey houses, people and livestock entered from the rear and access was either through a close, wynd or a courtyard, as depicted in Pringle's painting, but the form is still found in many small towns today. Tenement plans developed as towns and cities grew, often affected by Burgh Police Acts, Police Acts, bye-laws and Building Regulations.

As demand for housing grew, tenements rose higher, usually with stone construction at ground floor level and timber framed upper floors. Access would then be from a *turnpike* stair (at front or rear) into a rather narrow passage to serve each house. The plan form might have had several rooms, although there were no toilets, as a water supply was by a carrier and drainage was rudimentary, sometimes draining to the street or placed in a dungstead. When tenements were timber, turnpike stairs were also made from wood, although none of these remain today.

Some of the old stone turnpike stairs can still be seen in Edinburgh, Stirling, Dundee, Montrose, Brechin, Perth, Paisley and other towns. Early turnpike stairs would initially have been square edged steps but the sharp underside arris might be dressed back to provide headroom.

Front turnpike stair in Stirling

23 See section on 'Roofs' Chapter 5.

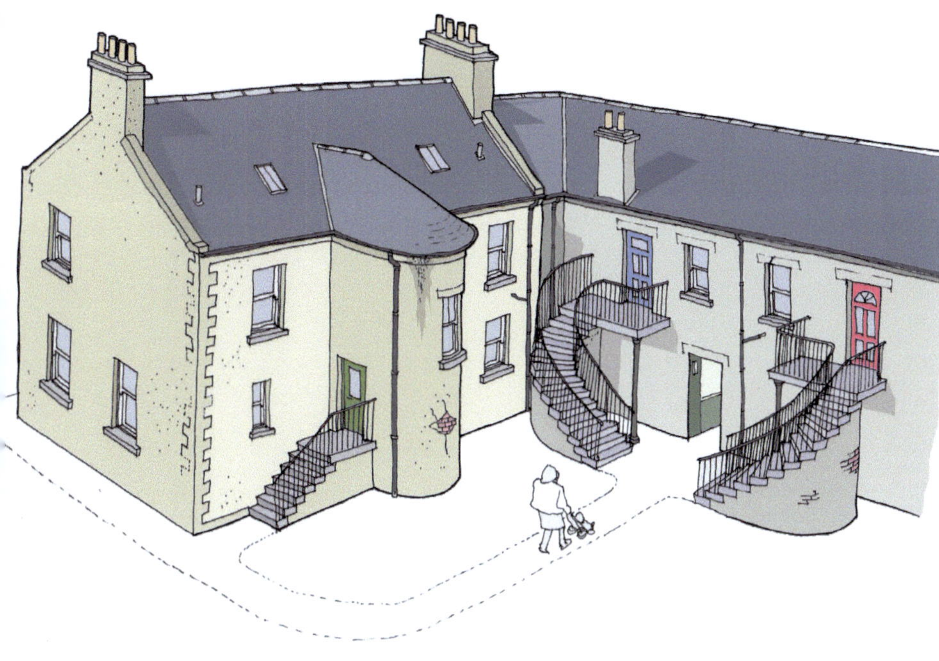

[above] Photograph of rear access stairs in Jedburgh

[upper left] Courtyard of White Horse Inn, Canongate

[middle left] John Pringle painting of old houses in Parkhead, Glasgow 1893

[lower left] Drawing of rear court in Coatbridge

Underside view

[above] The start of the spiral stairs

[right] How it looks in elevation, spiral stairs between two straight stairs.

[left] Front turnpike stair in Perth

[right] Front turnpike stair in Edinburgh

[above] Front turnpike stair in Oban

[upper right] Rear turnpike stair with Ionic columns

[right] Rear entrance in Stirling

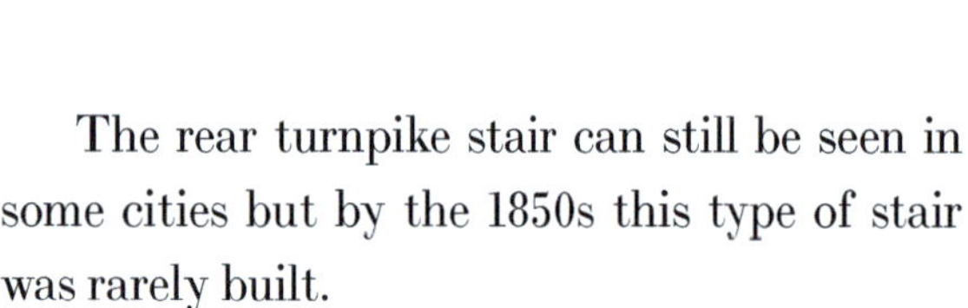

The rear turnpike stair can still be seen in some cities but by the 1850s this type of stair was rarely built.

Early turnpike stairs, as we have seen, were built at the front of tenements with access sometimes outside, sometimes within the close. However, the Georgians considered this untidy as it did not relate to any classical order that they wanted to emulate, so after about 1700 the turnpike stair was moved to the rear. There would be a close (or a pend) which led to the rear of the property where the turnpike stair gave access to the upper floors.[24] The turnpike

24 Edmund Burt, an army officer based in Inverness, wrote about building conditions in 1727–8: 'By the way, they call a "Floor" a house; the whole building is called "a Land"; an alley, is a "closs"; a round staircase, a "Tunpike"; and a square one goes by the name of a "Scale Stair". In this Town, the Houses are so differently modelled, they cannot be brought under any general Description; but commonly, the back Part, or one End, is turned towards the street, and you pass by it through a short Alley into a little courtyard, to ascend by stairs above the first storey. This lowest Stage of the Building has a Door toward the street, and serves for a shop or a warehouse, but has no communication with the rest.'

In the old Gorbals

stair, when moved to the back, often had an enhanced entrance accentuated with stone pillars or stone pilasters and entablatures to show the importance of the houses above.

These stone turnpike stairs gradually moved inside – to the rear of the tenement envelope, first as two straight flights with the return treads formed as radial treads (around 1850–60). Eventually the staircase became two straight flights and a half landing with treads spanning between the close wall and a central spine wall forming a skale stair.[25]

In the Victorian period most tenements placed the internal stairs at the rear with a stairhead window, unless it was a corner tenement when it would be lit from a skylight and the stair would have an open well to let in the light. In Edinburgh, as the plan form tended to favour houses facing one way, to the street and to the rear, the stair would be located centrally with a cupola or skylight to light the stairs.

The external walls of the turnpike stair were often lime harled. The treads would have been formed in individual stones, chamfered on the underside and forming, when stacked

25 A skale stair is a rectangular plan with two straight flights and either a half landing or angled treads at the turn.

[above] Turnpike stairs made in brick

[left] Not in or out, turnpike stair

[opposite] Stairs found in an early skale stair, the windings of the stairs

together, a central stone column, or newel and each tread built into the turnpike stair walls. In the 1850s they were also called wheeling stairs and when used in spires and turrets could be very tight.

The turnpike stair walls were originally built in stone, then eventually built in brick. For some early stone tenements, two turnpike stairs might be accessed through one close. Although the turnpike stair had ceased to be common after 1850, it lingered on in Glasgow and Dundee and was eventually used as the access for external gallery decks ('platties') when the law made it possible.

The plans below show how the turnpike stair developed after it was moved to the rear, gradually coming inside the envelope of the tenement, then with radial treads returning to join two straight flights with a central spine wall, and finally with two straight flights and a half landing forming an open well stair. The stair with two straight flights would commonly have a central stone spine wall which supported the individual stone treads. The return treads were initially formed as angles or wheeled treads, similar to the older turnpike treads. These stairs had no need of a handrail or banisters except on the top floor landing. In order to get more light into the close, open well stairs were developed.

The stair at the rear was gradually moved inside the close

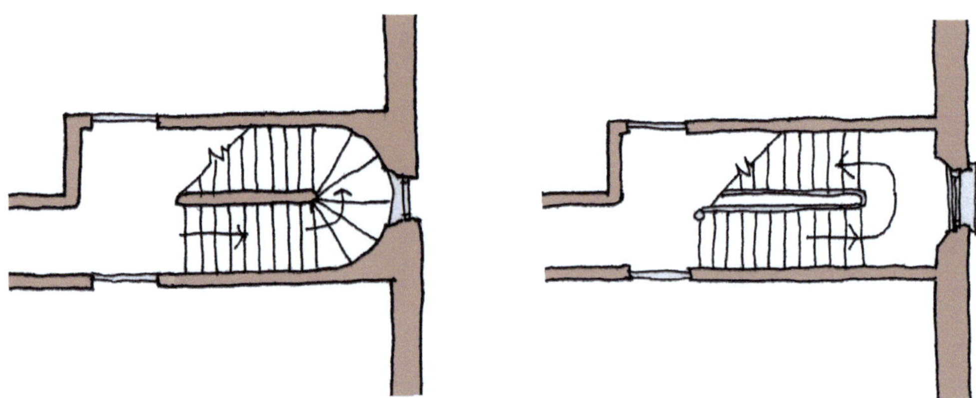

How the rear turnpike stair gradually moved into the close. By 1860 it was inside, but the faceted treads still formed the return

A set of plans

A set of building plans are provided to illustrate each of the core development stages which the tenement has undergone over time. This material acts as the book's visual grounding, before the detailed examining of different technical aspect of tenement construction, and their subsequent development and refinement over time is undertaken, element by element. We start by examining the external built form, the wall construction and stone that was used, its structure and the different roof forms that developed, before later moving to the close, the doors and windows and other significant features.

Because of the demand for housing, tenements might be built to a variety of plans. In early periods, some buildings would have several apartments to cater for richer tenants, but as the demand for workers grew in the industrial revolution, the demand changed to providing smaller houses with one or two apartments. The typical tenement layout developed after 1880 was to create a rectangular layout with four corners and a backcourt for each tenement. Sometimes these backcourts were developed with workshops and tenements were built around them.

Tenements eventually morphed into two two-apartment flats and a single end on each floor, although without toilets for the working classes. After 1892, the number of houses off a common stair was limited and water was provided. Some had internal and shared water closets. This, along with associated changes, such as the phasing out of wooden box-beds and common access corridors, and the provision of a separate scullery in many small flats from the 1880s diminished, but did not eliminate the shared use of domestic space.

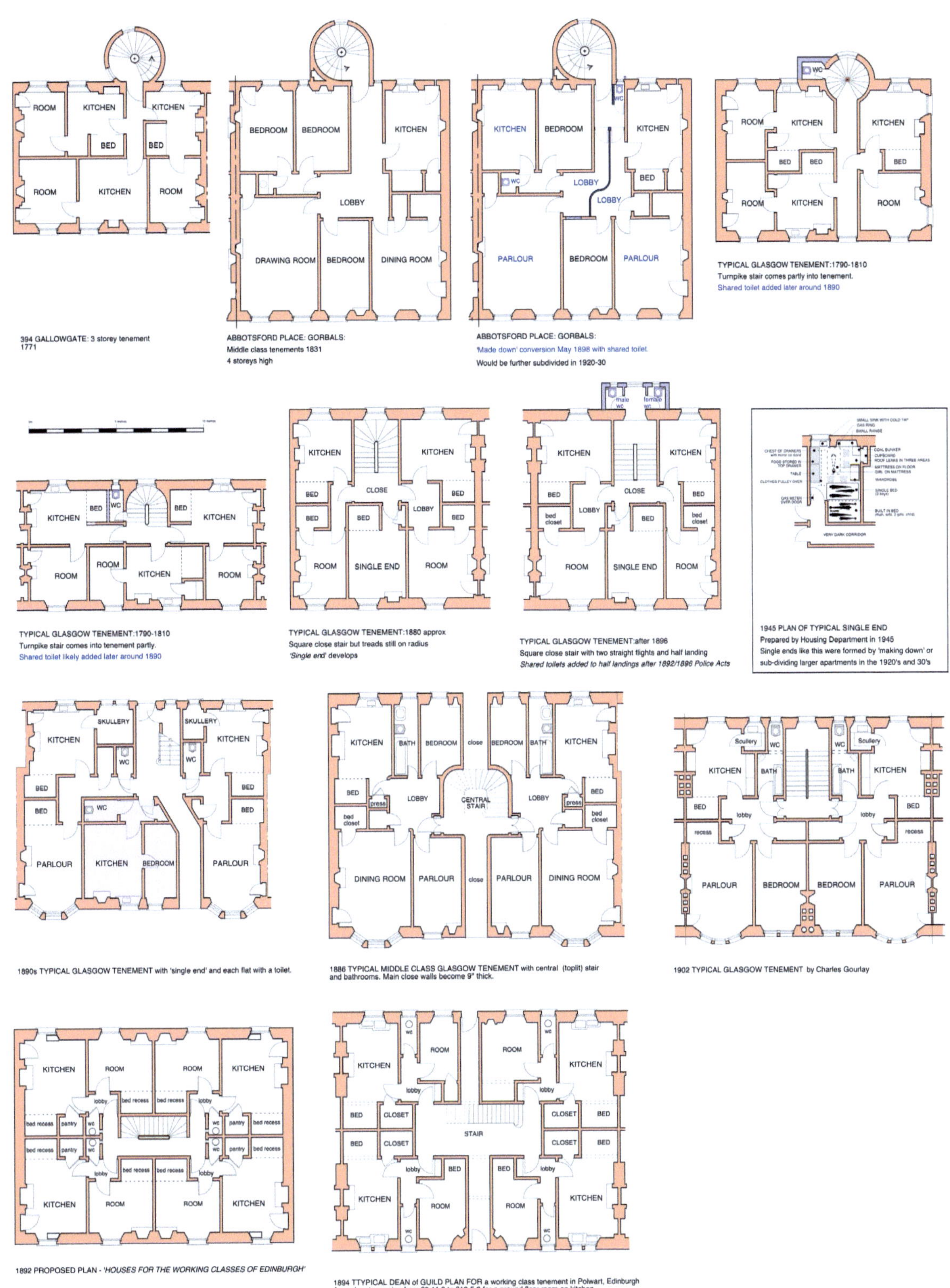

Change of tenement layout over time

Tenement under construction

Mullions being delivered to site

Chapter 3

External stone walls, chimneys and floor and ceiling construction

External stone wall construction: rubble walls

In early medieval and 18th-century times, tenements were largely built of rubble stone and then harled with a lime (or clay) mortar. In towns, where a tenement could be extended, a timber platform might be added to project over the street, the walls might then be infilled with wattle (interwoven twigs) and daub (clay mixed with straw). The rubble stone would be from a local quarry where possible. Different forms of window and door openings might be formed, sometimes with incised margins, sometimes simply with lintels and sills. (See window openings below.)

Whilst the outer surface of the walls were made from blocks of ashlar or rubble, the whole wall was about 2ft 6in. (550–600mm) thick and formed as a compound wall rather than a solid stone wall, simply described as two skins of masonry with rubble fill. The external face was usually a stone that was rendered; only much later would a broach stone be used on the inner wall (of uncoursed rubble) to tie the two walls together.

If the outer walls were built without any kind of lateral restraint, the higher they rose, the more unstable the wall would become. In earlier construction, joists would have spanned the shortest distance between load-bearing walls, usually parallel to the street frontage, using large balk timber joists, covered with thick floorboards. In the 16th century, a series of beams would support floor boards and there would have been no ceiling except the underside of the beams and floorboards (planks) which in upper class houses might have been decorated[1] on the underside.

In the 17th and 18th centuries room depths were constrained, but eventually plots were developed to add to the rear of properties. Additional joists would be used supported on the original rear walls. As timber construction above the ground floor stonework was common, timber joists would extend over the stone wall to create a bigger room on the upper floors; this was called a jettied floor.

Photos taken in Victorian times show that builders operated as small teams, that no

1 See Floors and Ceilings section later in this chapter.

[above] 17th-century floor, Provand's Lordship, Glasgow

[right] Tenement under construction

scaffolding was used (where possible) and that stones were pre-cut and delivered by a horse and cart to each site.

Harling

When walls were built using rubble stone, a clay or lime mortar might be used. The rubble stone was then harled. Lime harling is *'thrown'*[2] onto the rubble wall. It provides a weather protective and decorative coating and usually has a rough finish. They often used hot-mix lime mortars which provided better adhesion to stone walls. In the 1890s the Victorians tended to remove the harling, as they preferred to see the stone. After 1910, ordinary Portland cement was often used in the harling mix. It helped the set, and it was thought to be more weatherproof. However, as the mortar was largely impervious and much stronger than the backing, the render often cracked allowing moisture in. The moisture would freeze and cause more damage, throwing the render off, and as the stonework behind was unable to dry out, damaged the stone. Sulphate attack, differential movement, and high suction from the backing all lead to cracking.

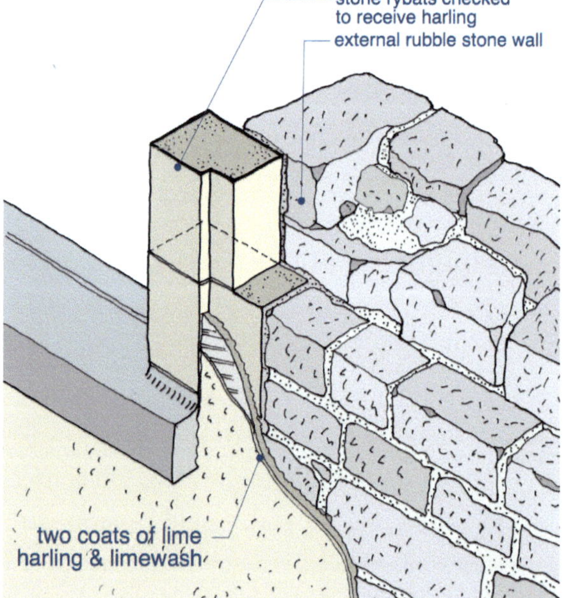

Stone rybats with a check to allow the lime harling to be dressed up to the stone surround

2 Thrown or hurled, hence 'harling'. In Ireland it is called 'wet dashing' and in England it is known as 'roughcast'.

The lime harling would also be lime washed to provide additional protection, but lime wash needed to be reapplied about every five years.

The advantage of harling is that it is a lime mortar (ideally hot-lime) thrown on and not trowel applied, so this provides better adhesion with the stonework. The lime mix[3] would be applied in one or two coats to a thickness of about 10mm. An initial straightening coat would be applied first, then scratched before the finishing coat is applied. The mix would consist of quicklime mixed with coarse varied aggregates, slaked with water and used whilst still warm.[4]

The important principle is that the external coating should be weaker than the background material, so weaker lime mortars are usually best at ensuring there is vapour transfusion through the wall. Stronger lime mortars may be specified in areas of high exposure.

Limewashing

Bo'ness lime harling and lime wash

When the existing rubble wall stone is friable, then it may be best to give the wall a few coats of lime wash before applying the harling.

Depending on how exposed a wall is will determine how many coats of lime wash should be applied. The lime wash acts as a sacrificial layer to the lime harling. Some lime washes may require as much as six coats, although this depends on the *limewash*. It is better to add several thin coats than a thicker coat. A typical mix might be one part lime putty to three to six parts of water, and the wall should be pre-wetted before lime wash is applied.

The lime wash might be mixed with natural earth pigments to provide a colour. The yellow/orange lime wash is usually created by adding ferrous sulphate to the lime putty and water mix. Venetian Red limewashes that were used in the 18th and 19th centuries simply required adding Venetian Red pigment to the limewash. To increase durability and water resistance, additives can be incorporated into the limewash such as linseed oil, tallow, casein or special proprietary products.

Ashlar walls

In Georgian and Victorian tenements, stone walls are about 2ft 6in. (500 to 600mm) thick, with an inner face and an outer face. These 'solid stone' walls are not really solid walls, they are composite walls. Any voids between the stones were filled with stone chippings, brick fragments, small pieces of rubble and lime mortar. The mortar used was lime mortar which beds the two stone skins so that individual stones are not touching one another, but simply held

3 Typically the mix would be 1 part lime to 3 parts sharp sand. It is often advisable to discuss any lime mix with the Scottish Lime Centre.

4 Referred to as a 'hot lime' mortar.

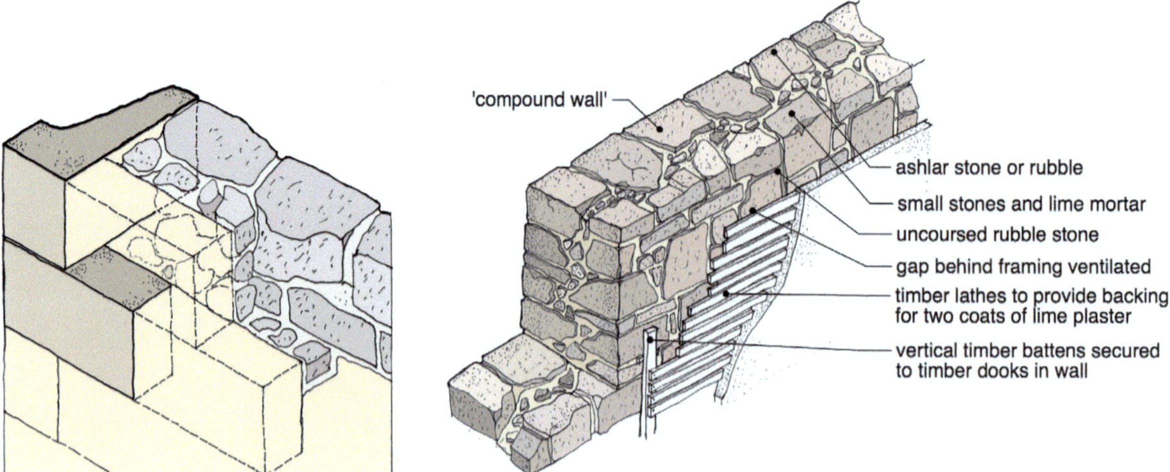

[left] Bonding stones were often used at window openings as well as through the wall

[right] A typical compound stone wall, ashlar to the front and rubble inside

in place. However, the rubble filling may have voids and the smaller rubble stones may touch one another, particularly when the lime mortar is roughly placed after the two skins are laid.

The outer face to the front is usually smooth ashlar and the inside face is rubble stone. At the rear, the outer face may be a coursed rubble stone with the inside face again a rubble stone.

These 'composite' stone walls absorb moisture, depending on the porosity of the stone and the degree of weathering. The stone should dry out naturally, but if the stone is pointed with cement mortar, or coated in a cementitious layer, or sealed in any way, then it cannot dry out. Frost can then attack the moisture laden stone.

Bonding stones should be installed about every other course to ensure the outer stone is tied into the inner leaf. These occur mostly at the window openings where stones are cut to allow for a check to accept the sash and case windows – the larger stones (as seen from the outside) are called '*outbands*' and the smaller stones are called '*inbands*'[5] (see Chapter 7 on windows). The wall thickness below the windows reduces to about 10–12in. (200–300mm) and would originally have been given a coat of lime plaster before any linings were installed.

Joists are bedded into pockets formed in the outer wall and may be set on embedded timber wall plates. In later (and better) tenements, terracotta pockets are sometimes used. Window openings are often formed with timber (safe lintels) to support the load of the masonry above (steel lintels would have been used after about 1910). Embedded timber joists and timber (safe lintels)[6] in external stone walls are always at risk from increased moisture levels, particularly if the stone wall cannot dry out. If left, it can lead to structural failure.

Stone quarries used to vary from city to city, and initially blocks of stone might have been cut into ashlar blocks by masons at each quarry. Stone blocks, lintels, sills and mullions would be transported by horse and cart to the site although, if the quarry was distant, ships via sea routes and barges using canals would be used. By the 1830s, granite from Aberdeen

5 Inbands and outbands are not to be confused with quoins which are staggered corner stones which usually project to become flush with any external harling.

6 Safe lintels: usually formed of one or two sections of pitch or red pine timbers bedded into the wall. Known as safe lintels because they 'saved' the other woodwork which formed the lintel of the door or window below.

[left] Victorian parapet in Glasgow

[right] Georgian parapet in the old town

quarries was being shipped not just to England but to places all over the world. It was not until Scotland had a reliable railway system after 1860 that stone could be transported from places like Dumfries, into Glasgow, although a horse and cart would still be used to get to the building site.

Ashlar stones were dressed to a fine finish and polished by masons in dressing sheds.[7] Rubble stone was more roughly dressed and used for garden walls, internal faces of external tenement walls and the rear face and gables of tenements. The quarry at Hailes, in Edinburgh, was known to produce high-quality rubble sandstone.

Precision was required to ensure the ashlar stones were cut evenly so they would bed correctly. Usually, it is just the front face that is accurately dressed: the stone behind will be much more uneven and any lime mortar bedding the stones will be of a much greater thickness inside than at the exposed face. The ashlar stones are pointed with a finely ground white lime putty, applied after the blocks are laid. It should be noted that lime mortar is designed to keep the stone apart from its neighbours, and not just to bond the stones together. Cement tends to glue the stones together and detrimentally prevents any moisture in the stone from evaporating. These compound walls are generally two foot thick, although at bays and oriels, the outer walls are much thinner.[8]

Stone parapets and stone cornice gutters were often formed to suit the Georgian taste, sometimes with stone balusters which were then capped with stone copings. Balusters might be secured into place with dowelled wrought iron rods and the coping stones were often tied together with inset wrought iron 'dogs' (gunmetal or bronze staples were used around 1900, eventually replaced with steel which rusted) which were then leaded into place. Slate bedded in mortar was also used to join stone elements. The Victorians also used parapets, usually with lead valleys (or dressed in asphalt) behind them.

7 Stone cutting and polishing machines were in greater use after 1880.

8 Alexander 'Greek' Thomson strongly argues for solid stone walls and not compound walls. He thinks there are tenements in older parts of the city, four and five storeys high, which are built of ten-inch stones, without showing any leaning.

When stone cornice gutters were formed, the stone sections would be sealed together with asphalt; in some cases the stone gutter may have been asphalted as well. The detail is prone to failure and most stone gutters no longer have the capacity to take the amount of water that falls on Scottish roofs.[9]

Scots baronial style

This developed from 1850 and was in use up to about 1930. Originally it may have been influenced by Flemish masons brought to Scotland to work on royal palaces in medieval times but the style was revived in 1850 and was soon adopted as a unique Scottish style. The stonework was usually coursed rubble with ashlar quoins and window dressings. Corbelled turrets[10] and crow stepped gables were commonly used, also with steep mansard slate roofs with dormers which echoed the style of the 17th century.

In Victorian construction, joists span in one piece front to back and are bedded into pockets set in the stone wall. The joists help to tie the front and back walls together. Embedded timbers such as joists, wall plates and safe lintel can be at risk from rot if the stone wall is constantly wet.

The head of the wall is usually finished with a capping stone which may be about 100mm thick and projects to carry the front gutter. On top of this stone is placed a timber *wall plate* to provide a seat for the rafters and ceiling joists.[11]

The rear walls usually have a similar capping stone, but it may only project slightly. Weakness of outer walls often occurs at chimney heads as flue bridges often provide a poor bond to the inner and outer leaves of stone, so they can separate (see chimneys in this chapter).

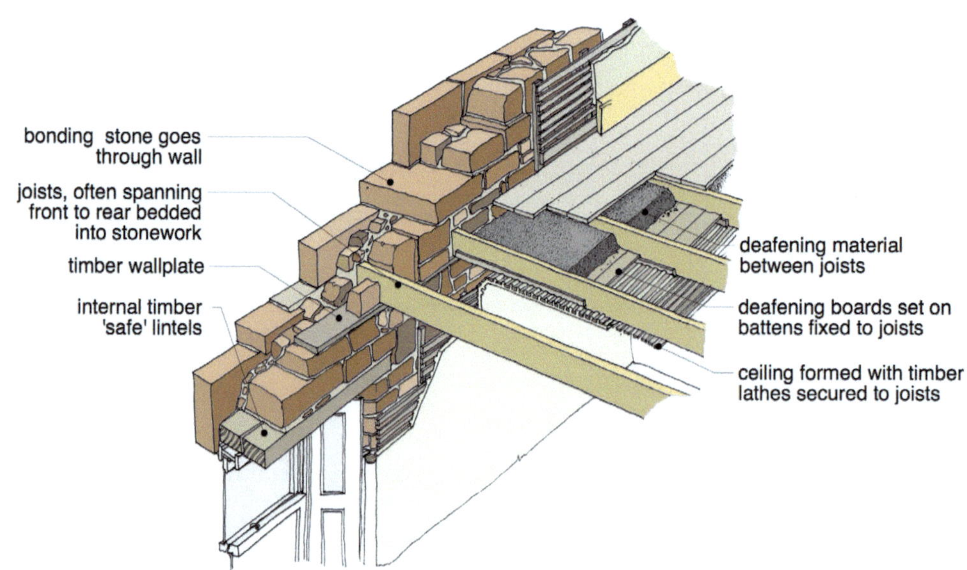

Floor joists are bedded into the front and rear walls

9 Such gutters can be increased in capacity by installing a raised section (metal or timber) and leading over the gutter, using a T-Pren expansion joint. However, such expansion joints have a limited life.

10 Also called *'tourelles'* or *'bartizans'*.

11 See Chapter 5 Roofs.

Window openings

Openings are formed in the external stone walls to allow windows and doors to be installed. Each side of the window opening (or window reveal) is formed of individual stones cut to bond into the outer walls. The sill usually projects at the base and allows water to be shed away from the stone, and the lintel (usually stone) carries the load of the wall above, although some of the load is also carried by any internal safe lintel. In tenement construction, the external walls with openings would be formed as the wall was built up, usually without external scaffolding, and then when the roof was on and the floors down, the windows, which would be made in a workshop and sized to suit, were then brought to site and installed. The construction of window openings has changed with styles of architecture, as so much of the construction was to create a certain appearance, although the 1892 Police Acts did require a certain minimum area of daylighting in a room.

In medieval times, the walls would have been either rubble stone or timber frame with wattle and daub infill. The stone lintel and sills would often be flush with the stone and the whole external surface would be lime harled.

In Renaissance and early Georgian periods stone rubble was still used to build the walls which would then be lime harled, with the harling taken into the stone reveals. Lime wash

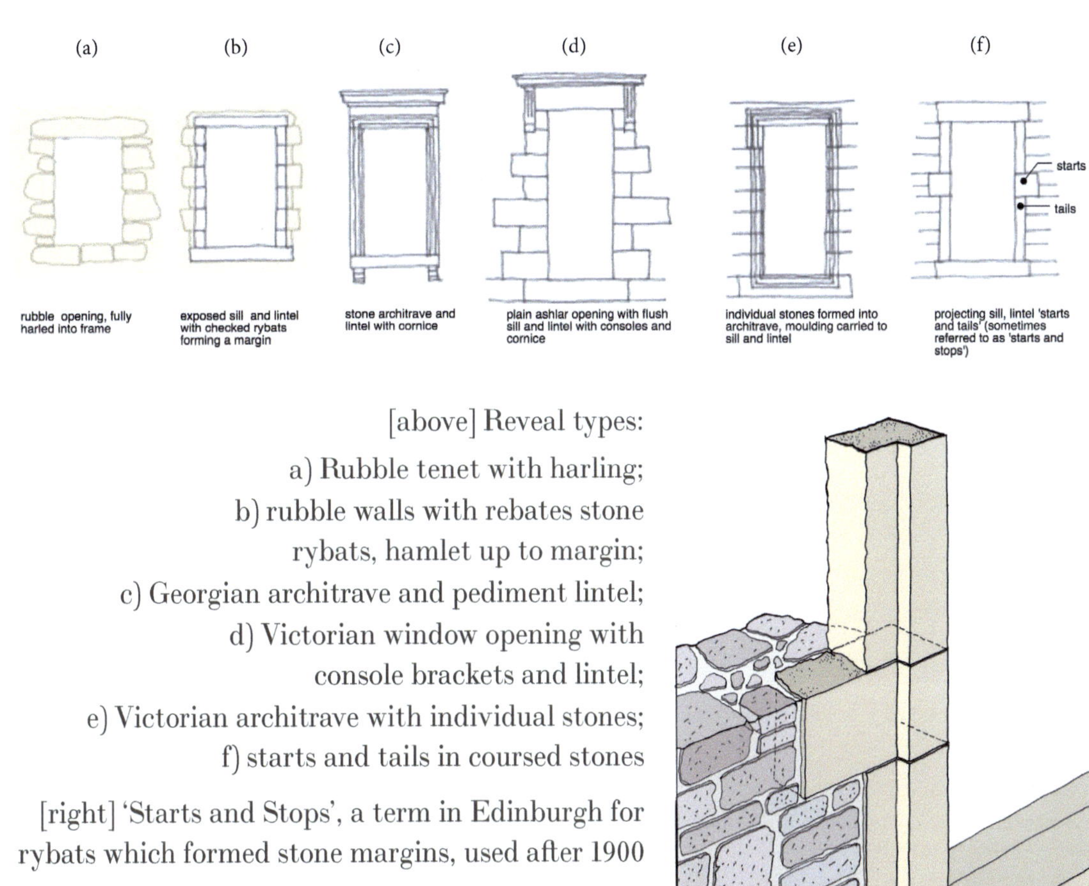

[above] Reveal types:
a) Rubble tenet with harling;
b) rubble walls with rebates stone rybats, hamlet up to margin;
c) Georgian architrave and pediment lintel;
d) Victorian window opening with console brackets and lintel;
e) Victorian architrave with individual stones;
f) starts and tails in coursed stones

[right] 'Starts and Stops', a term in Edinburgh for rybats which formed stone margins, used after 1900

was applied over sill, lintel and rybats[12] to create a very monolithic wall surface. In later periods, openings would be formed with stone margin, often incised to meet the harling formed around the window openings. This allowed an edge for the harling to dress into. As tenements grew in height, lintels were then set slightly proud of the stone, and sills projected past the harling to provide better weathering.

With the Enlightenment came the desire to improve the street appearance and to use ashlar stone on frontages. Some owners removed the harling to expose the rubble stone and these are still with us, for example, in Buccleuch Street, Edinburgh.

In Scotland, windows were always set back to protect them from the weather, unlike English construction, where the windows were usually flush with the external face.

Initially stone surrounds were simple and part of the stone frontage, as was the stone lintel. Only the stone sill projected, sometimes forming a string course on the first floor. The sides of the opening were formed from '*inbands*' and '*outbands*', which were set into the stone coursing. The inband[13] was the full depth of the stone wall so it tied the composite wall together; the *outband*[14], usually only about 7in. thick. In Edinburgh, the basement floor wall area became rock faced ashlar, the ground floor was often rusticated with V jointed ashlar and the first and second floor openings were framed with stone architraves[15], at openings and at lintels. Some openings had a pediment over the lintel. In grand houses, usually on the first floor, the openings were framed with stone architraves, capped with a frieze and pediment which was supported on stone console brackets. In Georgian times there was quite a variety of finishes around window and door openings, including arched window heads.

Sills

Stone sills would extend either side of the opening and a raised *stool*[16] at the ends would ensure water was thrown away from the joint with the stone architraves or scuntions.[17]

Stone sills are usually squared with a gentle slope to shed moisture, and projecting about 2.5in. (60mm) from the face. At frontages they may be profiled

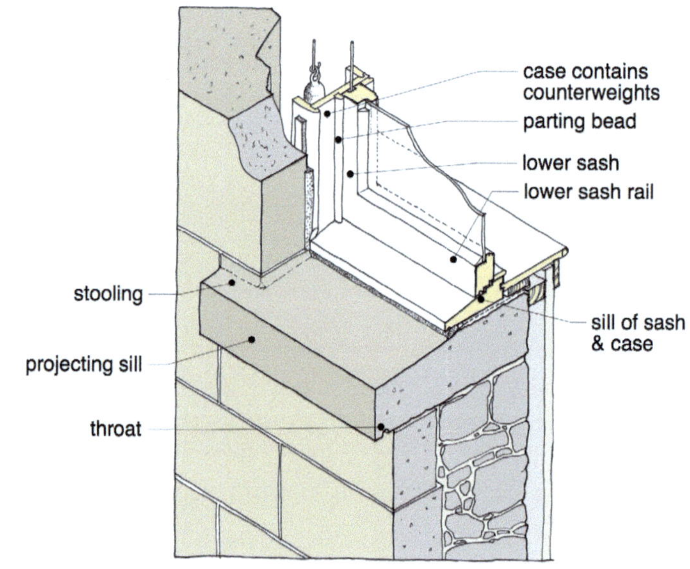

A 'stooled' sill

12 Illustration shows 'Starts and Stops'. The *Start* is the taller vertical stone and the *Stop* is the stone between, which is bedded into the wall.

13 Also called an *inbound rybat*. *Rybats* are the hewn stone at the side of a door or window.

14 Also called an *outbound rybat*. Rybats are the hewn stone at the side of a door or window.

15 Although these tall architraves should be formed from naturally bedded stone, often they had to be formed from face bedded stone.

16 Such sills are called 'stooled sills'.

17 Scuntions were the sides of door and window openings, commonly termed window reveals.

and match in with '*string courses*' which are often found at the first floor and in line with the top floor window sills. String courses help to throw rain off the face of the stone, although they often form a border between different stone finishes (such as a reticulated or rusticated base).

However, in poorer areas the openings remained a lot simpler. The architraves were simple although string coursing remained on ground and top floors. In the 1900s the stone *scuntions* were often formed in two pieces called '*stops*' and '*starts*' the starts being the central stone which was keyed into the external stonework, the stops being the thinner vertical stones.

Window openings on the first and second floors began to be emphasised with stone architraves and hoods.

Projecting sill

A projecting sill would sometimes tie in the elevations. It was formed primarily to get rid of lashings of rain, but also might ensure stonework was square and thus correctly built. The sill might have been cast as bonding stones, as they travelled through the wall to check the width of the wall.

Projecting sills are also found at door openings and sometimes higher, to tie in with a cornice above some shops.

Lintels at window openings

An exposed cast iron lintel on the ground floor of an Edinburgh tenement

Most window openings are not much wider than three feet. Increasing standardisation allowed the external lintels to be shaped and cut to size in a mason's yard, then transported to site. The internal 'safe lintels' were usually two or three *pitch pine beams*, although internal steel beams started to be used after 1900. When lintels were required at bays and oriels[18], they would be formed with a joint and supported on a vertical stone mullion. Because of the risk of movement at these joints, wrought iron cramps were used to tie them together (although maybe steel was used in some cases). Because of wetter weather, these cramps are corroding and the stone is vulnerable to packing because of expansion of the iron.

Stone lintels are usually about 12in. (305mm) deep and 13in. (330mm) wide, but on the ground floor, they carry the full load of stonework above, so they can be more vulnerable to cracking, particularly at ground floor lintels where the load is greatest.

Builders used a rule of thumb that said the thickness of the safe lintel should be one twelfth of the span.

18 See section on bays and oriels in Chapter 7.

Around 1896–8 in Edinburgh, cast iron lintels were sometimes used at openings instead of stone lintels. Such cast iron lintels are usually only found at ground floor openings and have lasted well, despite some surface corrosion. Some engineers thought this might be because the builder suspected that the ground was vulnerable to movement, or simply soft bearings. An alternative explanation could be because a stonemason's strike, to demand an eight-hour day, caused a temporary shortage of stone.

Lintels on the internal half brick thick walls were rarely used; instead the timber door frame would support the brick above which might be formed with a brick relieving arch above it. Stone lintels, usually Arbroath stone, are found at fireplace openings and these would also have a brick relieving arch over them.

Iron and steel lintels

Cast iron as a building material was in use from 1800 and wrought iron came a few years later. In the early days, they considered the material to be fireproof, which it wasn't. Although there are three main types of cast iron – grey, white and mottled – grey cast iron was normally specified, although white cast iron was less vulnerable to corrosion. The Carron Ironworks was founded in 1759 and led to a number of other Scottish Ironworks developing, initially casting items like railings, gates and iron products. Cast-iron columns and beams were in use until about 1900 although cast-iron columns were still being made into the 1930s. Cast-iron beams are most commonly found at close entrances, in shops as standards and columns, and as cast-iron plates to restrain oriels.

Wrought iron is almost pure iron, so it is good at avoiding corrosion. It is made by the puddling process from cast iron. Rolled sections were soon available and by 1879 beams were available from 3in. deep to 14in. deep. Belgium iron masters excelled at forging and it was then possible to import wrought iron cheaper from abroad than that made in the UK.

Steel came in around 1850 and had replaced wrought and cast iron by 1900.

It rapidly became the material of choice after 1860 when Henry Bessemer invented his converter. In 1879 William Beardmore built three open hearth furnaces at Parkhead in Glasgow, and although other furnaces opened in Scotland, Beardmore's predominated.

Steel beams and rolled steel joists (RSJs), were commonly used in tenements from 1900 onwards, although internal red pine safe lintels were still commonly used at window openings.

The structure of tenements

Tenements were built without engineers; the master mason or builder would bring their own experience, and traditions developed which only changed when there were major disasters such as fires or collapses.

The external solid walls are really compound walls made up from different sizes of stone and sometimes brick. To ensure the outer stonework was tied back to the inner rubble stone, some larger stones fully penetrated the wall thickness. This was most important around window and door openings.

Flues which went up gable walls were vulnerable since the flues could cause the outer leaf of stone to bulge outwards, particularly if the flue bridges were not built correctly.

The outer stone walls were high and were largely constrained by the floor joists that were

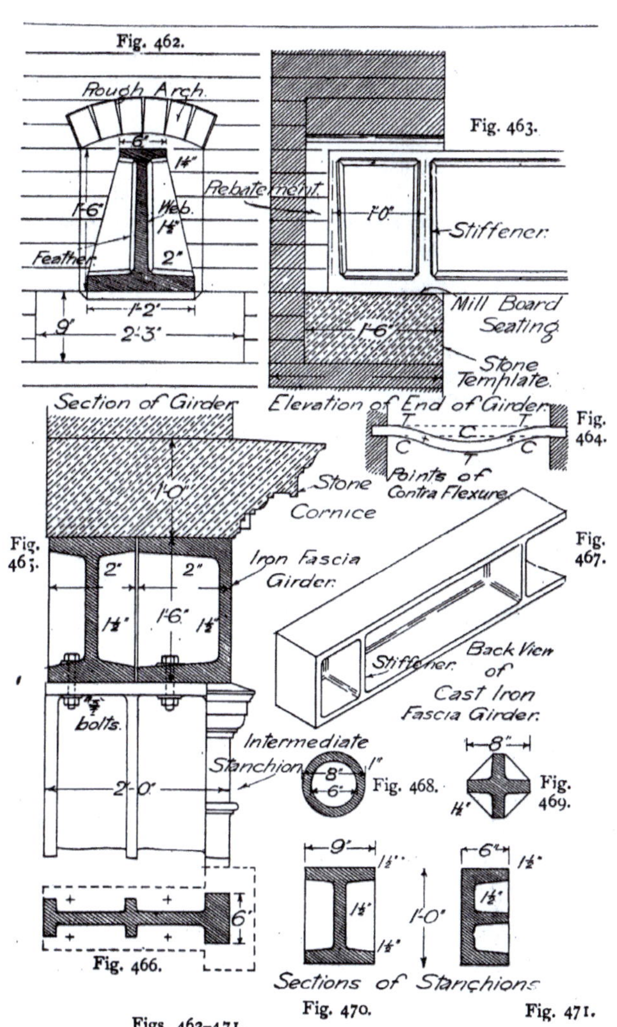

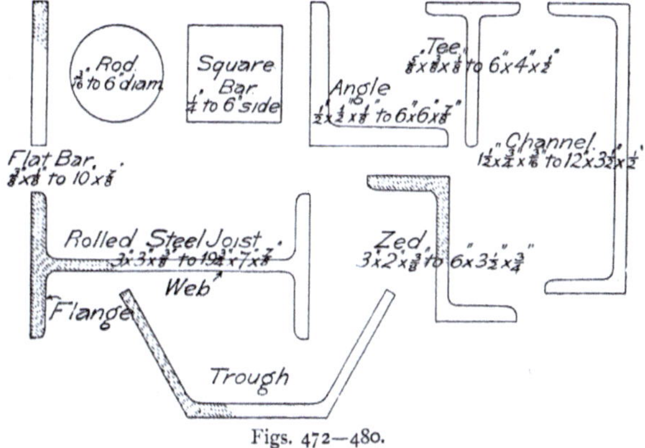

[above] Cast-iron lintels and steel section

[left] Cast-iron lintels and steel section

pocketed into the walls. Any timber embedded in an external stone wall is vulnerable to moisture and then rot. When the joist ends become rotten, the tying action is compromised, and the wall can bulge outwards.

The close walls, which are often made from a slender brick skin, are rarely well tied into the outer walls, so cracking and movement occurs at junctions between outer and inner walls (and as previously noted the foundations for internal and external walls may vary, as does their load, so ground movement can affect some walls).

The internal walls, whether they are made from timber studs or whether the studs have been infilled with bricks, are mostly load bearing. These internal walls can carry the weight of the wall above, or, if on the top floor, the roof structure. In addition, many of them will help to carry the floor load.

The roof structure retains the top wall head; however, the junction between the rafters and the ceiling joists is quite critical as it is close to the gutter. When gutters overflow, the water can go inside and start rotting the roof timbers. If this joint decays, then the roof can slump and push out the wall head, even causing it to collapse.

Tenements are basically cellular structures, so removing any internal wall can lead to problems. If this needs to be done, you are advised to hire a structural engineer and will also need to apply for a building warrant. Cracks can occur at many different locations. Cracks in outer stone lintels may suggest that there is rot in the internal timber safe lintels. If they have become rotten, the load of the thick wall will transfer to the outer lintel and crack it, so what may first appear as minor can lead to extensive repair work.

Cracks at sills, mullions and at the sides of windows can suggest there are stresses in the building which may require strengthening work as well as a simple sill replacement or repair.

Foundations, basements and solums

Early four-storey tenements in Glasgow were built in 1591–96 at Trongate and High Street; foundations then would have been wooden, set down by the builder. The only evidence we have of foundations under tenements is from tenements built in 1804 at Carlton Place, where brick strip footings were interspersed with timber sleepers on softer ground.[19]

Charcoal held within a timber grid was found to have been used in Edinburgh's New Town in areas, although generally simple strip foundations in Georgian Edinburgh were formed about 700mm below the lowest ground level and consisted of large flat stones[20] about 250mm thick and wider than the wall by about 50–100 mm each side (termed the 'scarcement'). This was common in the 18th century, although in sites where the ground was softer, a wider stone foundation would be used (for example in 1856, Glasgow's Queens Park tenements had sandstone

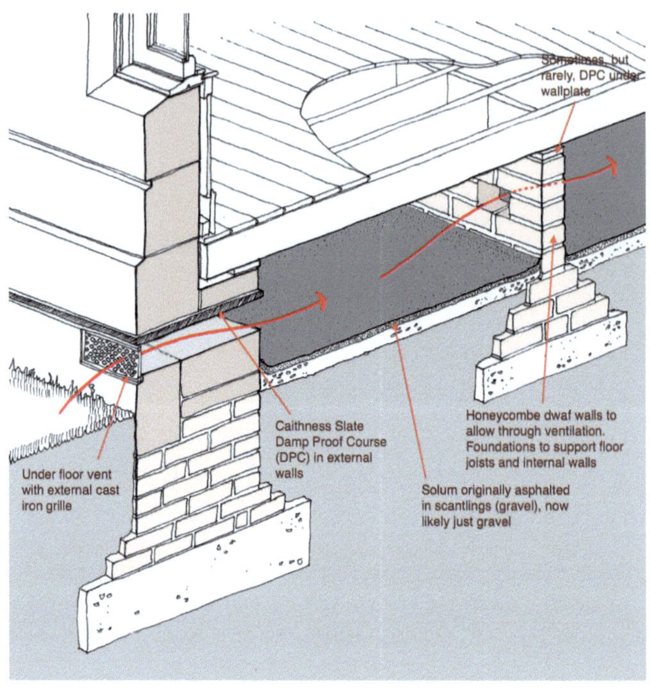

Typical stone slab at solum level

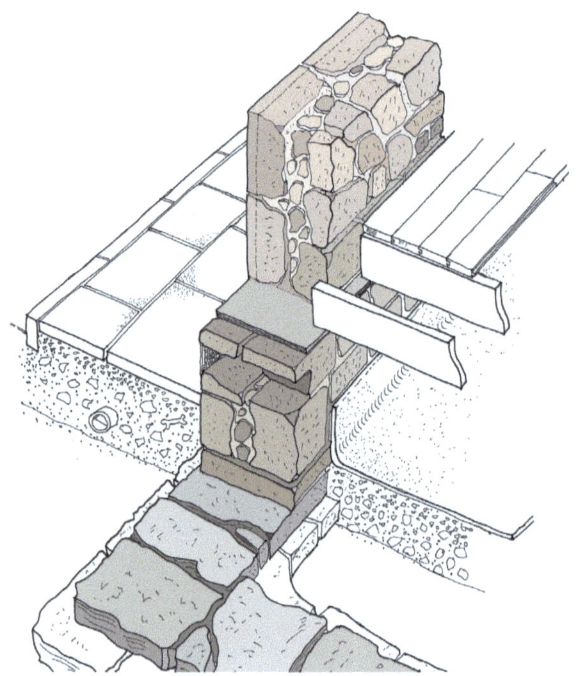

Foundation around 1900; concrete and brick footings showing slate DPC and venting in solum

19 Timber baulks as a foundation were found in a tenement in Bankhall Street when it was demolished in the late 1970s.

20 'The foundation having been dug out, we begin our wall by laying large flat stones of the roughest sort, from three or four feet in breadth, and, if there is no sunk floor, upon these are set the shop piers, of solid or cube stones, three or four in the height of the pier.' Alexander Thompson, Light of Truth and Beauty, lecture on masonry and how it may be improved, page 44.

foundations which were 1.3m across and 650mm deep, all about 500mm below lowest ground level). Although concrete foundations were used as early as 1820, they only became more common in tenement building around 1870. In later periods, depending on ground conditions, a brick footing might be built up with a much wider base to a point where the external stone walls started. These brick footings would have very little mortar between the bricks. It was common to have a stone plinth course and then a 50mm-thick stone (Caithness stone) damp-proof course (DPC), which was often set back for the face of the stone and then the gap pointed in a lime mortar. Such DPCs started to appear after 1875 and were common by 1890.

After 1900 it was common to have concrete strip foundations, which Gourlay[21] describes as being 12in. thick and at least 2ft below the ground surface (or lowest level). Also 4ft wide so that would give a *scarcement* of 1ft on a 2ft wide tenement stone wall. All vegetable matter would be removed and usually the base of the solum would be covered in a layer of '*shivers*' or broken bricks 'about 5in. thick" then a 1in. layer of asphalt applied on top and sometimes up the exposed foundation walls. Generally, the Dean of Guild[22] insisted that the internal level of the solum should be as near as possible to the external ground level but this was not always possible. Basement areas when they were built might have thick walls but were rarely 'tanked' against moisture.

Ground conditions

Some tenements were built over soft or unstable ground, sometimes above coal mines.

Steel channel to stabilise tenement in Govanhill, 1979

In Edinburgh, 1896–99, this might lead the builder to install cast-iron lintels at window and close door openings on the ground floor, with the thought that this would help prevent cracking to the structure down lower (but see p39 re. stonemasons' strike). In Glasgow, even timber foundations were used over soft ground. When a four-storey tenement in Glasgow's Bankhall Street was demolished, the foundations were revealed to have been made from large baulks of timber. Surprisingly they had not rotted, and the tenement was not demolished because of subsidence, but because of general disrepair.

However, those tenements built above 'stoop and room' coal workings did suffer later, as cracks began to develop in walls, as the voids underneath gradually collapsed. The answer to this in the 1970s was to insert steel channels around the tenements to hold them together. Later, it was decided to use a cement grout to fill the collapsing voids (see Chapter 11: Repair & Renewal 1970–90).

The internal walls were built from smaller brick footings, again with a wider underground level, and in the *solum* appearing as half brick width *honeycombed walls* to allow a flow of

21 Gourlay, C. (1903), *Elementary Building Construction and Drawing*, page 14.

22 Dean of Guild, under Scots Law was one of a group of burgh magistrates who, in later years, had the care of buildings.

ventilation through the tenement solum, front to rear. Cast-iron vents were installed in the outer walls to ensure the flow of air. Internal walls sometimes had a damp-proof course (DPC) either in slate or, later, bitumen, but often they did not and simply relied on ventilation to allow moisture to dissipate. Under bye-law 21 of the Glasgow Building Regulations Act of 1892: '*The damp-course in dwarf-walls may be of large and squared slates laid in two layers and properly bedded in neat cement*'. The same Act also required the ground floor of any house to be at least a foot higher than the external ground level, to ensure there would be ventilation, although this was not required for shops.

Subsidence sometimes causes differential settlement between the outer walls and the inner walls, possibly because the foundations for the outer walls were seen as more important since they took much of the load. However, if the internal brick close walls did not have good foundations, or were built on softer ground, then differential settlement can occur later, most noticeable in the corners of close stairs as the brick close walls are rarely well bonded into the outer stone walls.

At the base of the outer walls there would usually be a half-brick-wide offset to support a timber wallplate which would carry the ground floor joists. All joists should have been bedded onto a wall plate ideally with a damp-proof course below it.

Timber piles were in use around 1800, but mostly used to build retaining walls although timber piles were used to consolidate the earth before brick and stone foundations were laid.

Engineers were only involved in designing foundations for large prestigious properties, otherwise the tenement builder would use a rule of thumb that said if the weight of soil excavated was equivalent to the weight of masonry built onto the ground, then the structure would perform satisfactorily. This is the reason why some tenements in areas of poor soil conditions were built with basements (and not simply to store material or barrels of beer for the shop or pub above!).

Basements

Early Georgian tenements were often built with basements (usually for servants' quarters), in which case a 'dunny'[23] would be formed below ground level by a stone retaining wall with railings above. Steps would lead down to the basement level. Basements were also formed when the topography demanded it. Windows were often secured with steel bars embedded into the lintel and sills. Some basements were simply voids or large solums, sometimes used as coal holes. Steep changes in ground level could be catered for if a basement was built; these sometimes contained stores under the pavement and also coal holes. Basements for shop units were also popular to store goods. Georgian tenements did not have damp-proof courses; these were only introduced in Victorian times, and then inconsistently. Floors in basements that were occupied would ideally be raised above the level of the dunny floor outside. Timber floors might be laid on joists with a very shallow solum, often poorly ventilated, leading to rot in timbers. Otherwise, basement floors might be solid Caithness flagstones laid onto a layer of ashes over the limestone bedding.

The stairs that led to the main ground floor of the tenement would span over the 'dunny'; in Georgian times these would be stone arches and in Victorian times cast-iron support rails would be used on which the stone treads would be laid.

23 'Dunny' – an underground passage or cellar. In the West of Scotland 'dunny' used for basement area.

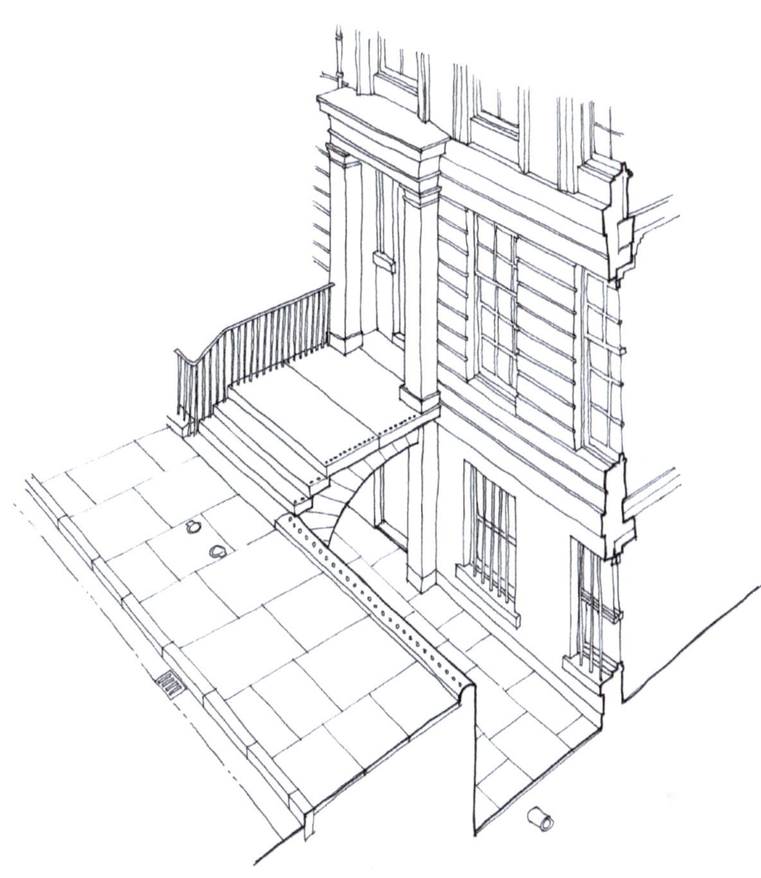

[above] Typical Georgian tenement with elevated ground floor and basement

[upper left] Dunny area below street level

[middle left] Construction of base to support treads

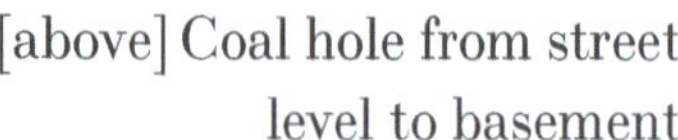

[above] Coal hole from street level to basement

[upper right] Access to basement for loading kegs etc.

[right] Air raid notice in basement advising of marked brickwork to allow escape

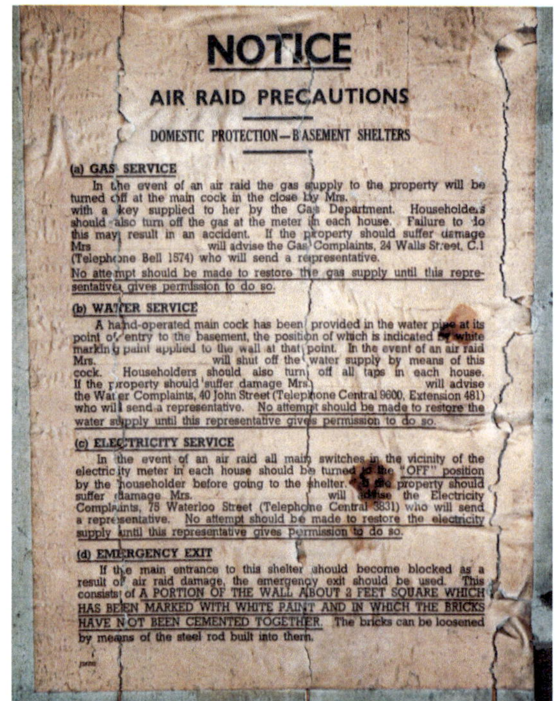

NOTICE

AIR RAID PRECAUTIONS

DOMESTIC PROTECTION—BASEMENT SHELTERS

(a) GAS SERVICE

In the event of an air raid the gas supply to the property will be turned off at the main cock in the close by Mrs. with a key supplied to her by the Gas Department. Householders should also turn off the gas at the meter in each house. Failure to do this may result in an accident. If the property should suffer damage Mrs will advise the Gas Complaints, 24 Walls Street, C.1 (Telephone Bell 1574) who will send a representative.
No attempt should be made to restore the gas supply until this representative gives permission to do so.

(b) WATER SERVICE

A hand-operated main cock has been provided in the water pipe at its point of entry to the basement, the position of which is indicated by white marking paint applied to the wall at that point. In the event of an air raid Mrs. will shut off the water supply by means of this cock. Householders should also turn off all taps in each house. If the property should suffer damage Mrs. will advise the Water Complaints, 40 John Street (Telephone Central 9600, Extension 481) who will send a representative. No attempt should be made to restore the water supply until this representative gives permission to do so.

(c) ELECTRICITY SERVICE

In the event of an air raid all main switches in the vicinity of the electricity meter in each house should be turned to the "OFF" position by the householder before going to the shelter. If the property should suffer damage Mrs. will advise the Electricity Complaints, 75 Waterloo Street (Telephone Central 3831) who will send a representative. No attempt should be made to restore the electricity supply until this representative gives permission to do so.

(d) EMERGENCY EXIT

If the main entrance to this shelter should become blocked as a result of air raid damage, the emergency exit should be used. This consists of A PORTION OF THE WALL ABOUT 2 FEET SQUARE WHICH HAS BEEN MARKED WITH WHITE PAINT AND IN WHICH THE BRICKS HAVE NOT BEEN CEMENTED TOGETHER. The bricks can be loosened by means of the steel rod built into them.

Basements would also house coal holes, allowing coal to be easily shovelled into basement coal cellars. Sometimes the coal door would be on the pavement with a chute to the coal cellar and sometimes in an opening in the wall, although these are mostly found in houses rather than tenements.

Pubs in tenements would often require a basement to store the beer and kegs would be dropped through a keg door in the pavement, or in a wall, or sometimes both.

During the Second World War, tenement basements were often used as air raid shelters.

Bombing was frequent over Clydebank, Greenock and Glasgow[24]. The air raid precautions in a tenement basement would list the name of the responsible person (designated a Mrs…) who had to ensure the gas, water and electricity supplies were turned off and in the event of the emergency exit being blocked: '*A portion of the wall about 2 feet square which has been marked with white paint and in which the bricks have not been cemented together*' could be broken out to reach another basement.

24 In total, there were more than 500 German air raids on Scotland – ranging from single aircraft hit-and-runs, to mass bombings by 240 planes.
During the air war in Scotland, 2,500 people died and 8,000 were injured.
With 4,000 properties destroyed and a further 4,500 severely damaged, the Clydebank blitz saw more than 35,000 people made homeless as a result.
Greenock had 5,000 homes demolished.

Solums

Solums are found under most timber joisted ground floors. They can be low so if inspecting or installing insulation, access is problematic. The solum area should be vented[25] from both front to rear. As the floor joists are smaller at ground floors, additional brick honeycombed support walls were built with timber wall plates or, alternatively, larger timber support beams would provide midway support for the ground floor joists. After 1892 these small walls, like the external walls, should have had a slate damp-proof course installed. There was often a requirement that the ground level in a solum would be no lower than the external ground level, to ensure there was no flooding. This is rarely the case in most tenements. However, once a layer of stone shivers or bricks was laid over the solum, it would be coated with a thick layer of asphalt.

Damp-proof courses (DPCs)

Slate damp-proof course circa 1900, often covered with earth or mortar

After 1900, lead DPCs came into use

The earliest reference to an asphalt DPC comes from 1860 when Alexander Porteous, a contractor who laid cement pavements used asphalt on Leith Walk. He then offered his services which *'Affords a complete protection from damp... When spread on foundation walls and other sleepers...'*[26]. Asphalt was often used in the solum base over *scalpings*,[27] but slate DPCs in external stone walls only came into use around 1875, and were common by 1890, although it is always best to check this, as many tenements gradually sank, or ground and pavement levels increased over time. Most internal brick walls (and vented dwarf walls which provide intermediate support for the ground floor joists) did not have DPCs but relied on the solum space being well ventilated.

In most Victorian tenements the outer walls had a Caithness flagstone bedded between mortar beds and often set slightly back from the face of the stone when it was then covered with a coloured mortar to match the colour of the stone. The Caithness stone was usually an inch thick.

Bedding stone correctly

Sandstones vary greatly in texture, porosity and colour. As sandstone is formed from layers of sand continually under compression and forming a stratified rock, they usually have a natural bed. It is important to lay the stone on its correct bed (which varies depending on how it is to be used).

25 Cast-iron vents were normally used.

26 From Slater, I. (1860) *Slater's Royal National Commercial Directory and Topography of Scotland*, London: Slater p138 but noted in Moses Jenkins PhD 'The Technical Development of Brickwork in Scotland 1700–1900'.

27 Scalpings, small stones, used to be called shivers in Victorian times.

The *freestone*, the best stone as it is homogeneous and without any clearly defined strata, was used for the public facing side of a building, to impress the public and visitors, whereas roughly dressed rubble stone was used for the rear elevations and non-public facing edges, such as servant entrances and backcourts.

Stone that is laid on 'the cant' or 'face bedded' is where the strata is laid parallel to the building face and often found on rubble stone at the rear of tenements. Stone that is 'edge-bedded' (or 'joint-bedded') is where the bedding planes are vertical and perpendicular to any exposed surface. Not all stones should be naturally bedded: stones for cornices, lintels, sills and string courses may be edge-bedded to prevent erosion at the exposed edges.

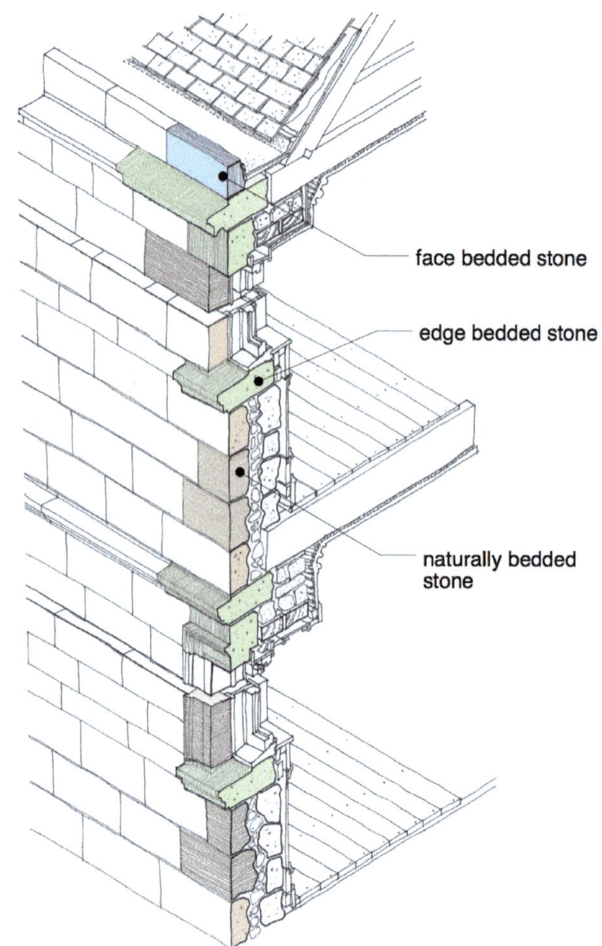

Diagram showing stone laid with natural bed (pink), face bedded (blue) and edge bedded (green)

Although columns, architraves and mullions should be naturally bedded[28], the builder might find it difficult to get the right depth of stone, or for cost reasons decide to face bed the stone. Stone architraves[29] at window opening are sometimes formed in sections, with a mid-section, to allow smaller lengths of stone to be naturally bedded. Columns could be formed in one piece. The stone ashlar used in Georgian times would be hand chiselled by masons and sanded.

In Victorian times, the saw cutting of stone blocks facilitated production, and rotating rubbing tables allowed the faces to be sanded smooth, although all the ashlar stones, cornices and mouldings would still need to be hand chiselled to apply a finish. Rubble stones would be roughly hand dressed once they had been roughly shaped for bedding.

The drawing above shows how the stone would be bedded in a typical stone tenement. Lintels, ashlar facing stones, would be naturally bedded, although carved architrave lintels would be edge bedded. A projecting cornice would be edge bedded well and the blocking course above it could be edge bedded but usually face bedded. The sill would also be edge bedded. Mullions should be naturally bedded but are often face bedded because of the difficulty of getting such long sections naturally bedded, but if face bedded, they are vulnerable to delamination.

28 To obtain a long length of natural bedded stone would mean digging deeply into the rock to fully expose a long length, hence it was easier to form long lengths from face bedded stone which did not require so much excavation.

29 Also called 'rybats' and 'scutcheons'.

Mortar used for stonework

Most of the mortar used in Georgian and Victorian tenements is a lime mortar (see Chapter 4 on Lime Mortars). The lime mortar would have been composed of fresh burnt lime (*Quicklime*), slaked[30] and exposed to the air until hydrated, then mixed with a clean sharp sand or grit, or with hard burnt clay or freestone, usually in the proportions of one of lime to three of the other material. Around 1890, ordinary Portland cement came into common use, particularly for brickwork. The Glasgow Building Regulations of 1892 mention both lime and cement mixes[31] to be used, although we suspect that masons would have stuck to the traditional lime mixes.

However, if lime mortar remains wet during the building process, it can calcify and remains as a sort of putty, with little strength.

Mud mortars were used as *daub* coatings onto timber laths or *wattles* and in solid wall construction. Sand is often added to the mud and quicklime is sometimes added to harden the mortar more rapidly.

Plasterwork inside face of external walls

As a solid stone wall will get damp (and eventually dry out), the internal face was usually formed by installing small vertical battens fixed to timber dooks[32] plugged into the mortar courses of the external wall. Riven split timber laths were then nailed to the vertical battens. *Laths*[33] were originally made from oak, but in Victorian times beech or Douglas fir was used; when nailed to these battens, a small gap of about 9mm was left between each lath. Three coats of plaster[34] were formed over the laths. The first coat is called contained ox hair or goat hair[35] to give added strength to the plaster as it was keyed into the timber laths. This was then scratched to form a base for the second levelling coat, finally a finishing coat of stucco plaster was added[36].

When brick walls were plastered 'on the hard' it means they were 'rendered' with a lime plaster. This would often occur on walls facing the close stairs. Over time, the plaster became detached from the backing, possibly because of movement in drying out.

---

30 Quicklime, which reacts with water to form lime. The process is called 'slaking'.

31 Gourlay notes that bye-law 10 of the 1889 Regulations states that 'Mortar made from cement or hydraulic lime must be mixed with clean sharp sand or grit… in the proportion of one of cement or lime to two of the other material.' Gourlay, C. *Elements of Building Construction and Drawing*, page 1.

32 Timber wedges driven into mortar joints to provide a fixing for the vertical battens to support the timber laths, which formed a backing for the lime plaster.

33 Today, new laths can be bought pre-split, usually in chestnut or Scots pine (30mm wide, 5–6mm thick).

34 Gourlay describes the plaster mix as follows: 'The lime for the first two coats is generally specified to be of Scotch lime shells, mixed with pure water, clean sharp sand, and long fresh hair. The proportions are about one part of lime to two or two and half parts of sand by measure, with 1lb of hair to 3 cubic feet of mortar. A larger proportion of hair is used for ceilings than for walls.' *Elements of Building Construction and Drawing*, Charles Gourlay, page 17.

35 Gourlay again: 'The hair formerly used for plaster was clipped from the hides of oxen, but in the present day methods of tanning, the hair is subject to chemical processes, so that when the hair comes into contact with the lime in plasterwork it decays in a very short space of time…a mixture, half of goats' hair and half of Manilla fibre, is what is commonly used for good work.' *Elements of Building Construction and Drawing*, Charles Gourlay, page 17.

36 Greek Thomson stated that: 'I would recommend a lining of brick inside the walls, with a space for air between it and the stone, in lieu of lath and straps – the dooking for which is a frequent cause of mischief in wedging walls off the perpendicular.' Alexander Thompson, *The Light of Truth and Beauty*, on masonry and how it may be improved, page 47.

Riven timber laths form a basis for lime plaster

Dados would be formed in a stronger setting plaster of Keene's cement[37] which was hard-wearing; otherwise tiles would be applied to the dado.

At certain other areas, a single parging coat of lime plaster may be applied to the inside face of the stone. This would normally be found at the ingoes (the recesses for windows) and applied before the window or any timber lining or shutters were formed. It was thought that any dampness or 'sweating' of the stone, would be safely absorbed by the plaster. Indeed, the inside face of the window reveals (the *checks*) would be plastered along with the stone sill and whilst wet the window case would be bedded into place. The external gap between window case and stonework would then be pointed with a linseed oil mastic.

Plaster was also applied at the rear of presses, on gable walls, and in presses before any timber lining or shelving was fitted.

The inside of the close wall, as well as the internal brick walls, would be finished in three coats of lime plaster. In the close, the lower dado section would be rendered in a harder mortar such as Keene's cement, which was then gloss painted. The lime plaster above and the close ceiling would be lime washed.

For tiled dados, the tiles would be bedded in cement and edges closely butted together.

Internal wall construction

From 1700–1860, internal walls were sometimes stone and sometimes a timber framed structure with vertical planks to form partitions. In Georgian and early Victorian times, internal walls were also built using timber studs and lath and plaster. These stud walls, built with 4-inch studs and stiffened with horizontal *noggins or dwangs*, although built in timber, were often carrying a structural load. It is not unusual to find the load-bearing stud partitions do not line up above each other. The wall's surface would be formed by nailing thin timber riven laths to each stud and then applying two or three coats of plaster.

The timber studs to the partitions supported the long joists that travelled front to back (although in earlier Georgian times, in Edinburgh, larger timber beams (about 170 × 335mm) were installed to support the joists, which were notched into the beam midway. This probably accounts for why the Edinburgh New Town floors are suspended on timber hangers, rather than secured directly to the ceiling joists). This allowed the builder to use the floor plate as a working platform. After 1860, internal brickwork was used more extensively, often only half brick thick (6in. with plaster), but the studs remained at door frames and to support the smaller mid span beams across the bed recess, to allow the joists and floor plate to be formed. The space between the studs would then be backfilled with bricks and plastered over. These walls are termed 'chivers' or sometimes 'brick noggins'. At certain locations, timber laths

37 Keene's cement is actually a gypsum plaster, like Plaster of Paris which is then steeped in a solution of alum and subjected to intense heat, ground to a powder, and sifted. Invented around 1840.

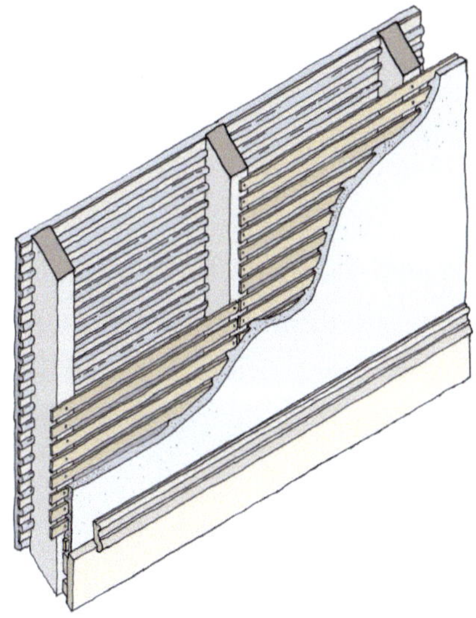

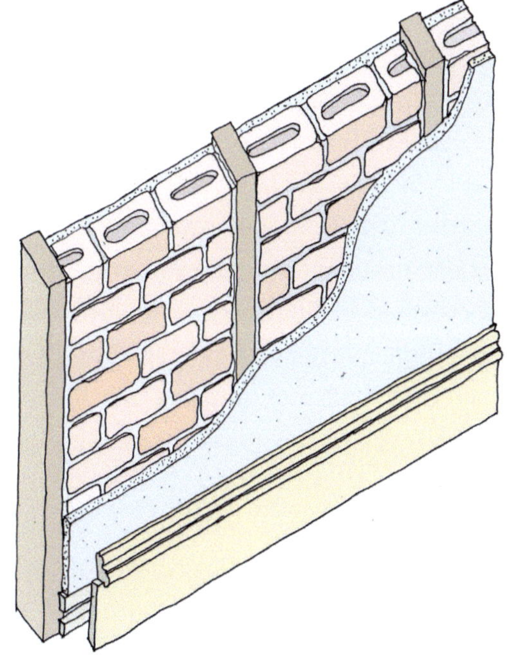

[above] Stud wall partition plain faced. [upper right] Stud wall shivered with brick noggins

[right] Timber stud partitions, often load bearing, were more common in early tenements

might be added to provide a better grounding for the plaster (see 'the Humble Door Frame' in Chapter 7).

Partitions were also made with locally made clay bricks bedded in lime mortar and usually about 9in. thick (with plaster, about 11in.). In some better tenements, the close and party wall might be made from solid brick construction using the Scottish bond, also plastered on both sides.

The void between the head of the door and the top frame might have an additional *transom* and be infilled with glass to form a fanlight to light the hallway. The top of the door frame would have a timber wall plate on which the joists would rest and, if the leg of the door frame landed between joists, a timber board would be recessed into the adjacent joists to secure the legs of the door frame. Door frames were usually secured into brick-sized timber dooks[38] (four for each post) that were bedded into brick walls (after 1900 often became brick or concrete plugs).

38 Also termed 'bilgates'. Usually about 6in. × 3in. × 1in. thick blocks of wood to fit into the brickwork coursing and to which the door frame is secured. It was best not to bed vulgates in with mortar, but keep the blocks tight to the bricks.

Perhaps because of fire risk, internal partition walls are commonly formed from 4½in. thick brick, plastered both sides. In Victorian times and probably as a cost saving to speed erection times, it was common to build a timber frame to support the floor joists at each level, with bed recess beams. The door frames were often designed as 'H frames' to rise and support the floor joists above. This allowed a temporary floor plate to be formed to raise the tenement higher. At the same time, bricklayers would be building up the new walls between the H frames and temporary studs. The space between the timber studs and sides of door frames would be built up in brick bedded in lime mortar[39] and coated both sides with three coats of lime plaster added to the face of each wall. Originally the first '*scratch*' coat and the second coat would contain goat hair fibres (or ox hair) to give them strength.[40]

Half-brick thick internal walls were often stiffened on the top floors because there was less weight on the internal wall to ensure its stability. This was done by installing timber wall plates.

The rapidity of building tenements led to the process of 'nogging' where brickwork was infilled between timber studs, although it started much earlier (in the 1780s in Brechin). Sometimes vertical cracking will occur above a door; usually this is where the H frame rose to the ceiling and the brickwork is not continuous.

However, the close walls were always built in brick. Usually, but not always, half-brick thick.[41] This gave some degree of fire protection to the close staircase and allowed the individual stone treads to be set into the brick close walls.[42]

Once the floor-plates were built and floor joists installed (to act as a temporary scaffold), half-brick walls would infill between the timber framing from the dwarf walls in the solum. Although the construction of internal walls varies depending on the likely market a house was designed for, the mix of timber and brick is common in tenement construction.[43]

Brick walls

Bricks in the period before 1800 were on average 213mm long, 91mm broad and 58mm high. Between 1800 and 1850 the brick size increased to 228mm long by 114 broad and 71mm high. In the period from1850 to 1900 Scottish brick increased in size again to 231mm long, 110mm broad and 78mm high, all of which reflects the change of mechanisation processes during these periods. In comparison, today's brick is sized at 215mm long, 102.5mm broad and 65mm high, much smaller than traditional Georgian and Victorian bricks. (See Chapter 4 for more details about bricks and brickmaking.)

Glazed bricks were introduced from 1870 to 1900. Facing bricks (higher and medium quality) were fired to allow them to resist frost and first used on rear tenement walls around 1870. Common bricks were used internally or, if used externally, harled to protect them.

39 A process known as 'nogging'.

40 If you are ever stripping plaster from an internal wall, best to only do one side at a time, as the plaster helps to give the wall some stiffness. Also, where there are embedded timbers, cracks can develop as the brick coursing is not continuous.

41 Some middle-class tenements have brick thick close walls around the stairs and in the front close, but reducing to half-brick thick in the passage to the rear door.

42 Which is one reason why you can often hear people's footsteps on the close stairs.

43 Fanlights above doors are often sealed in. If the door has to be moved, the whole frame has to be taken out and a new precast lintel inserted with the brickwork above toothed into the existing brickwork.

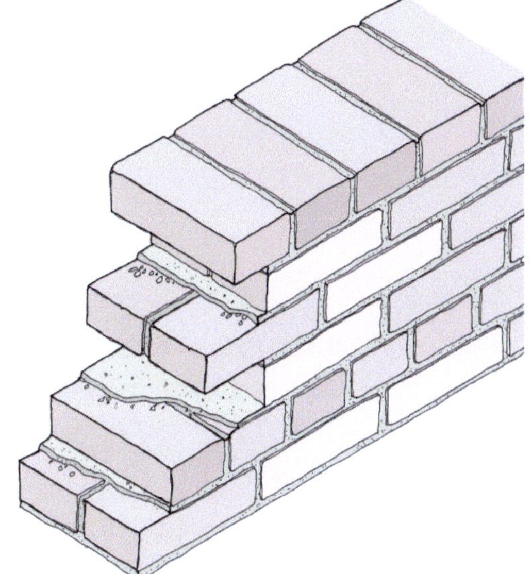

[above] Scottish bond often used on gables and sometimes party walls. The headers would have three stretcher courses between them, sometimes as much as six courses

[left] Chain timbers or wall plates to stiffen slender brick walls on top floors

Scottish Bond was used to make full brick thick walls; it was developed as a variation of English and Flemish Bonds. It comprised one header course and three, four or five stretcher courses between each header course. It was simply quicker to build, thus cheaper, than English or Flemish Bond. Eventually the external compound walls of tenements would be built from brick, bonded into the stonework.

The *gauge*[44] of bricks is the distance, with mortar beds, between four courses of brick. Typically, in tenements this would be 13in. or 330mm which would correspond with the height of an ashlar stone block; this would allow keying of gables and party walls into the external stone walls. However, internal brick walls which were mostly a half-brick thick (4½in.) were only keyed into the stone walls every fourth brick, if that. In close walls, this is the reason why vertical cracking appears in the outer corners of a stairwell.[45] Gourlay advises that it is much better to provide projecting stones from the outer walls to then bond into the brickwork of the close stair walls.

How the bay and oriel developed

The great thing about bays and oriels is that they allow much more light into the principal room. The sill levels are low and would not be allowed today, but generally the windows are quite safe with the right fittings. The amount of daylight is significantly reduced when PVC windows are installed into existing timber cases. (See Chapter 7 on windows.)

44 The gauge could vary between 12in. and 15in. depending on the elevation and size of masonry.

45 Cracking may have started because of differential settlement between the outer (heavier) stone wall and the internal brick (lighter) walls, leading to a vertical crack in the corner of the close wall.

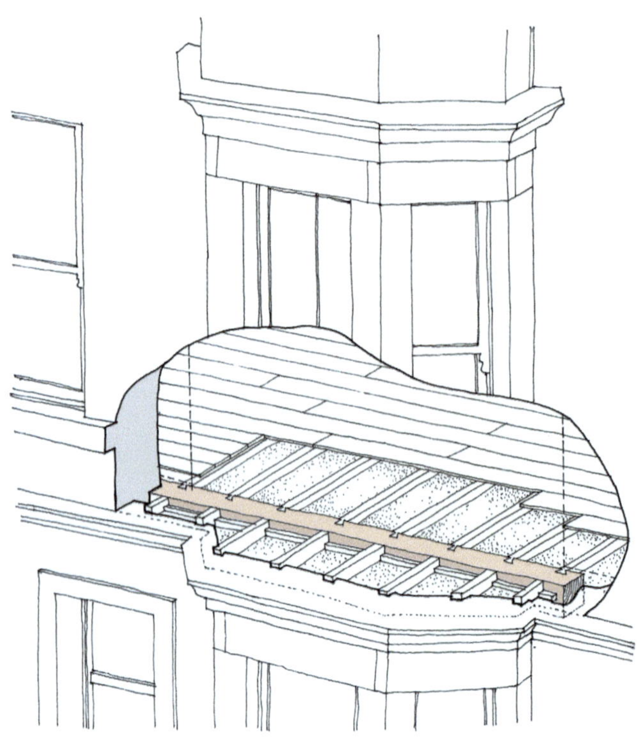

[left] The large timber bressumer beam support and [above] the bressumer beam provides strengthening for the joists

In the late Georgian era, bays and oriels developed at the front of tenements and sometimes at the rear. When the principal room (sometimes the parlour) would originally have had two or three windows in Georgian times, the Victorians changed the spacing of the windows to have twinned windows with a central stone mullion, or sometimes to tripartite windows, with two small windows on either side of a central window.

In areas where the Feu requirements specified a front garden or a basement to cope with levels, front bay windows developed, as in 18th-century Edinburgh in North Castle Street, with bow fronted windows.[46] However, oriel windows also appeared around 1850 in Glasgow, at Claremont Gardens, with first floor oriels, although they only came into common use around 1890 when tenements were built up to the pavement edge, without front gardens.

The wider feu arrangement which resulted in a wider plot, allowed for bays to be developed as flats got bigger than two apartments, mainly for white collar workers.

Bay window construction

Initially stone fronted bays were built in Georgian times to have three principal windows. This soon changed into the three-sided bays with two windows at the canted walls and separated by slender stone mullions. Such bay windows were commonly built in Edinburgh's middle-class areas, as well as in Glasgow, possibly after Ruskin's comment '*that every one of your principal rooms shall have a bow (bay) window, either large or small*'. Bay, and oriel window walls were much thinner than the main external wall, so the inside face would often receive a *parge* coat of lime mortar beforehand and timber framing was formed at the sides or under the sill. Gourlay explains that this was to prevent 'sweating'[47] of the wall.

46 Some New Town properties have rear stone bows which usually have a single window.
47 Sweating – probably referring to the possibility of surface condensation on thin walls.

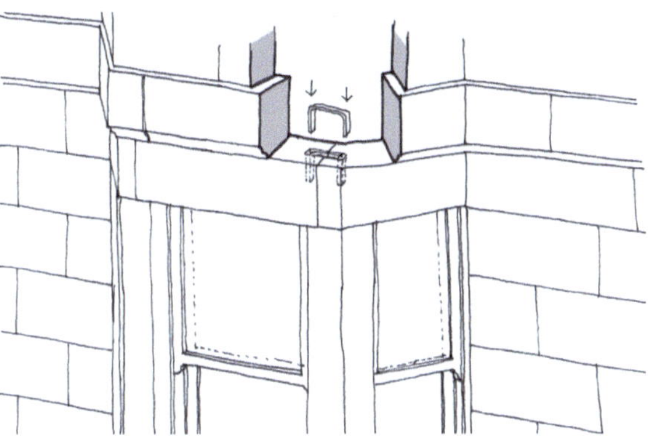

[far left] A new stone mullion is used [upper left] iron staples used together at bays

[left] iron staples used to tie lintels together at bays

A large timber beam[48] would span between the sides of the opening where the bay protrudes and the main floor joists on the upper floors. The upper floor joists would be tongued or keyed into this timber beam. Smaller joists would extend from the beam to the stonework of the bay window. The load of the roof and floor joists is carried on the beam and the load transferred to the external wall, rather than placing the load on the bay or oriel structure. This works well unless the stonework becomes damp and rot develops within this main timber beam. The timber beams were replaced by steel H beams[49] from 1892, if the width of the bay or oriel exceeded[50] 5ft. (RSJs and cast-iron beams were also used.)

The slender vertical stone *mullions* should be formed so that the natural bed is at right angles to the height of the mullion. This was increasingly difficult to achieve, so many mullions were supplied with the natural bed vertical.

The stone lintels at bays and oriels were shaped and sometimes dowelled into the vertical stone mullions with wrought iron cramps[51], called *'dogs'*, were set into pockets in the stone, and leaded into place. The lintels would also be tied together with wrought iron cramps.

Oriel construction

As oriels sprang from the first floor, the construction was similar, with a timber beam spanning the opening although in some areas with shops a cast-iron lintel would carry the

48 Called a *Bressumer* beam or Baulk; bressumer comes from '*Breast Summer*' which is derived from the French '*sommier*' meaning a pack horse. Thus, such a beam is bearing a great burden or weight.

49 The H beam is shaped like the letter H. I beams were also common. They are known as 'Rolled Steel Joists' or RSJs. The horizontal sections are the 'flanges' and the vertical section is the 'web'. They came into common use around 1890 although were manufactured around 1850.

50 This was a requirement of the 1892 Glasgow Building Regulations Act that batts or oriels exceeding 5ft had to be supported with incombustible lintels.

51 Bronze cramps were better but steel was also sometimes used.

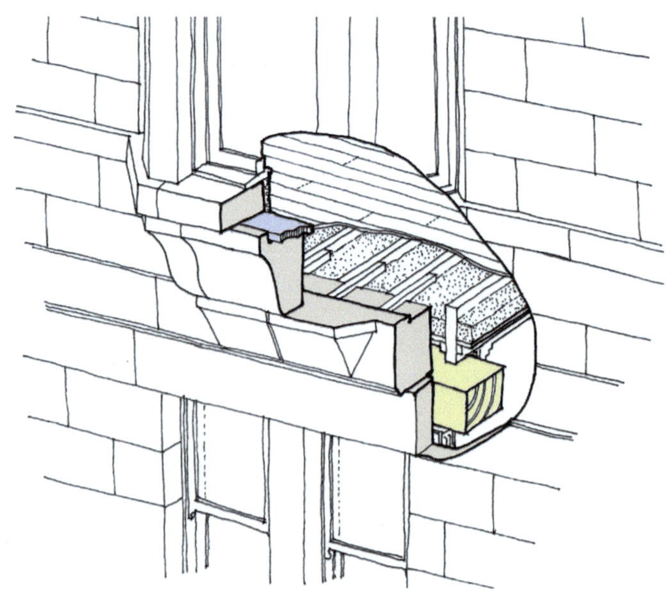

[above] Glasgow oriel, view cast-iron plate to restrain oriel

[left] Claremont Gardens

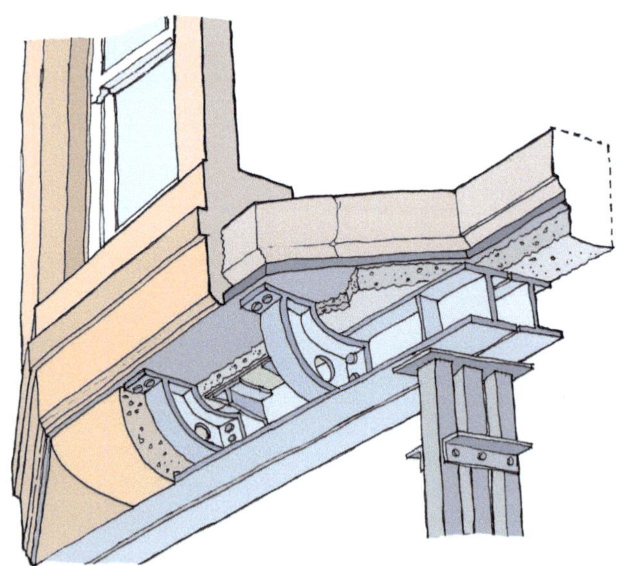

[left] Oriels above shops with cast-iron corbel as -

[above] Steel angles used to support the oriel plate when retrofitting a shopfront

main floor load. Corbelling[52] of the oriel varied greatly; sometimes it was formed entirely in stone, sometimes with stone brackets and sometimes with cast-iron corbel brackets formed as part of the main cast-iron beam. The earliest corbels in Glasgow probably date from 1850 and can be seen in Claremont Gardens (1857).

In Glasgow, after about 1900, large, shaped, corbel stones would be set into the lower stone lintels and then a 9in. × 2in. (229 × 50mm) thick *cast-iron plate*, shaped in plan to the bay, would be bedded onto the stone and the stonework above cut to ensure the cast iron was concealed. This cast-iron plate acted to restrain the oriel. The stonework of the oriel was usually, but not always, checked over the plate to hide it. On the second and all upper floors, a beam would support the floor joists. In earlier construction before 1890 the beam would usually be timber and in later construction, steel RSJs would be used. Beams are usually set within the depth of the floor but not always.

In Edinburgh around 1890, the use of cast-iron beams for ground floor shops (and in some cases houses) became common, often developed with cast-iron corbel brackets to support the oriel above.

When ground floor houses were converted into shop units below an oriel, new steel beams would provide the opening for the new shop window and cast-iron brackets would be added to the flanges of the beams to support the oriel above. In this case, the brackets support the original cast-iron plate at the base of the oriel.

Problems in bays and oriels

The main problem has been the increasingly wet weather as the stonework has little time to dry out and so any embedded timbers or cast iron or indeed steel, can be affected by moisture, particularly in Glasgow with a much higher incidence of rain due to climate change. The construction of bays and oriels is also much more slender than the main external walls, so dampness can penetrate them sooner and, because of wet weather, they do not have time to dry out. Glasgow suffers from more annual rainfall than Edinburgh[53] receives.

Small angular cracks that are found in many bay and oriel lintels are often caused by these embedded metal cramps rusting and so expanding and causing cracking of the stone.

In the post 1890 red sandstone tenements, stonework under oriels is starting to crack, as the embedded cast-iron plate corrodes where it is close to the external stone surface. The same seems to be occurring at some close entrances where cast-iron plates or steel beams have been used.

There are often twinned windows immediately below the oriel; this can stress the ground floor mullion and stone lintels under the oriel. Any movement at this point will tend to allow the oriel corbel stones to rotate creating more cracks. Usually that is the time when support shoring is required.

Where mullions have been face bedded, the stone can decay more easily. In such cases it may be best to replace the mullion completely, ideally in matching stone.

Another problem is the risk of rot in the timber bressumer beams that span between

52 A corbel is a projecting stone which carries some of the weight of the stonework above.

53 Glasgow's annual average is 49in. and Edinburgh's is 28.7in. Met Office.

[left] Cracking of stone lintel caused by an ironstaple

[right] Similar cracking at stone corbel joint which was constrained by an iron staple

[upper left] Thick cast-iron plate used under oriels circa 1890, corrosion to iron causing damage to stone

[above] Timber bressumer beam with rot at ends

[left] construction of bressumer at bays bedded into stone

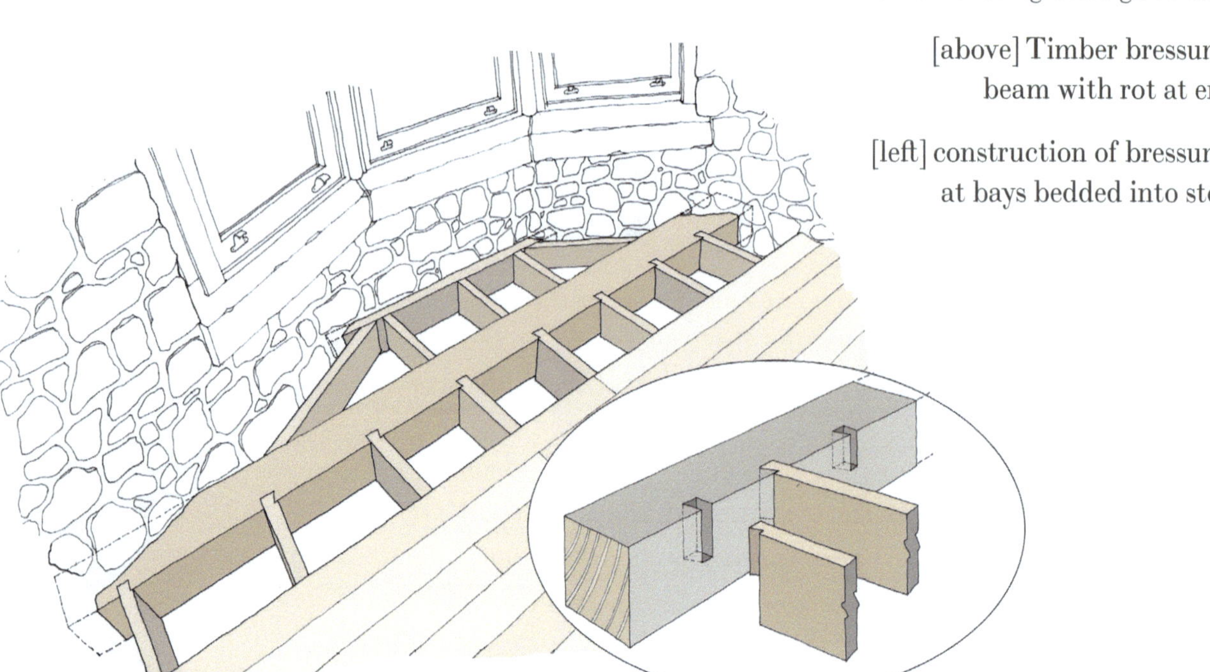

the bay or oriel opening. As we get wetter weather, the stone fabric has less time to dry out and the moisture, together with a lack of ventilation, can lead to a rot outbreak in these beams.

In some cases this can be repaired with resin repair techniques, but if the beam is badly affected by dry rot, it may require the beam to be completely removed and replaced with new steel beams. It may also require floor joists to be spliced.

Chimney construction

As coal fires were essential in the past in heating houses, tenements were built with fireplaces in the main rooms with flues into mutual chimneys, wallhead chimneys and individual chimneys. After 1750 and up to 1840, coal became cheaper so more chimneys were needed to vent a fireplace in every room. Each chimney stack might contain six or more flues, each taking the smoke from individual fireplaces or stoves. A typical flat in a small tenement would have at least three flues. With eight flats on a stair that would amount to twenty-four individual flues.

Mutual chimneys built between tenements would contain flues from the adjoining tenements, so a typical party wall flue could have about 16 individual flues. When developers built a tenement, they would build the adjoining flues at the same time, even though they might not be building the adjoining tenement.

Under the 1892 Glasgow Building Regulation Act, chimneys had to be at least three feet above the ridge and the flues would have a slight bend to prevent rainfall going down a chimney. Bedrooms on the ground floor at gables would sometimes have problems drawing

Mutual chimney serving 10 flues

[left] Decayed chimney cut away to show construction

[right] Wallhead dismantled, note brick flue bridges

the smoke up because the stone at ground level might be damp. In these case, glazed flue terracotta liners were sometimes used.

Flues would be built inside the party walls, either *double banked* or *single banked*. In the 18th and mid 19th century, the divisions, feathers or 'flue bridges' between the flues would be built from a thin stone, usually set on cant and liable to laminate. Brickwork then became more common; to build the divisions between each flue, they should be tied to the outer stonework and fully pointed so there is no leakage between flues.

As chimneys were swept by a sweep dropping a flue brush down the flues with a weight on the end, the bridges were often damaged, although salts from the soot would also damage the stone. Water penetration at flues located on external walls can deteriorate more rapidly as moisture combines with sulphates in the flue gases to form sulphur dioxide which corrodes the masonry inside the flues.

Where chimneys are located at gables or front and back walls, if the flue bridges separate, the outer skin of the wall can peel away as there is little to restrain it. This is more likely to occur in larger chimneys with many flues.

Above the roof line, the stack would be built in stone, but below the roof, brickwork was always used, unless it was on a gable wall.

The brick flue bridges were rarely well bonded into the adjoining stone and would often collapse, causing a flue blockage. In later Victorian and Edwardian construction, flues were sometimes lined with glazed terracotta pipes which generally worked well. Chimney flues had to be swept regularly as chimneys were prone to go on fire when soot built up. The Police and Improvement (Scotland) Act of 1850 set a penalty for any person occupying or using the premises who wilfully set, or caused a fire in a chimney within the premises to a penalty not exceeding ten shillings. Up until the 1660s chimney heads were finished with a stone cope only. From about 1700, chimney copes were finished with clay chimney pots. Some of the clay pots used in Edinburgh can be 6ft high. Victorians introduced squared and hexagonal pots with friezes and decoration. Georgian chimney copes tended to be

[upper left] Front chimney exposed as flue bridges cannot restrain outer ashlar stones, although this was mainly a problem at the cornice level

[upper right] Individual flues would snake up the party wall, to vent through individual pots

[left] Radar testing of an elevation

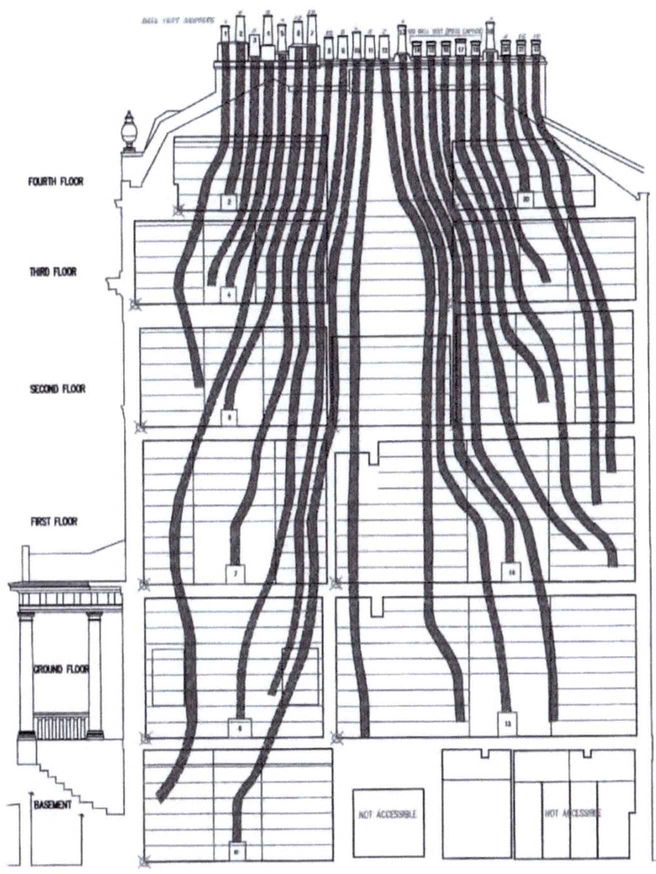

KEY
RADAR AND FEMR SURVEY

- Survey datum
- Radar survey position
- Well resolved flue
- Poorly resolved or inferred flue

BALL TEST RESULT

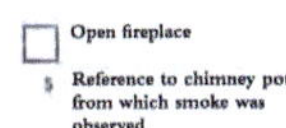

- Maximum distance that 200mm ball could be lowered down flue measured in metres from pot
- Ball would not fit into chimney pot - no measurement possible

SMOKE TEST RESULT

- Open fireplace
- Reference to chimney pot from which smoke was observed

Fig 1 Results of a radar investigation to identify flues and voids in the wall construction

Edinburgh closes with
the flues shown

[upper left] Standard Edinburgh pot, Victorian hexagonal pots and square pot

[above] Iron cramp to hold projecting cornice has corroded, damaging stone

[right] Slated 'saddle' at rear of chimney in Edinburgh

Brick flue extended from lower roof to higher roof, after tenement built next to it

simple cylinders with double beads although square pots were also in fashion. Victorian pots tend to be more decorative, and often change in different locations.

Victorian chimney copings were finished with a slightly projecting string course under them, then a projecting moulded and weathered stone coping and finally topped with a stone capping, which acted as a counterweight to the projecting coping. The top stone coping had recesses to house the terracotta chimney pots into, which were then haunched into place with a mortar.

The cornice and copings stones would act to hold the outer walls of the chimney in place. After 1900 the Victorians liked to use large projecting cornices which would be tied back with wrought iron dog cramps before the top capping was set. In time, these iron cramps corrode, and can crack the stone coping.

As many chimney stacks have been rebuilt over the years, copings are often replaced with concrete copings. These are often made from *'in-situ'* concrete mix poured into a timber shutter. They rarely have a drip, and as they are cast in one piece, can crack in time, particularly if formed as a large coping. Also, pots are sometimes cast into place, making them difficult to replace.

Front wallhead chimneys had to ensure the pots vented above the height of the ridge which is why there are tall chimneys at the front and back of tenements. In Glasgow, many of these have been reduced in height as coal fires were no longer in use and planning controls were weak. Sometimes chimneys have been entirely removed or replaced with stainless metal flues, all of which detracts from the streetscape.

Where tall Victorian tenements were built adjacent to lower Georgian properties, the flues on the lower building had to be raised to the height of the higher chimneys to ensure there was no blow back of flue gasses. Flues for any future adjacent tenement would be built at the same time with the builder intending to recoup some funds when the next tenement was built, by themselves or by others.

In Edinburgh, wallhead chimneys have a small slated roof behind them. These are called 'shoulders' and are flashed into the slate roof and the chimney. In Glasgow, the chimney abuts the slate roof and a small lead valley seals the joint between chimney and slates. The chimney stalk would also have a wrought iron ladder stay which helped to support the ladder and provide access for the chimney sweep.

Fireplaces

Fireplaces were usually in two parts, the mantelpiece which could be in marble or pine forming the frame, and a grate, which could be a raised metal grate with the fireback formed in firebricks or a cast-iron register grate to form the backing, flue opening and grate. Sometimes air flow might be controlled by flap at the throat of the flue. The cast-iron register grate was initially quite simple but in Victorian times it became embellished with decorative tiles which were inset into side panels. The photo shows an added cast-iron hood to contain the smoke, with the fire converted to gas using an open mantle.

Kitchen ranges were large as they usually had a bread oven as well as a stove with a hob for cooking. Sometimes there might be a small boiler for hot water, although later ranges could act as back boilers to heat hot water in a separate copper cylinder.

Each opening for a fireplace was usually formed from stone *jambs*, usually with a brick relieving arch above it, set into the wall with a stone lintel 12 in. × 14 in. (300 × 355 mm) in section, although in later construction concrete jambs and lintels were used. The back wall of the fireplace should be a half brick thick and when sited 'back to back' with a fireplace on the other side of the party wall, was normally plastered before the cast-iron grate or fire bricks were installed.

The flue vents from the fire were about 10in. square. The lintel that supports the opening is shaped at the back to provide a smooth angled passage to allow entry to the chimney stalk (flue). This area of the lintel is termed an oncome.

Hearths

In Georgian Edinburgh, the flue openings were formed with rough stone lintels and stone jambs. The back of the fireplace opening would be plastered before any fireclay brick lining was installed.

Hearths were formed in Arbroath pavement stones set onto a lime ash bedding onto timber deafening boards supported between joist and wall. The fireplace was added later as it could be marble, timber or cast iron.

Edinburgh's floor construction used intermediate support beams to carry the floor joists, so joists were rarely close to the fireplace wall.

In Glasgow, the floor construction was more problematic. This was because in earlier tenements of 1850–90, the main floor joists, often running close to the fireplace wall, were cut down to a very slender height to allow space for a thick stone hearth to run flush with the floor. As in Edinburgh, the stone would be bedded onto lime ash set onto timber deafening boards.

After a report by the fire authorities, it was found that this led to joists catching fire and then causing dangerous tenement fires. A new bye-law was introduced in 1892 which required any floor joist to be at least 18in. away from the chimney breast, stating that:

> *'Every fire-place opening shall have a sufficient hearth or horizontal slab of durable and incombustible material, at a level of the floor, extending the length and depth of such opening, and to a distance of 18 inches beyond the face of the chimney breast and six inches beyond each side of the opening.'*[54]

54 1892 The Glasgow Building Regulations Act.

[left] Fireplace with open gas fire

[right] Fireplace with stove

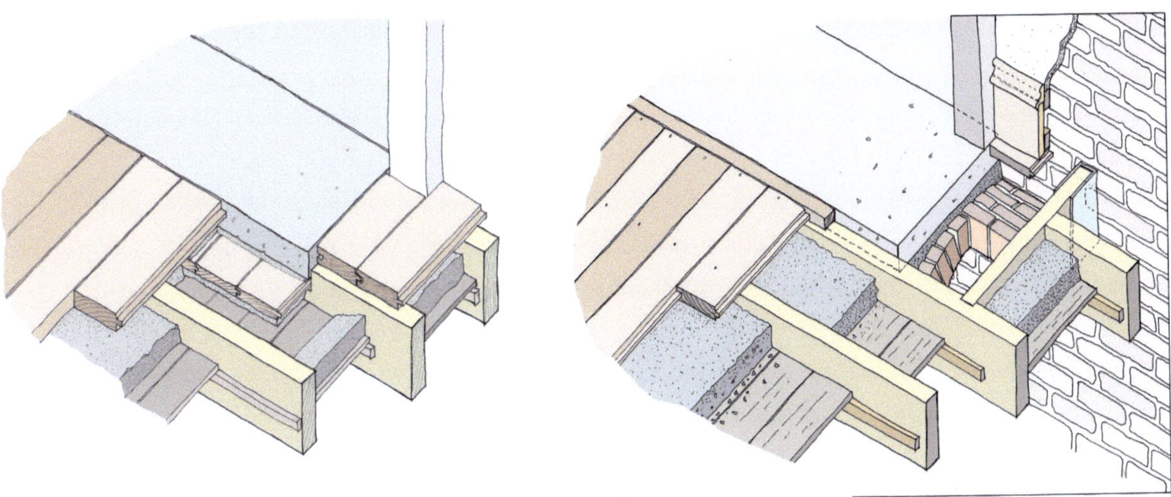

[left] Early hearth set onto checked floor joists which resulted in fires

[right] Later regulations required brick arches and trimmer joists to be kept away from the hearth

This required the joists supporting the stone hearths to be bridled by thicker joists. The two small bridling joists were bedded into cast-iron shoes in the fireplace wall to ensure the bridled joists were 6 inches away from the flue opening. A stone hearth would then be set onto a brick arch that spanned between the wall and the bridling joist. In later tenements after about 1900, concrete hearths would be used, about 6 inches thick and sent into a timber base plate. The top surface would be tiled.

A coal delivery, in jute bags

The hearth depth could be about 24–30 inches and comprised a front hearth and a back hearth. Arbroath stone pavement was normally used which was about 75 mm to 100 mm thick.

The coal bunker was usually formed in the kitchen, into a recess with a timber front side and top. Coal would be delivered in jute bags. In some richer areas, basements would be used to store the coal and coal chutes from the pavement allowed the coal to be funnelled down into the basement coal cellar.

Floor construction

In medieval Scotland, joists would be formed from roughly square baulks of timber[55] (oak or Douglas Fir) placed about 1 ft 7 in. (500mm) apart and spanning about 14 ft 7 in. (4.5 metres) between the cross walls. Joists were often parallel to the street frontage. Inch-thick floorboards were laid over the joists, so there would be no ceiling apart from the underside of the floorboards. Noise transfer would have been a problem and they would have presented a serious fire risk.

In Georgian Edinburgh, the floor construction made use of imported timbers, but as the tenement plans were deeper, and because of fire regulations, floor joists were mostly perpendicular to the front and back walls and were supported on internal timber beams spanning between internal brick walls. Joists might be about 2ft 7in. × 6ft 7in. (67 × 167mm) at 500mm centres and be dovetailed into the main beams 6ft 8in, x13ft 9in. (170 × 350mm) which ran parallel to the frontage.

One advantage of this type of construction is that the shorter lengths and main beams were able to also support Caithness stone slabs (90 mm thick) which would be bedded on mortar over rough timber boards laid over the joists. Tenements in Victorian Edinburgh followed the same tradition of flooring construction.

The joists of a Glasgow Victorian tenement always run from front to back, with a rest of about 9in. The joists were in one piece, usually about 30–36 feet (10 to 11 metres), but the actual span is broken up by the internal walls and bed recess beams (usually 6in. × 5in. (152 × 127mm) but more slender beams are often found in cheaper or earlier construction).

A typical 10in. × 2¼in. (254 × 57mm) joist could span 18 feet (5.5 metres) with joists at 14–15in. (35–38mm) centres. To stiffen the joists, solid timbers called dwangs[56] were inserted between the joists, two rows for a span of 18ft (4.5 metres).

Bye-law 28 of the 1892 Glasgow Building Regulation Act set down permissible sections of timber for different spans for joists at 15in. centres:

Span of eight feet would require 6in. × 2in. joists

Span of 16 feet would require 9in. × 2¼in. joists

Span of 20 feet would require 11in. × 2½in. joists

In all cases, as these long joists were prone to twist, dwangs were required to ensure they stayed in a vertical position and thus prevented buckling. When spans exceed 12 feet a dwang was required midpoint. When the span exceeded 18 feet, two dwangs were required.

55 Approximately 6in. × 6in. in section. Also termed ‘balks’.

56 Dwangs ensure the joists do not twist and help to stiffen the floor. Solid dwangs are commonly used although diagonal dwanging can be found.

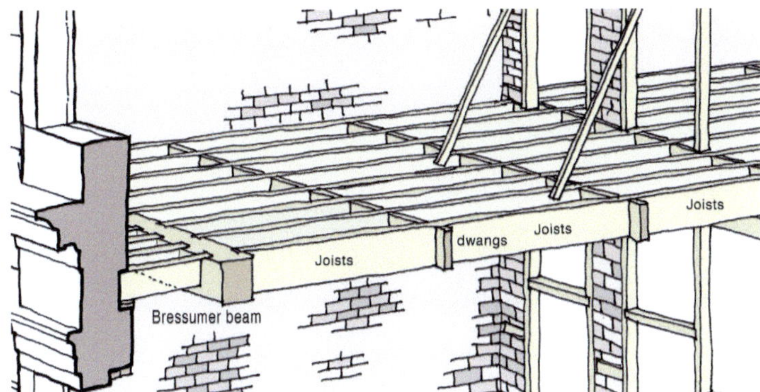

Joists span front to back, at bays supported on bressumer beams, dwangs between to stiffen floor

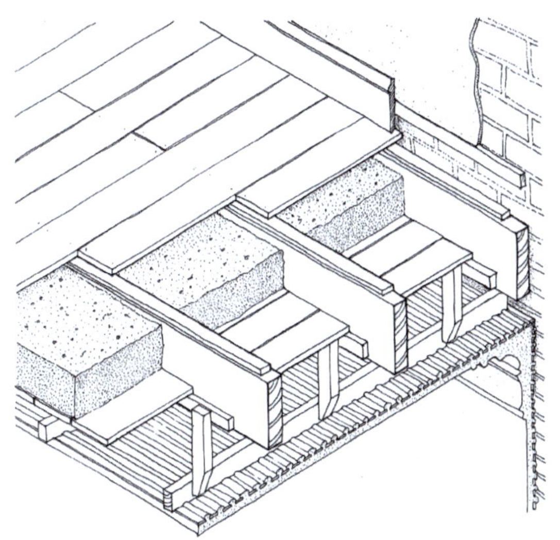

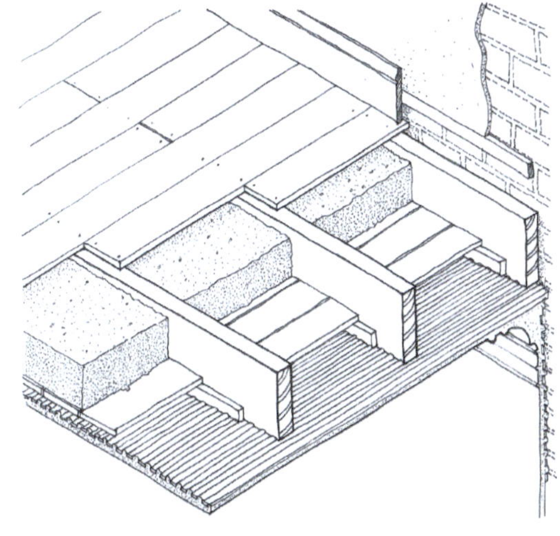

[left] Edinburgh floor construction

[right] Glasgow floor construction

Bed recess beams

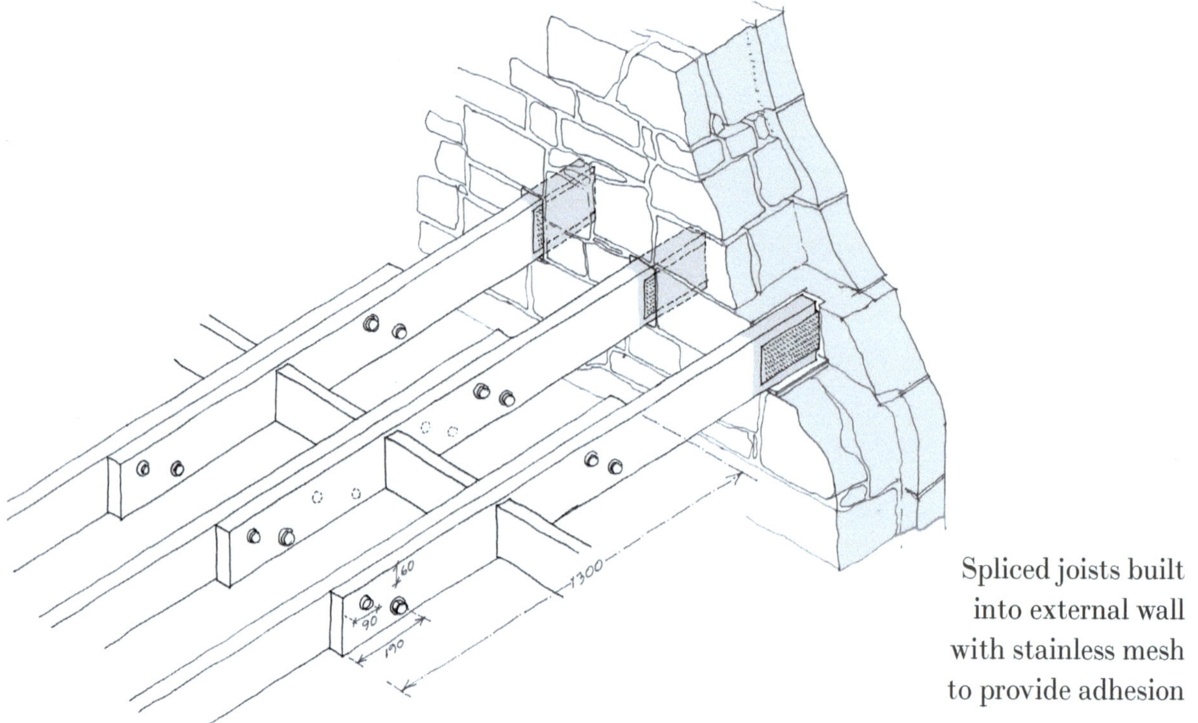

Spliced joists built into external wall with stainless mesh to provide adhesion

The joists are usually built into the outer walls and mortar filled. Sometimes 'ceramic shoes' are bedded into the stonework to hold the joists. However, it is important that the joists are well 'bedded' into the outer walls as they help to restrain the outer walls – this is usually done by securing a stainless-steel mesh to each joist and bedding the joist into the wall with a lime mortar.

Large bridling joists can be found at fireplaces to support hearths.[57]

Ground floor joists

These were called 'sleeper plates' as they were thinner than joists on the upper floors and spanned between small half-brick sleeper walls which were designed to allow through ventilation of the solum, front to back. The joists are typically about 6½in. × 2½in.

Deafening

Deafening is the material which is laid between each joist and which helps to reduce sound transfer between flats. It is designed to provide a mass to the floor to prevent the transfer of sound to the flat below. Deafening did start to be used in medieval times although the deafening boards were visible and the battens that support the deafening boards were neatly chamfered – this can be seen in Glasgow's Provand's Lordship.

Timber deafening boards, about ½in. thick (12–14mm) are installed to fit tightly between the joists and are set onto timber fillets nailed to the sides of each floor joist. Originally there would have been small gaps between the deafening boards, so the builders would lay a layer of coarse lime plaster to seal up any gaps, then, when dry, fill the void with about 2in. of dry ashes.[58] and

57 See Chapter 3 Fireplaces and hearths.

58 Engine ashes or other suitable material.

Deafening material placed on deafening boards set between joists

a final coat of lime, sand and ashes about 1½in. thick.[59] However, most of the deafening is a waste product from the building site and thicknesses of deafening may vary, but are usually about 70 mm. In later periods after 1900, foamed slag was used.

One disadvantage of deafening is that if there is moisture in the floor through flooding or a plumbing leak, dry rot can develop in unventilated spaces between the joists as the deafening holds the moisture and prevents drying of the timbers.

Where joists are tight to a wall, the space between wall and joist must also be filled with deafening.

When flats get flooded from plumbing defects, the deafening material will get soaked so, to prevent rot developing, the floor and deafening may need to be removed to dry out. Plumbers or DIY enthusiasts also often remove deafening to make space for pipework – in both cases, the deafening should be replaced. If the original deafening material is not available, then a bagged material of limestone chips can be used. Glass wool is often used as a replacement, but it has no mass so it is generally ineffective as an acoustic barrier.

When wet rot occurs at the embedded joist end, it can turn into dry rot if the conditions are right and travel, unseen, under the floorboards to affect other timbers. The deafening material can hold moisture and the space is usually unventilated, so this is a major risk in older tenements.

Floorboards

Timber tongued and grooved boards are laid across the joists and fixed with two flooring brads[60] at each joist, one as a secret fixing through the groove and the other through the board. The flooring boards are commonly 1⅛ inches thick, 5 or 6 inches broad. Any joints in a floorboard must be made at a joist position, ideally chamfered, to allow two nails through both boards.

The floorboards should be kept away from the walls to allow for movement and to prevent contact with any dampness in the wall. Each floorboard is double nailed to the joist below, one nail through the face of the board and the other set under the groove.

Concrete floors and ceilings

Concrete floors and ceilings started to come into use around 1900, really for the purposes of fireproofing rather than for structural reasons. The five-storey tenements at Camphill Gate in Glasgow were all built using concrete floors, including the flat roof. The concrete was

59 During repair and renovation works, the old deafening may have been removed and filled with glass fibre quilt. Unfortunately, this is not effective as an acoustic barrier. Modern alternatives to deafening are limestone chips, trade name "Quietex' which are pre-dried and bagged.

60 A cut nail with four sharp sides that tapers in width.

laid in-situ, about 5 inches thick and corrugated metal was used to increase the transverse strength of the floors. Concrete reinforcement was just coming into use, although concrete floors in tenements and high-rise buildings existed in America well before 1870; in Glasgow, concrete floors for fireproofing were seen as essential for five-storey buildings by 1890.

Traditional ceiling construction

Ceiling construction techniques vary between Edinburgh and Glasgow. In Edinburgh's Georgian tenements, the ceiling would be suspended from the floor joists and a grid of timber branders (approx. 1¼in. × 3in. (32 × 80mm) at 13¼in. (340mm centres)) used, supported on timber hangers fixed to the floor joists.

In Glasgow's Victorian tenements, the timber laths would be secured directly to the underside of the flooring joists, so there was a much closer distance between the ceiling and the floor above, which made sound transfer slightly more problematic than Edinburgh's floors.

Timber laths would be fixed to the brander framing or floor joists. The first 'scratch' coat of plaster was pushed through and over the laths and the gaps to form '*rivets*' or '*keys*' of plaster which acted to hold the plaster firmly to the backing. The 'scratch coat' contained ox hair, goat hair, or manilla fibre, but for ceilings more hair was used to give additional tensile strength to the plaster. This first coat was then scratched to form a key for the second straightening coat and then followed by the final finishing coat.

Any cornices would be formed over with a lath framework built up at wall junctions and plaster '*run*' in situ using a metal template formed to the cornice profile. Decorative cornices were only found in the main public rooms, rarely in bedrooms or kitchens which would have plain run mouldings. Any decorative additional features would be cast in a workshop then fixed to the cornice and bedded into the plaster backing when it was still setting.

Ceiling roses would also be made in a workshop and then secured with wires to the central ceiling joists. Fibrous plaster was introduced after 1856 and it affected the plastering trade as plaster cornices were no longer run by hand except in select buildings.

Over the years, many original lime plaster ceilings have cracked and been replaced. This is usually done with plasterboard then given a coat of *bonding plaster* and a *skim coat* of *gypsum plaster*.[61] However, if the laths are still in place, a better result will be made using a lime plaster to the original mix.

61 At least 30mm of plasterboard as well as plaster skim will be required to provide adequate fire protection.

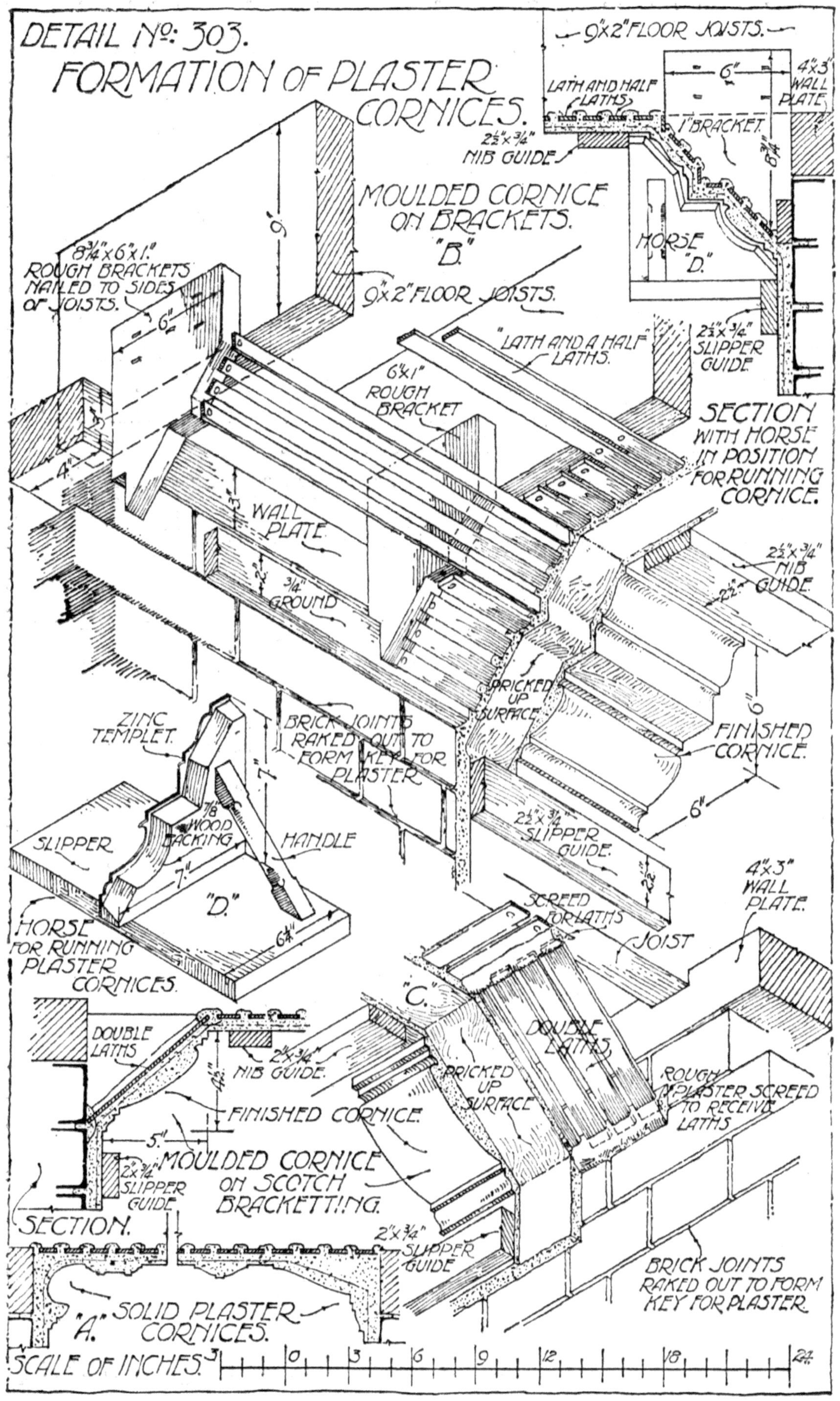

Formation of plaster cornice in-situ

Chapter 4

Stone that built our tenements

What makes Scottish tenements unique is that the main outer walls were made from stone, often sandstone and initially extracted from local quarries. The majority of tenements throughout Scottish cities are made of it, as are larger public buildings. This may have been because at one time quarries were plentiful, over 1,200 in Scotland, and they may have been local to where the building was. However, it's likely that the skills of the mason had been developed from building castles and defensive buildings, as stone was considered to be stronger, and masons became builders for the tenements. Originally rubble stone was used and protected with a lime harling. Once the Georgian period started, harling fell out of fashion and ashlar was more common. Two colours of sandstone can be seen in the buildings of Glasgow: red sandstone and buff sandstone (cream/yellow in colour).

From early days up to the 1880s, buff sandstone from Giffnock and Bishopbriggs, for Glasgow, and from Craigleith and Hailes, for Edinburgh, as well as other local quarries, was transported initially by horse and cart, then eventually by barges on canals. After 1880 when a railway network started, quarries from further away became much more affordable. In some areas near the coast, stone was transported by sea. Red sandstone from Dumfries only started to be used extensively once railways could be used to transport stone effectively after 1890.

Sedimentary, igneous or metamorphic rocks

Sedimentary rocks are sandstone, red or buff. Sandstone can be hard or soft depending on the sediments it contains. Sandstone was common in the central belt so that accounts for its use in Glasgow and Edinburgh. Flagstone is also a sedimentary rock which is split into different thicknesses, quarried in Caithness, Angus, Arbroath to form flagstones and hearths. *Greywacke*, a dark grey hard sandstone with quartz and *feldspar*, is quite hard to cut, found and used as a building stone in Southern Scotland.

Limestone is formed from an accumulation of shells or calcareous hard parts of marine organisms laid down in the Carboniferous period. Portland Stone is a typical limestone. Most limestone is found in England and Southern Scotland, where it is used to make lime and lime mortar rather than quarried as a stone, although some limestone buildings do exist in Scotland.

Igneous rocks are formed by hot molten magma from the earth's crust. They tend to be basalt or dolerite, dark grey and black. *Andesite* found in the Pentlands can be pale pink and red.

Granite and black gabbro found in Aberdeen have a wide range of fine, medium and course grained rock, often including feldspars. Granite setts were extensively used to form streets.

Metamorphic rocks are formed by high pressure within the earth's crust. Slate is metamorphic and extensively quarried at Easdale and Ballachulish. Although west of Scotland mostly used the thicker West Highland slate, Welsh slates were used in Scotland as quarries in England could offer the smaller and variegated slates which were unsuitable for the English market, to Scotland, because all slate sizes could be used on a sarked[1] roof.

Marble is also a metamorphic rock and quarried at Ledmore in the Highlands (although rarely used in tenements).

Whinstone is a generic name, used to describe any dark coloured rock. It could include igneous rocks such as basalt or dolerite, although it's mostly used to describe the sedimentary rock, greywacke, from the Southern Uplands. Whinstone is mostly used for road chippings as it is hard to shape.

Sandstone

Sandstone is the material mostly used across Scotland, apart from granite (mostly used in Aberdeen).

Early tenements and residential buildings were built using stone in a rubble form, with rebates at openings and quoins (coins) at corners. They were then given two coats of lime harling and lime washed. After 1750, harling fell out of fashion, initially with the wealthy, then with everyone, and stone with an ashlar finish was considered more pleasing and easier to maintain. (Harling needed to be lime-washed about every five years.)

The buff sandstone is generally from the Carboniferous period (approx. 320 million years ago), a period which also produced the coal seams found in and around Glasgow. The buff sandstones were laid down in vast river systems, evidence of which comes from finding pebbly layers or even bits of plant debris within this rock. In the 18th and 19th centuries, this sandstone was quarried from a number of sites around the city centre, Glasgow Queen Street station is built on the space left after one such quarry. There were quarries at Cowcaddens and around Kelvingrove and Partick as well as other areas of the city. Once the railway network was established and grew, by the 1890s stone could be brought in from further afield and much of that brought in to Queen Street station was from the east, towards the Stirling area.

The red sandstone comes mainly from Dumfries and Ayrshire and is from the Permian period (approx. 270 million years ago). During this time, a vast and expansive desert stretched across Scotland, resulting in massive dunes and arid conditions. Today the evidence of this desert can be found in the red sandstones used in Glasgow, in the form of cross bedding (evidence of where desert sand formed dunes) and by the red colour itself, representing an iron-rich coating of the sand grains, a phenomenon which can still be seen today in the Sahara.

1 Sarking was timber boards laid across the roof joists (derived from 'sark' a shirt or sail as in *Cutty Sark*).

Greywacke with sandstone quoins

Raploch Quarry. Taken by Stewart McKenzie, Image courtesy of Smith Art Gallery and Museum. Note proximity to housing!

Although the original sandstone quarried in Glasgow was of the finely bedded variety, once transport improved it was much more desirable to use more thickly bedded stone called 'Freestone' or 'Liver-Rock', which makes for a smooth, clean-looking building stone. Freestone is the best stone as it is homogeneous and without any clearly defined strata.

Scotland still has a number of quarries, although the list is dwindling. When working with stone or lime, it is best to get a sample analysed first to ascertain what type of stone or lime it is and what its characteristics are, as it is best to try and match it, not just for colour and texture, but for how it will perform when placed beside other existing stones. Such analysis is done by the *British Geological Survey*, who have a database of samples from Scottish quarries, or the *Scottish Lime Centre* and is usually a requirement if you are working on a listed building.

Most of the stones used in Scottish tenement construction were sandstone, although granite was used in the Aberdeen area for building tenements and whinstone for smaller properties in country areas.

Granite

In Aberdeen, the stone used was granite, and although the stone had a sparkle, the tenement facades tended to be repetitive and basic as granite was a lot harder to cut. Parapets also tended to be simple without cornices and/or projecting courses. Granite stone has lasted, although, because of how impervious the stone is, it does not absorb moisture. Instead, the moisture can be driven through poorly mortared joints. Good lime mortar pointing is still essential for such stonework.

Granite was shipped over the world. In 1796 there were 600 workers employed in the granite industry making setts to pave London streets and when lathes were able to make granite columns after 1870, columns were made for banks and sometimes polished smooth after 1886 when Jenny Lind polishers were introduced.

Granite was difficult to hand cut, so it was often left squared with little decoration and in the 1830s was '*cut by an iron blade swinging backwards and forwards, grinding at each downward stroke a mixture of sharp sand and water into the granite*.'[2] After cutting it might then be hand dressed. If there was any decoration, the stone might be incised with a pattern. Aberdeen bond was often used on gables and frontages in Aberdeen, where large blocks of ashlar had

[left] Aberdeen bond

[right] Galetting of mortar joints

2 *Sixty Years in an Aberdeen Stone Yard* by John McLaren page 6.

Workshop in Aberdeen

three or four smaller blocks in between them. This made best use of the cuttings. The bonding pattern was sometimes used for decorative effect as in George Square in Edinburgh. Galleting[3] occurred in Aberdeen around the 17th century.

By 1890 machines were available to turn, cut and polish the granite. Columns could be turned on lathes, and by 1900 air powered chisels were used to incise and shape the granite.

The bonding patterns

All stonework needs to be stagger bonded in some manner, most visible on the front walls. Where there are long vertical joints, sometimes between buildings, these are termed Risband Joints or 'Racebond Joints' and should be avoided. Ashlar stones are laid in courses, averaging at about 300mm in height; the depth of ashlar masonry should be about 175mm.

One area where *risband* joint occurs is inside a close, where the brick close walls meet the outer stone wall. Usually there is a brick tie every fourth course, but this is rarely effective and cracks develop in the vertical outer corners of close stairs. Party walls which are sometimes built of brick and meet outer walls of stone are also at risk of developing disband joints. If cracks or movement occur here, remedial work may be required in the form of steel straps and rods to secure the inner walls to the outer wall.

Uncoursed random rubble

Rubble stones which are not in any consistent shape require skill to place the stones to ensure good bonding. Uncoursed random rubble is the cheapest form of walling and may be used in garden walls, although in earlier times (1700–1800) random rubble would be used as a base course for lime mortar harling. Smooth ashlar quoins and sills would eventually be used at openings.

Coursed random rubble

This is where the random rubble is roughly cut and laid in levelled courses. Regular coursed squared rubble are courses laid level but vary in height. In Glasgow tenements, the rubble stone is roughly *stugged* or *hammer dressed* so that the blocks are fairly square. This facilitated speed of erection.

Uncoursed square rubble

This is where the squared rubble stones are laid to level courses, with larger stones or quoins formed between smaller stones. Uncoursed stones would still be laid level but to a certain pattern which might contain smaller stones and snecks.

For rubble work the finishes would usually be:

Scabbled rubble

Pitched faced rubble

3 A 'galet' was French for a pebble. Small stones would be formed into the mortar beds during construction.

[above] Uncoursed random rubble wall

[upper left] Rubble wall

[lower left] Scabbled random rubble wall with stone margins

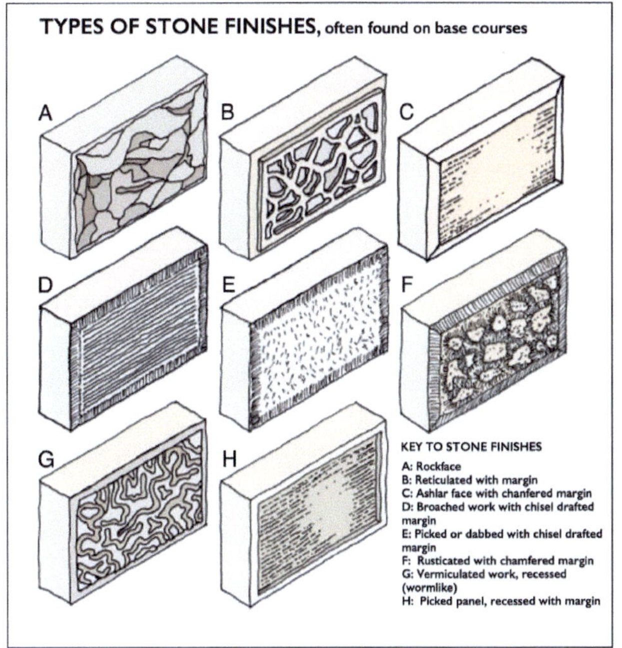

[left] *Droved ashlar wall*

Broached ashlar finish (horizontal lines)

Stugged or *boasted* or dabbed ashlar: formed by a mel (mallet) and point chisel to break up the surface, and sometimes pointwork in ashlar

[right] Stone types

Stone finishes

Different finishes are possible in ashlar stones as well as rubble work, as shown in the attached drawings.

For ashlar work the finishes would be:

Polished ashlar. Originally this would be done by hand, with masons first broaching the stone to shape, then droving it with a chisel and finally polishing it by sanding. With the advent of mechanical sawing around 1900, smooth ashlar can be readily achieved but sanding was still necessary to avoid saw marks.

Droved ashlar (rows of vertical lines).

Quarries

Quarrying was originally done by quarrymen using jumpers to drill holes in the stone, then using plugs and feathers (wedges) to split the stone. Dynamite was only used sparingly after 1873. Machines were available to cut and polish stones, but most ashlar and even rubble stones had to be hand dressed using droving chisels and mallets.

Many of the quarries that supplied stone for our tenements are now closed and compatible stones from other quarries (often from England rather than Scotland) are required for any stone replacement. This will be done after analysis of the existing stone is carried out and the stone type is matched (often carried out by The Scottish Lime Centre or British Geological Society).

The main quarries are listed in Appendix 5.

Lime

Lime is made from limestone, a sedimentary rock formed from chalk and shell laid down in prehistoric times. When limestone is burned in a kiln at high temperatures, this produces lime. Lime was mostly used as a fertiliser in farming, but it was also used as a binder to form lime mortar which was used for pointing, bedding of stone, harling and lime wash as well as internal plasterwork.

Lime mortar[4] is composed of lime, an aggregate and water. The aggregates may contain sand, fine gravel and even clay. The advantages of lime mortar – its ability to breathe and move – are unfortunately offset by the length of time required by the mortar to achieve resistance to rain, frost or rapid drying out[5]. It was in common use up to 1900 but eventually Portland cement began to be used and lime. Portland cement was available from the 1840s and had the advantage of having a faster setting time, so it was often used in foundations. Lime was not used after 1950, although it's coming back after the 1970s.

However, lime mortar had a lot of advantages when building in stone, as it was more readily available and easy to work with. After 1900, cement mortar harling was sometimes used in exposed areas as it was considered to wear better in coastal areas. Indeed, some of these early cement renders have lasted well. However, as builders began to use harder mixes, problems arose.

The advantage of lime mortar in stone is that it allows the wall to be vapour permeable so it can dry out. Much of the drying out occurs at mortar joints. If the joints are repointed with a hard cement mortar, then this prevents the stone from drying out and leads to decay of the stone. Cement repairs to stonework surfaces, and to individual stone, can lead to complete decay of the stone underneath.

Our advice is that existing lime mortar should be analysed first to assess its make up and only then can a choice be made on selecting an appropriate match to the lime.[6]

Scottish lime, particularly in the Western Isles, was produced from seashells. When burnt in a kiln these produced a high-quality lime that was suitable for internal plastering.

Types of lime

There are two main types of building limes: hydraulic lime and non-hydraulic lime. For repairing traditional tenements and other buildings built with lime mortar, current day 'natural hydraulic limes' (NHLs) and 'air limes' are available.

NHLs are typically used externally for repointing and harling. They are produced from limestone which contains additional minerals such as silica and alumina – when burnt this allows the lime to set in the presence of water although carbonation is still important in achieving a good set. They set faster and harder than air limes. NHLs are classified by strength; NHL2, NHL3.5 and NHL5 (the number is the compressive strength in N/mm^2 at 28 days, so NHL5 is typically the strongest and fastest setting, but also the least permeable

4 Bye-law 10 of Section 177 page 752 of the 1892 Glasgow Building Act stated that 'Mortar must be made from quicklime carefully slaked....mixed with clean sharp sand or grit.... proportion of one of lime to 3 of any other material.'

5 Additional Lime mortars: Pat Gibbons: *Material and Traditions in Scottish Building.*

6 The Lime Centre in Scotland carry out such tests and should be able to provide advice on lime types.

and least flexible). NHLs come in the form of pre-bagged dry hydrate (powder), to which aggregate and water are added on site.

Air limes are typically used externally for repointing of fine ashlar joints and internally for plastering. They are produced from pure limestone and set fully by carbonation (absorbing carbon dioxide from the air, hence the name 'air limes'). Air limes will not set in wet or damp conditions, and they set more slowly than NHLs, though they are the most permeable and flexible of all lime mortars. Air limes come in the form of quicklime (calcium oxide), lime putty or dry hydrate (both calcium hydroxide). The latter is also known as builders' lime and is generally only suitable as a plasticiser in a cement mortar and is not suitable for making a lime mortar.

Lime putty is mixed with a coarse or fine sand to make a coarse stuff or fine stuff (available pre-mixed) for ashlar mortars and internal plasters. Lime putty is what is often used to point thin beds of ashlar stone. It is produced as a result of the slaking process. Once the lime has been fully slaked and allowed to mature ('fatten up'), it makes putty that is soft and permeable, and contains finer lime particles.

Quicklime can be used (by contractors experienced in their preparation) for making a hot lime or 'hot-mixed lime' mortar by slaking with aggregate and water (this reaction creates heat and steam, hence the 'hot' reference). Many historic mortars were made in this way.

Aggregates: sand, ideally sharp sand and fine gravel make up common aggregates in lime mortars. It is often best to have an analysis of the lime[7] (and stone) before specifying repair work on tenements (and usually required in all listed buildings and Conservation Areas). This will tell you what lime mix was used originally, the type and size of any aggregate, the lime type, the void ratio and the clay content in the lime. The Lime Centre in Scotland can help with such analysis.

Cement

Ordinary Portland cement was patented by Joseph Aspdin in 1824 and developed by his son, William Aspdin in the 1840s. It was recommended for parapets, chimney heads, plinths and sills in 1898 and Gourlay recommended it for half-brick internal partitions as 'cement mortar' is stronger than lime mortar, and sets more quickly.

Roman cement (1:2.5 clay and chalk) was imported to Glasgow in 1854. It was not used in tenements but Roman cement and other proprietary products were developed and used from the late 18th century.

Cement (ordinary Portland cement) was more commonly used to make concrete and used in foundations where it would set in a few hours even in the presence of water. In the post-war period, cement was often used (and continues to be used) to repair and point stonework. This has the effect of preventing the stonework from drying out leading to decay of the stone. Sometimes a cement lime mix is used to help with plasticity, but only lime mortar should be used when pointing stone. Modern-day cements tend to be much stronger, harder and less permeable than most of the 19th century cements.

Tenements were built from no fines concrete in Court Street, Dundee in 1874, well before no fines was used by the SSHA to build houses in the post Second World War period. The buildings in Court Street still remain.

7 See HES Guide No 6 'Lime Mortars in Traditional Buildings'.

Court Street, Dundee, no fines construction 1874

Bricks and brickmaking

With the exception of a few isolated examples from the Roman period and the 17th century, the use of brick in Scotland began in earnest after 1700. Bricks from the period 1700 were generally smaller than the imperial-sized bricks used in the latter part of the 19th century following the advent of mechanisation. A tax on bricks was introduced in 1784,[8] which in some cases saw larger bricks manufactured, although this was the exception to the rule. The standard imperial-sized brick, which was larger than today's bricks, came into common use in the latter part of the 19th century.

Despite the common conception that Scotland's tenements were built of stone, this is only partially true. Bricks form important parts of many tenements in Scotland, particularly in the 19th century. A technique known as 'compound masonry' or 'brick ashlar' was commonly employed. This saw walls with a backing of brick, but a front facing of either rubble or ashlar stone. Compound walls which bonded brick to stone were common and brick bonding patterns developed to suit the needs of tenement builders.

Scotland had a number of distinct craft practices in respect to building in brick in the 19th century, most significantly in the use of Scottish Bond (three, four or five courses of stretchers between heading courses) and a distinct Scottish gauge which saw four courses of brickwork rise to heights of between 13 and 15 inches.[9]

Brick was not, however, commonly used for the construction of tenements in their entirety. The earliest reference to a tenement where brick was the principal construction

8 The brick tax lasted until 1850. It was originally set up to help pay for the wars in the American Colonies.

9 Jenkins M. (2022) gives details of Scottish bond and gauge, pages 151–70.

material comes from 1832, when a group of six three-storied tenements, known as Thomson's Buildings, at 162 Orr Street in Glasgow, were constructed.[10] The earliest surviving example of a brick-built tenement is the Rosemount Building in Edinburgh, built in 1857. This was an innovative building, constructed using an early form of cavity-wall construction, and is also the earliest use of polychromatic brickwork in Scotland. Later examples include brick fronted tenements, as in 1892 at Cathedral Court[11] in Glasgow and Greenhead Court in 1900. Later still, brick walls were roughcast as in 1904 in Perth at St Johnstoun's Buildings.

Quarries for brick clay were found throughout Scotland in the 18th and 19th centuries. These ranged in geographical location, from Cruden Bay in the north-east, through Perthshire and Angus; and many in the central belt. The clay used for brickmaking in Scotland varied considerably; much of it, however, contained ferric oxide, which gave rise to the red and orange colour of most Scottish bricks.

The methods of manufacturing brick changed considerably in the 19th century. In the 18th century and the earlier part of the 19th bricks were moulded by hand and fired in small kilns. Handmade bricks continued to be made up to the 1950s, but generally mechanisation developed after 1836 with machines which extruded the clay in a rectangular shape which were then wire cut. Brick was used to build factories in the 1820s and 1830s but rarely used exclusively for tenement construction except for housing built by factory owners for their workers as in Calton in Glasgow. Along with compound walls, chimneys, party walls, gables and close walls, it would be used to build turnpike stairs. After 1850 when the brick tax was withdrawn, bricks were used to build wash houses and spine walls and basement footings. In the mid 19th century this made bricks cheaper to produce and saw an increasing use of brick being used in tenements.[12]

The mid 19th century saw the use of various mechanised methods of forming bricks, an early Scottish example being the Tweeddale brick and tile machine. The advent of continuous kilns for firing bricks also greatly increased brick production in Scotland.

The use of shale, often a by-product of the coalmining industry, was used in Scottish brickmaking from the latter part of the 19th century. This led to the manufacture of colliery or composition bricks from the shale.[13] Steam-powered crushing machines allowed the shale to be ground and plasticised, so that it could be used in brickmaking. Although some of these bricks were unsuitable for use internally, many had sufficient durability and strength to be used externally.

These advances in methods of manufacturing had a stark impact on the number of bricks manufactured in Scotland. In 1802 15.25 million bricks were manufactured, rising to 47.75 million in 1840. There were hundreds of brickworks operational in Scotland throughout the 19th century. By 1901 in Glasgow, 40 million bricks a year were being produced with 32 brickworks in Glasgow alone.[14] Now there is one, Raeburn, with most bricks coming from England and Germany.

10 Worsdall, F. (1979) *The Tenement: a way of life*, page 83

11 This also had rear access by gallery decks.

12 Allan & Mann Works in Rutherglen in the 1860s had two Clayton brickmaking machines to make bricks.

13 'Many bricks now used in Glasgow are called composition bricks because they are made from pit refuse, which is clay shale and is called blaize or blaes.' Gourlay, C. (1903) *Elementary Building Construction and Drawings*, page 1.

14 Jenkins M. (2018) gives background on these figures and advances in *Scottish Brickmaking*, page 42.

Brick was commonly used for internal walls in tenements, a practice which can be seen from the early 19th century. Brick internal partitions were used extensively and were often only half a brick thick. In some tenements, the rear and side elevations are of brick, with the front façade built of stone. When it is considered, however, that the rear and side elevations plus internal partitions were constructed of brick, this means a seemingly stone-built tenement may in reality be formed with up to 80 per cent brickwork. Brick was also used for specific purposes, such as the formation of stair towers, an early example of which was 37 Eglinton Street, Glasgow, built in the 1830s. Glazed bricks can also often be found in tenement construction, on internal walls or rear and side elevations. These fulfilled several purposes, being easier to clean and maximising light.

Although stone was always the preferred material for principal elevations, the use of brick in Scottish tenements should not be underestimated. Although often hidden from view, brick has been a vital part of Scottish tenements, particularly from the second half of the 19th century. This remains an important and often underestimated part of not only the heritage of Scottish tenements but also their repair and maintenance needs.[15]

The industrial revolution increased brickmaking: in 1802, 15.25 million bricks were manufactured, rising to 47.75 million in 1840. By 1869 there were 122 brick and tile manufacturers in Scotland, rising to 270 operational brickworks in the 19th century.

By 1901 in Glasgow, 40 million bricks a year were being produced with 32 brickworks in Glasgow alone, although Lanarkshire had as many as 52.

Although stone was still the preferred material for frontages, brick rear elevations began to appear after this date, often combined with gallery access flats and model lodging houses.

Stones and quarries used in different towns and cities

The following list describes the different tenemental areas in some of our major towns and cities and attempts to name the stones that were used and the quarries that provided the stone.

Glasgow and Paisley:

Main areas: Govan, Ibrox, Hyndland, Govanhill, Strathbungo, Shawlands; Pollokshields, Mount Florida, Cathcart, Bridgeton, Denniston, Woodlands, Partick, Maryhill, Dowanhill, Broomhill, Woodlands, Anniesland. Paisley Causeyside, Oakshaw and Central area.

Notable tenements: Provand's Lordship; Charing Cross Mansions; Caledonian Mansions; Fotheringay Road; Govan: British Linen Bank; Walmer Crescent; 335–45 Argyle Street (1870 tenement); Beaumont Gate; StVincent Crescent; Strathbungo

Mostly using buff sandstone from local quarries, tenements in the west end might have used stone from local quarries such as Woodside quarry between 1790 and 1850, although there were local quarries in the Glasgow area. Later, when railways provided ease of transportation, buff sandstone from Bishopbriggs and red sandstone from Giffnock was transported. Once the railways had developed, by 1890 stone could come from further afield, such as red sandstone from Ayrshire

15 Jenkins, M. (2014) *Short Guide: Scottish Traditional Brickwork 7*, Edinburgh: Historic Scotland, all pages.

[upper left] Inside a half brick close wall, joists for half landing and stone treads embedded in wall

[upper right] Exposed brick gable wall in Dunfermline

[left] Brick party wall in Glasgow

[upper left] Glasgow West End

[upper right] Paisley

[left] Edinburgh, North Bank Street

[right] Leith

(Ballochmyle) and Dumfriesshire (Locharbriggs and Closeburn). Sometimes stone types were mixed. In the north there was a quarry at Garscube near Anniesland Cross. After 1880 more quarries opened up in the central belt and at Earnock and Auchinheath, with lime coming from the Campsies.

Edinburgh and Leith

Main areas: Old Town for late medieval tenements and New Town for high-class Georgian tenements; also Victorian tenements in Gorgie and Dalry, Bruntsfield, Merchiston, Morningside, Polwarth, Slateford, Colinton, Marchmont, Portobello, Dalkeith and Musselburgh.

Notable tenements: Corner of York Place circa 1824 by David Paton; John Knox House; Thistle Court.

Predominantly using local buff sandstone from Craigleith and Hailes but also from the north of England. Since the early 16th century, stone had been shipped across from the Fife ports of Cullross and Burntisland into the late 1790s.

The Union Canal was used to ship stone from West Lothian Humbie and Binnie quarries until the railway network allowed moving stone from further afield in 1842. Stone from Polmaise, Plean and Dunmore came from the west and red sandstone came from Locharbriggs and Corncockle in Dumfriesshire. The Forth railway bridge in 1890 made it easier to transport stone from Fife (Grange and Fordell) as well as Cullalo in Fife and Longannet quarries. Lime was transported from Burdiehouse and Middleton as well as Charlestown. Welsh and West Highland slates were used.

Mostly tenements were built as four storeys, or sometimes five storeys if at a corner or overlooking an open area. They often had main door ground floor flats and close doors were common. Early tenements tended to have basements for the servants.

Early tenements were built with two pitched roofs, disguised by a front parapet in a classical Georgian style. Around 1890 roof styles developed into having a small pitch at the front and then a flat roof at the rear. Sometimes mansard roofs were developed with wallhead dormers and a lower pitched roof above. Generally close stairs were in the centre and top lit. The design for working-class tenements was four flats on a landing, all single aspect. Bay windows were common in better tenements.

Building tenements with red sandstone came in around 1890.

Stones for paving and steps in the New Town usually came from the Hailes Quarry.

Inverness[16]

Main areas: Medieval: Church Street. Georgian era: Church Street, Fort George 1748, High Street, Academy Street, Bank Street, Telford Street, Inglis Street, Bridge Street; Victorian era: Bridge Street, High Street, Station Square, Inglis Street, Academy Street, Queensgate, Union Street, Eastgate.

Notable tenements: Abertaff House 1590 – oldest house with turnpike stair and square cap house; Bow Court: Church Street 1722. Merchant's House dormers 1840; 22 Church Street 1898 Arts and Crafts four-storey red sandstone, blond sandstone,

16 Interestingly it was twinned with Augsburg in Germany in 1956.

Inverness

pilasters; town houses in Church Street 1690, steep roofs, stair to upper level; Leakey's bookshop end of Church Street 1790; 2–6 Eastgate 1795 Georgian three-storey scroll skew puts and harled.

Stone from Covesea and Tarradale, Inshes and Muckovie quarries and sometimes imported red sandstone from Dumfries and granite from Aberdeenshire. Lime from Keith. Welsh and West Highland slates used.

In Inverness, two-storey houses were often subdivided into flats, otherwise between 1830 and 1910, three and four storey tenements, often commercial at ground floor, mansard roofs with slated pitched roofs and pedimented dormers in timber with leaded roofs to dormers and slated flanks. Projecting cornices at eaves. Earlier tenements pre 1780 likely to be rubble stone and lime harled. Also with steep pitched roofs, now slated but would have been thatched originally. Simple stone margins around windows.

Aberdeen

Main areas: Georgian era: Urquhart Road, Summerfield Terrace, Rosemount St, Upper Kirkgate, Union St, Marischal St, Golden Square, Old Aberdeen. Victorian era: Rosemount Viaduct, Crown St, Bridge Street, King Street, Union Street, Market Street, Crown Street.

Notable tenements: Golden Square 1810-21. Window tax 1748; Bridge Place; Rose St; Chapel St; Bon Accord Terrace; Bon Accord Square; Queens Terrace; 106 Rosemount Viaduct/ Skene St; Rosemount Square.

Aberdeen

Local granite came mostly from Rubislaw Quarry with finely dressed work coming from Kemnay. Red granite came from quarries at Peterhead and even darker red from Deeside. Pine granite came from Corrennie of Deeside. Granite was a big industry in Aberdeen employing 9,000 men at one stage, so there were several granite yards cutting and shaping the stone. Lime came from Keith or Dufftown. Macduff slate from Aberdeenshire was mostly used but Welsh slates eventually imported by sea.

Large wallhead chimneys. Often with below ground level basements with pavement lights. Also two-storey Georgian properties with top attic floors and dormers. Four-storey tenements often with mansard roofs and top attic floors with dormers. Simple banded string courses and any detail incised into the stone. Properties in Old Aberdeen mostly two storeys with cherry caulked mortar beds.

Tenements with stairs at rear, usually two flats to a landing, timber stairs and wooden dados common, possibly because of difficulty in cutting granite. Ground floor flats entered off the close. Aberdeen bond developed to use smaller stones and provide pattern which would otherwise have appeared grey and flat, usually found on rear and gable walls.

Dundee

Main areas: Blackness Road, Perth Road, Springfield, Nethergate, Seagate, Panmure St.

Notable tenements: 56, 58 Arboath Road and 1 Kemback St; Gardynes House; Gray's Close.

Dundee

Local stone was good, particularly from Leoch Quarry, which produced a sandstone that was bluish-grey and fine-grained.[17] *Also the Carmyllie Quarries in Angus, described as 'greenish-brown and very hard'.*[18] *The quarry also supplied slates. The quarry reached a peak in 1890 when it employed 700 men, and exported south and to the continent.*[19] *Sandstone in Dundee also came from Kingoodie Quarry in Invergowrie, Pitairlie Quarry, and Cullalo Quarry near Aberdour in Fife. Lime from Charlestown and Cults. Mostly buff sandstone but also some red sandstone and brick was used in later periods.*

Platties developed before the *Burgh Police (Scotland) Act 1892* which allowed 24 houses off a single stair. Decks usually concrete in steel or cast-iron columns, staircase was a type of turnpike stair between landings at the rear, with an open close. Some mansard roofs with dormers and wallhead chimneys. Local slates and also Welsh and West Highland slates used. Flat roofs became common.

Paired doors in what appeared to be wide closes. House doors off the close, often with sidelights and fanlights. Tiled closes on ground level. Square stair in stone with central spine wall and rear close windows. Bays and oriels used. Some detailing looks like precast concrete, but is in fact stone.

Dunfermline

Main areas: Abbot Street, East Port, High Street, Kirkgate, mainly central area.

Quarries at the Grange, Burntisland, Newbigging, Fife and Brieryhill Quarry in Dunfermline, also Rosyth Quarry and Clunevar Quarry, west of Dunfermline. Yellowish grey sandstone from

17 Leoch stone was used on the Usher Hall in Edinburgh.

18 Edinburgh Geological Society (1980) *The Edinburgh Geologist* Issue 8, page 31.

19 Cologne Cathedral is floored with pavement stone from Carmyllie.

Dunfermline

Dumfries

Cullalo in Fife and Whinstone from Cruicks Quarry near Rosyth. Lime from Limekilns or Charlestown. Welsh and West Highland slates mostly used, although some scotch pantiles exist on earlier properties.

Early tenements likely to be rubble stone, harled and lime washed, often with stone wallhead dormers. Mostly two- and three-storey tenements often with dormers, and brick gables after 1900. Mostly commercial on the ground floor and by now converted to flats on upper floors although some four-storey tenements exist. Few bays or oriels.

Dumfries

Main areas: English St, Irish St, Church St, Buccleuch St, Friars Vennel.

Quarry at Locharbriggs providing red sandstone, also cove from same area. Corsehill and Corncockle for buff sandstone. Greywacke (a dark hard sandstone, like a whinstone) from Morrinton, Beatockhill. Roofs with Welsh and Cumbrian slates. Lime from Annan and Carlisle area.

Rear elevations sometimes in brick. Wide pens often lead through to rear courtyards. Some closes with twin doors, otherwise single panelled doors with fanlights. Close stairs have open well and cast-iron railings, straight stairs with half landings and close windows. Earlier tenements with stone margins and harled. Oriels above shop units. Tenements mostly three storey, but some four storey, often with fourth floor in a mansard with dormers set behind parapet. Parapets common. Quite a lot of two-storey housing, some with rear access to the upstairs flats. Some use of granite using Aberdeen bond, although whistle (Greywacke) also used and coursed with buff sandstone rybats.

Stirling

Stirling

Main areas: Victorian era/ late renaissance: Bow St; Spittal St; Georgian: Barnton St and Queen St, Baker St, Broad St; Georgian and Victorian: Cowane St; Upper Craigs Victorian: Friars St, Murray Place, Upper Craigs.

Stone from Auchinheath, a micaceous white and pale brown rock and Craigforth Quarry, granite with feldspar from Polmaise Quarry. A coarse-grained sandstone containing quartz pebbles from Raploch Quarry, and honey coloured sandstone from Thornydyke Quarry near Denny. Locharbriggs and Closeburn quarries in Dumfriesshire would have provided red sandstone. Brown sandstone blocks from Ballengeich Quarry. Lime from Charlestown. Brick also used. Welsh and West Highland slates.

Early tenements from the 16th and 17th centuries, would have used a local stone, possibly a whinstone or greywacke, with sandstone margins or rybats. Some Georgian two storey but mostly three and four storeys up Spittal Street. Some early 18th century lime harled buildings with crowstepped gables and cat slide dormers, with front turnpike stair access and basements. Later Georgian tenements with gable mounted turnpike stairs. Buff sandstone mostly for 1860s with dormers and wallhead chimneys. Red sandstone tenements with flat roof and rear deck access on Spittal St.

Greenock

Main areas: Victorian era: South St, Newton St, West Blackhall St, William St, Wellington St, Mudieston St, Dempster St, Greenock West, Gourock and Port Glasgow.

Clyde Sandstone and local sandstone, also possibly Mauchline and Ballochmyle Quarries for early tenements. Later imported red sandstone from Locharbriggs. Lime from Girvan, slate from Wales, West Highland or Cumbrian slates.

Originally mostly three-storey tenements, but also four storey often with rear turnpike stairs. Later tenements three storey with dormers in attic. Greenock was built on the sugar trade and also cotton, so significant links to slavery. Earlier part of Greenock in buff sandstone, sometimes with bays and basements to cater for hilly sites. Close stairs sometimes at the fronts, with front close doors with large fanlights. When three-storey, they usually have an attic storey with dormers. Older tenements with Dutch styled nepus gables and harled, windows with stone rybats. Also two-storey Georgian properties, often with dormers and basements. When red

Greenock

sandstone was introduced around 1880, four-storey tenements with grand corners and oriels, sometimes with buff sandstone at the rear. In Gourock, Sandringham Terrace built in 1900 in red sandstone by William Strachan of the Baltic Sawmills, a timber merchant, features wonderful close tiles by J. Dean of Glasgow.

Ayr and Kilmarnock

Main areas: John Finnie Street and Bank Street areas in Kilmarnock; Market St, New Market St, Canongate, High Street, Hope Street, New Bridge Street in Ayr.

Notable tenements: 61 Newmarket St, 4 Hope Street.

Buff freestone from Dean Quarry and red sandstone from Mauchline area after 1850. Much later red sandstone was imported from Locharbriggs and Corsehill. Lime from Girvan. Welsh and West Highland slates used. Brick used with sandstone.

Ayr

Kilmarnock three and four storey, both buff and red sandstone tenements. Earlier buff tenements mostly Georgian. In Ayr, some older tenements with nepus gables and harled walls.

Kilmarnock

Perth

Main areas: Kinnoul St, South St, King St, Tay St, South St, Marshall Place, George St, Tay St, Atholl Place.

Notable tenements: High St, George St corner. Riggs at rear; 30 St John St, four-storey tenement with dormers and five-storey at Marshall Place with tripartite windows.

Local stone, or later from Corsehill, Burghmuir and Newhouse quarries. Quarrymill quarry was a purplish red stone from the Scone Sandstone Formation, and the Witch quarry provided a local source of basalt. The local red-brown sandstone is from Scone. Lime from Charlestown and Cults. Roofing slates from Dunkeld and Logiealmond but also Welsh and West Highland slate.

Pre 1800 four-storey tenements with rear turnpike stairs and built from uncoursed rubble with stone margins, with nepus gables at the front, but also fully harled. Turnpike stairs with open well but lit by windows in stair wall. Some two-storey tenements with front turnpike stairs. Two-storey Georgian terraces with basements and dormers which may have been houses but are now mostly converted to flats. Three-storey sandstone tenements with nepus gables and dormers in a classical style but, sadly, painted.

Oban

Main areas: Nursery Lane, Hamilton Park Terrace, High Street, Ardconnell Road, George St.

Notable tenements: Oban Times Building on the Esplanade; Argyll Mansions 42 George St.

COUNTY CLASSICS
SALE
A.S.DEUCHAR

Perth

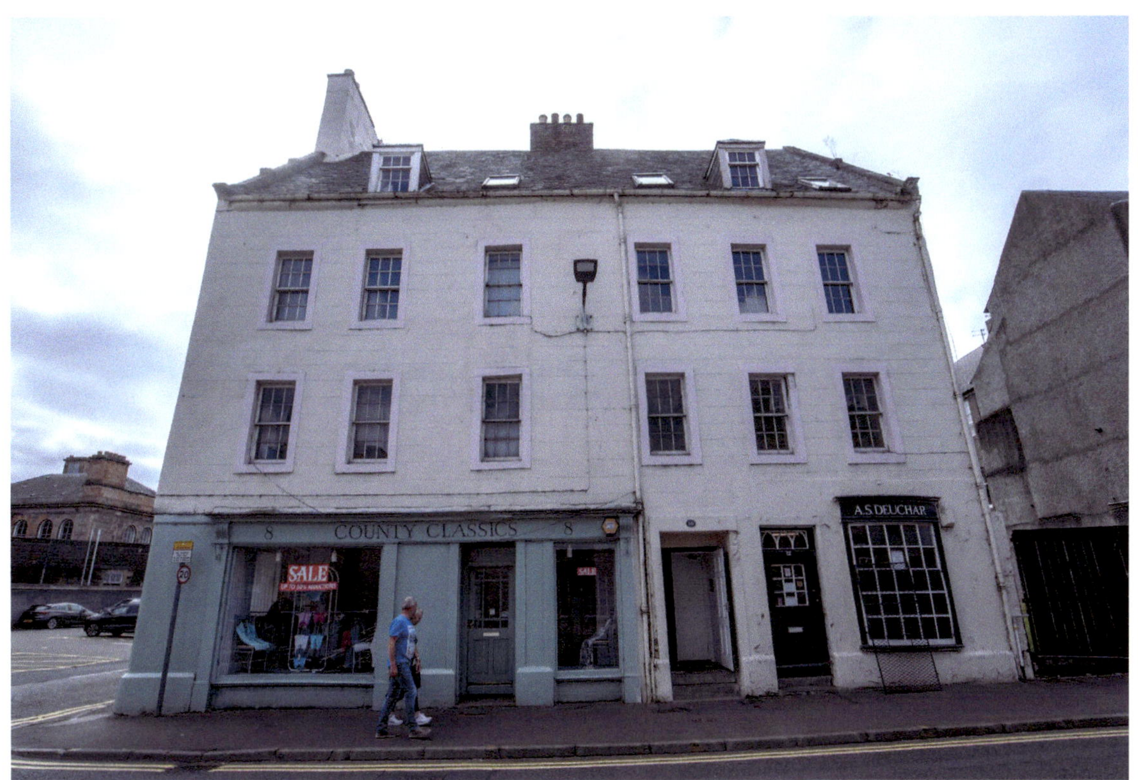

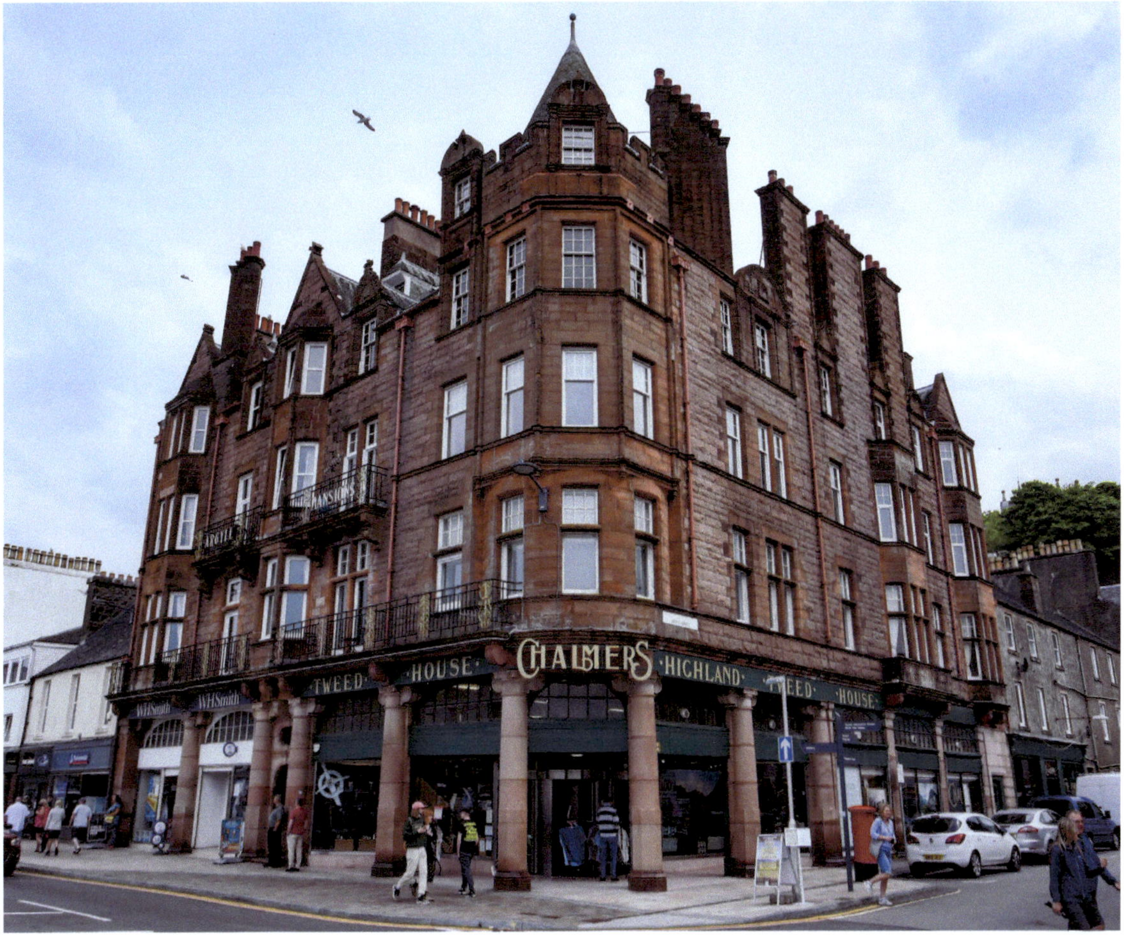
CHALMERS
TWEED
HOUSE
HIGHLAND
TWEED
HOUSE
WHSmith
WHSmith

Oban

Oban features quite a number of buildings of red sandstone which was likely imported from Ayrshire or Dumfries or St Bees although the quarry at Corrie or Arran may have provided some red sandstone. Quarries at Inninmore Bay and from Carsaig may have provided some of the buff sandstone although again it could have been transported from the Central belt. Sometimes granite was used, possibly from Cruachan.

I did not get to photograph every city or town, but here are a few. The location of work is listed under each photo:

Glasgow 1
Kelso_
White Swan
Arbroath1
Arbroath1-2
Inverness 1
Inverness 1-2
Paisley 1
Glasgow 1-2
Dunfermline 1
Jedburgh
Haddington_
Kirkintilloch_
PET SHOP
Stirling 1
Stirling_
Stirling 1-4
Portsoy_
Rothesay1
Rothesay1-2
Rothesay_
Paisley 1-2
Edinburgh 1
Paisley 1-3
Coatbridge_
Coatbridge_-2
Strathaven_
Edinburgh 1-2
Edinburgh 1-3
Paisley 1-4
Falkirk 1
Falkirk 1-2

Edinburgh 1-4
Leith 1
Dumfries 1
Dunfermline 1-2
Dunfermline 1-3
Inverness 1-4
Inverness 1-5
Inverness 1-6
Glasgow 1-3
Dunbar_
Dunbar_-2
Linlithgow_
Linlithgow_-2
Leith 1-2
Leith 1-3
Leith 1-4
Edinburgh _1-3
Dunfermline 1-4
Dunfermline 1-6
Glasgow 1-4
Glasgow 1-5
Dundee 1
Dundee 1-2
Dumfries 1-2
Dumfries 1-3
Dundee 1-3
Dundee 1-5
Dundee 1-6
Musselburgh
Greenock 1

Greenock 1-2 | Greenock 1-3 | Greenock 1-4 | Gourock_ | Greenock 1-6

Fort William 1 | Fort William+-2 | Fort William 1-4 | Fort William 1-5 | Fort William 1-6

Edinburgh__ 1-5 | Kilmarnock 1-3 | Kilmarnock 1-4 | Kilmarnock 1-5 | South Queensferry_

Dalkeith_ | Stirling 1-5 | Stirling 1-6 | Bridge of Allan_ | Oban 1-2

Oban 1-3 | Oban 1-4 | Oban 1-5 | Dumfries 1-6 | Annan_-3

Annan_-4 | Annan_-5 | Alloa1-5 | Perth1 | Perth1-2

Perth1-3 | Perth1-5 | Perth1-6 | Arbroath1-3 | Arbroath1-4

Forfar 1 | Forfar 1-2 | Brechin_-3 | Forfar 1-3 | Aberdeen1-2

Aberdeen1-3 | Aberdeen1-4 | Aberdeen1-5 | Aberdeen1-6 | Ayr-2

Ayr-4 | Ayr-5 | Ayr-6 | Ayr-7 | Ayr-10

Ayr-11 | Ayr-12 | Ayr-14 | Ayr-15 | Ayr-19

Ayr-20 | Ayr-21 | Ayr-22 | Ayr-23 | Ayr-24

Roof at Abbot House, Dunfermline

16th century. Illustration of crow steps and cat slide dormer.

Chapter 5
Roof construction, gutter, flashing and drainage

Early tenement roofs

17th and early 18th-century tenements would have had steeply pitched roofs of between 45° and 60°, often with the gable end facing the street and forming an attic space with dormer windows or catslide roofs.[1] Crow-stepped gables[2] were found in the 17th century, often with thatched roofs. Thatched roofs were common until about 1840 and remnants of thackstanes[3] can often remain in chimneys where the original pitch was steep. Later tenements with crow-stepped gables might have been slated or tiled with pantiles[4], or a bit of both, as in Cullross, but crow steps are rarely found in buildings after 1750 and not before 1840 when they reappeared when the Scots Baronial style (1840 to 1920) became fashionable. Gables used to face the street up until 1800, but after that tenements and houses were turned to have a frontage so that the pitched roofs ran parallel to the street.

Thatching

Thatching was still in use by the 1850s, but mainly in cottages and smaller buildings. Otherwise, thatching on tenements would reflect local conditions. Bracken, heather, bere straw, sods of turf, broom, fir, wheat and rye sheafs might all be used and clay was often intermixed with thatching in 1770.

Thatching was used on early tenements, but it soon became banned because of the serious fire risk in urban areas.

1 This is thought to be due to a shortage of timber.

2 Whilst crow-stepped gables had a series of squarish stones, which allowed a plank to be placed from gable to gable allowing access for re-thatching, the steps also allowed access to maintain a roof, although the form probably started because the stones were easier to form than cutting stones to the pitch of the roof.

3 Thackstanes were projecting stones at the base of chimneys which the thatch was dressed under to protect the joint.

4 Clay pantiles are mostly found on the east coast of Scotland, although originally produced in Fife in 1700, because of the coal trade. Used as a substitute for thatch, also found in Perth and Argyll. Flemish pantiles were imported back to Scotland as ballast in the ships, making home produced pantiles uneconomic. Scottish pantile profiles are different from Dutch pantiles, so cannot be intermixed.

Shaped clay pantiles in Crail and [right] slated edge detail found in Fife

Clay pantiles

Although ceramic roof tiles were found in Scotland in the late 12th century, the first use of plain terracotta tiles was not until 1669.[5] The form gradually changed into the interlocking wave form pantile[6] with a single overlap around 1830 as a substitute for thatch, although not because it was cheaper.

The Scottish pantile was made after 1714 and continued in use until imported clay pantiles became common around 1850 in Fife, likely because of the trade that developed between places like Anstruther, Crail, Pittenweem and St Monans, shipping coal to Denmark and the Baltic when returning with cargos from places like Antwerp with Flemish clay pantiles and bricks, as well as timber.

Development of tenement roofs

The diagrams below show different roof configurations that developed over time between the Renaissance period up to 1915. Clearly there were many variations to roofs, but the main change occurred after the medieval period when longer lengths of timber could be imported, allowing the rafters to have a long tie beam to constrain the rafters and permit a bigger roof span. From 1645 to 1715 woodland cover is estimated to have fallen to 10% of land in the whole of Britain and by 5% in the whole of Scotland. It was hardly surprising then that doors, posts and roof timber were so prized in Scotland that they were written into tenurial agreements. Oak was highly prized, but then timber from oak was imported from Norway.

The steep pitches of roofs in the medieval and Renaissance periods became lower as slating became more common with thatch being phased out due to the fire risk, also the difficulty of roof maintenance in tenements. The Georgians liked to emphasise the street frontage rather than the roof, so *cornices* and *parapets* took precedence over the roof which was kept low. This may have led to the M-Type roof found in Georgian Edinburgh, although it may have also been influenced by the length of timbers available at that time.[7] Dormers were added mostly

5 The Mussleburgh Tile Company started in 1780 and William Adam started a tileworks in 1714 in Kirkcaldy.

6 'Pantile' is derived from the Dutch word 'Dakpan'.

7 The length of timber shipped by boat would have been limited by the size of the hold. Fast clipper ships allowed timber to come from across the Atlantic, but it was not until the steel hulled ships were made that long lengths could be stored on deck.

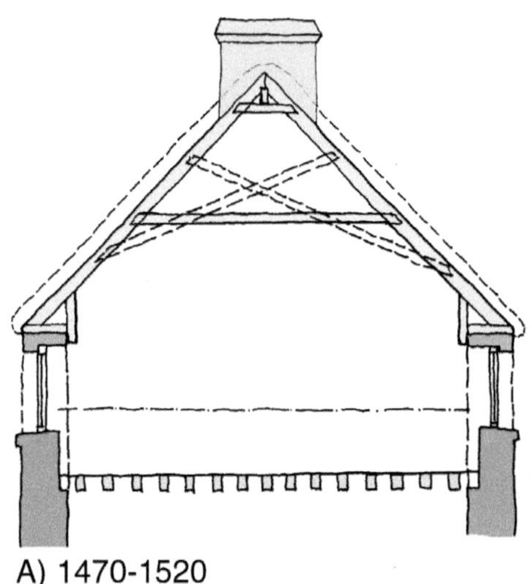

A) 1470-1520

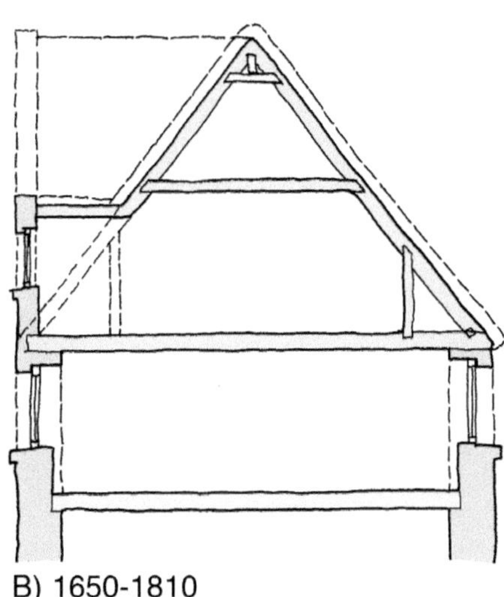

B) 1650-1810

Roof illustrations above depict medieval to Renaissance type roofs, which were normally quite steep

A Provand's Lordship in Glasgow, oxterpieces and collars providing most of the strength to the rafters *B* Wallhead dormers and *C* cat slide roofs. All these roof types might have been thatched at some time

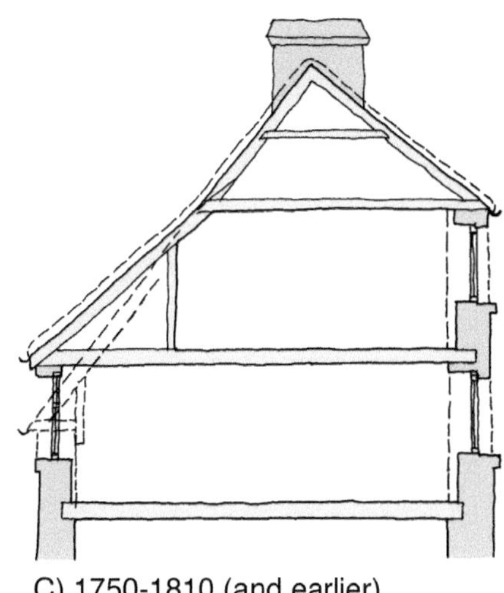

C) 1750-1810 (and earlier)

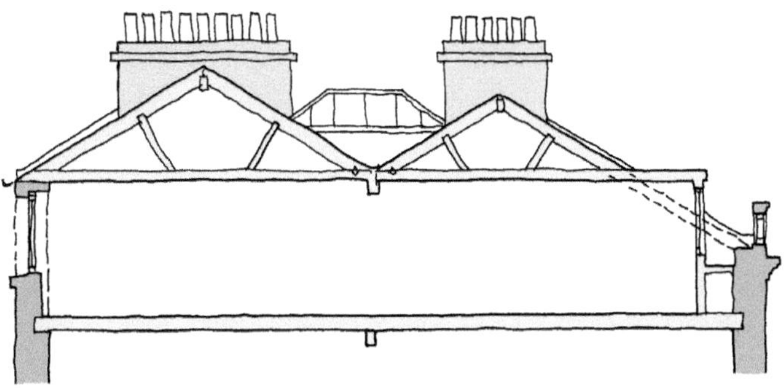

D) Glasgow 1820-1840 (terraced)

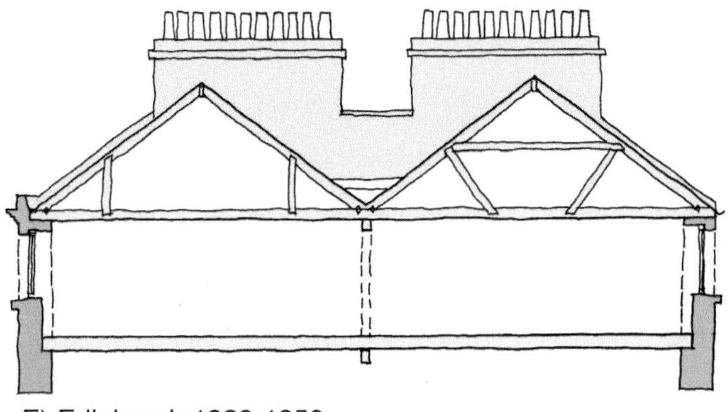

E) Edinburgh 1820-1850

D Georgian, but in a terrace in Glasgow, now converted into flats. Central cupola lighting the stairs. *E* Georgian New Town in Edinburgh. The M-type roof is typical and might have had a central cupola, although other versions developed

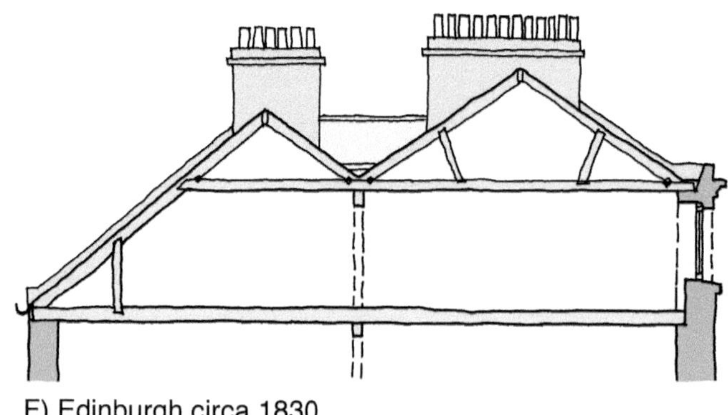

F) Edinburgh circa 1830

F Early Georgian in Rutland Square, sometimes with stairs at the front being top lit, but full height rooms at the other frontage.

G A typical Glasgow tenement roof from 1860 to 1910

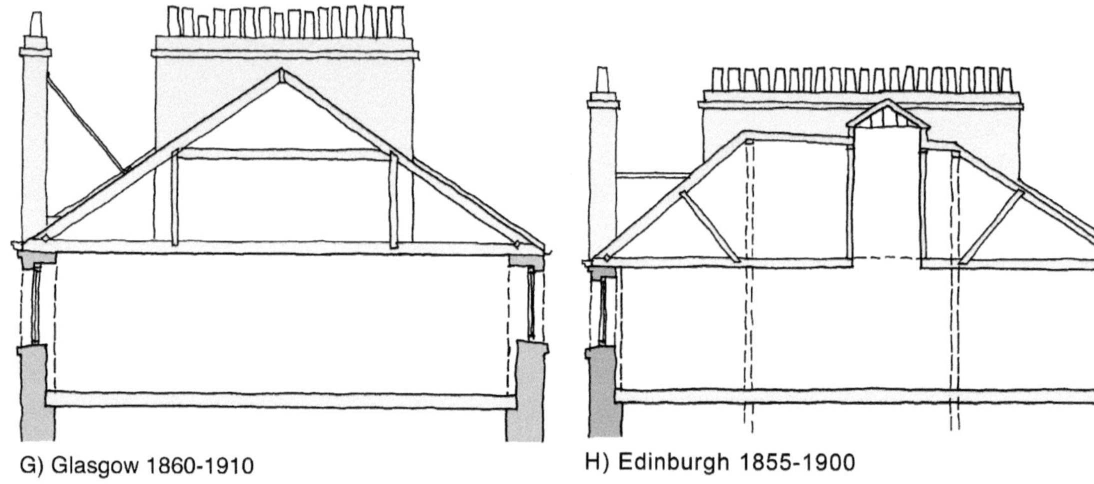

G) Glasgow 1860-1910

H) Edinburgh 1855-1900

H A development of the M-type roof but with a flat roof above two sloping slate roofs. It basically kept the same format which Georgian Edinburgh approved of.

I Shows Victorian influence with Mansards allowing better use of the top floors (sometimes five storeys). Wallhead dormers, two pitches of slate roof topped by a flat roof and central cupola

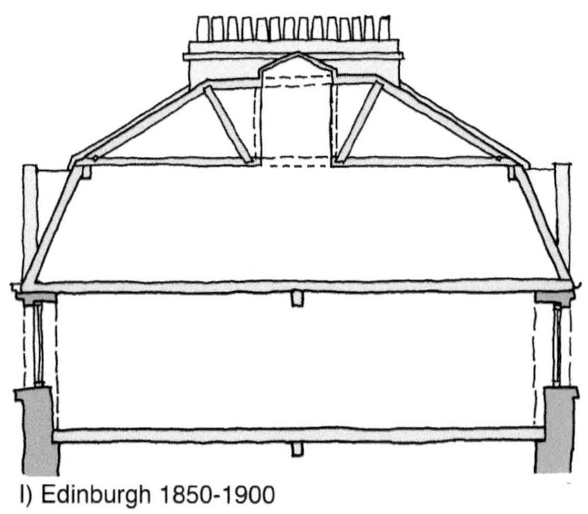

I) Edinburgh 1850-1900

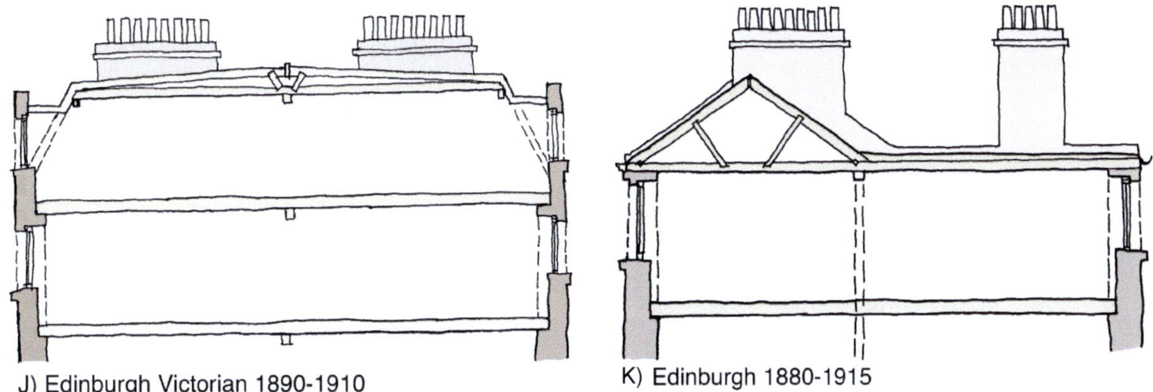

J is similar with mansards and wallhead dormers topped by a flat, usually felted roof.

K The later Edinburgh roofs continued to mimic the Georgian roof, with a small pitch at the front and a felted flat roof for most of the tenements

after 1840. Thistle Court in Edinburgh (1768) was the first building to be built in Edinburgh's New Town. It consisted of two houses (an early 'semi' detached building) built around a courtyard. The slated bay dormers were added later, probably around 1860.

Some details of rafter joints at eaves and wallplates.

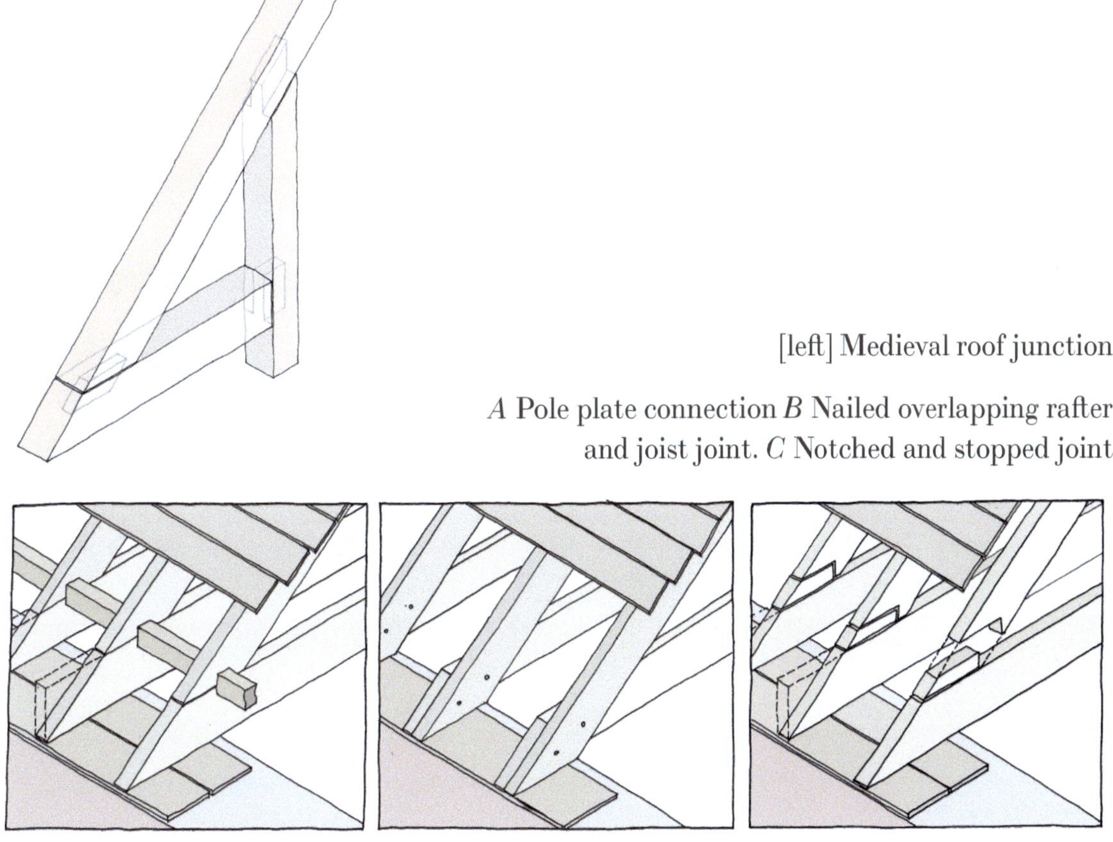

[left] Medieval roof junction

A Pole plate connection *B* Nailed overlapping rafter and joist joint. *C* Notched and stopped joint

Variety of Georgian roofs, M-type, mansard, platform roofs, bay roofs

Georgian tenement roofs

Edinburgh tenement roofs from 1800 might have had the rear wall a storey higher so that the top floor room would be lit by full height windows at the rear and dormer windows at the front. The ridge would be off-centre and the rear half might be flat. This mansard like elevation produced a lower frontage and a higher rear elevation.

However, in Phase 2 of Edinburgh New Town, M-type roofs[8] set a trend which may have come from London architects keen to produce a classical style with stone parapets that set off the classical frontage. The low ridges of these roofs made the roofscape less dominant. Between the two roofs a central lead valley was formed with a cupola to light the stairwell. These roof types are mostly found in Georgian Edinburgh and Georgian Glasgow,[9] but a small number exist in most Scottish cities (although mostly in early Georgian townhouse terraces).

8 M-type roofs consisted of two parallel pitches with a central lead clad valley to drain water, as well as front and back downpipes.

9 St Vincent Crescent and Queens Crescent in Glasgow both feature these M-type roofs.

In later years in Edinburgh, small front pitched roofs mimicked the original M-type design with a flat roof formed to the rear above (see image J on page 131). Flat roofs[10] would originally have been covered in lead laid on top of a thick timber sarking, although asphalt was often used after 1880 and bitumen felt[11] was in common use from about 1915.

Normally the timber rafters are set at a pitch of about 33º 6.5in. × 2.5in. at 18in. centres (250 × 50mm at 450mm centres) and ceiling joists tie them together with hangers and collars and sometimes additional purlins.

If there were mid supports, not unlike purlins at mansard roofs, the timbers were called binders or hangers and struts. They are fixed into a ridge board about 11in. × 1.5in. and their ends are notched and seated on square pole plates set into the ceiling joists, then bedded on timber wall plates set onto the external stone walls and onto a central internal support beam. The roofs are fully marked to receive the slates.

Victorian tenement roofs

Victorian tenements in Glasgow and other cities have a single dual pitched roof, using longer lengths of timber,[12] pitched at about 33°. Rafters are 6.5in. × 2.5in. at 18in. centres. Timber purlins or balks which are 5in. × 2in. may be used to stiffen the structure. Horizontal collars[13] (5in. × 2in.) are fixed between each set of rafters and vertical oxter pieces[14] (4in. × 2in.) also known as hangers are half checked and secured to each ceiling joist and rafter. Ceiling joists are 6½in. × 2½in. The rafters and ceiling joists were checked onto a square pole plate which ensured the rafter did not slip down and ensured it was aligned with the ceiling joists. In some tenements, the ceiling joist and rafters would be simply nailed together and rest on a timber wall plate.

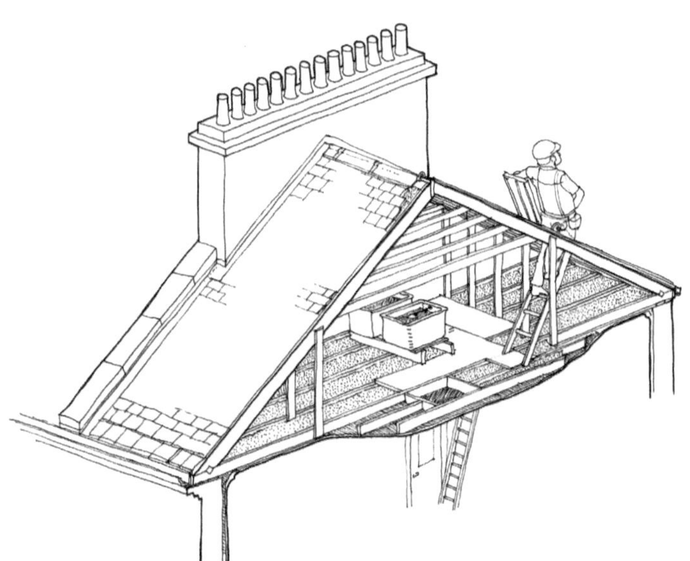

Space in the loft to house water tanks

A ridge board is placed at the highest point which all the rafters are fixed to. It rises above the sarking boards and allows a fixing for any ridging.

The sarking boards are ½–⅝in. (14–16mm) thick and 9in. (170mm) wide; where jointed at rafters they should be half checked but are usually butt jointed. Each roof would have had a sliding access hatch, lead covered. These still exist in Edinburgh but have largely been replaced by cast-iron skylights, or even more recently by plastic skylights. A roof ladder was

10 Also called 'platform' roofs.

11 See 'bitumen felt roofs' in Chapter 5.

12 Longer lengths of timber were imported from Russia, Norway and the Baltic ports.

13 Collars or ties in 1900 were then called balks, not to be confused with baulks.

14 Also called 'hangers'.

Typical Victorian tenement loft space; note lack of purlins

originally provided to get workers to the ridge. These are long since gone. In the 1980s, stainless steel anchors were fitted to ridges and hatches to allow a rope to be attached which could provide a degree of safety for any roofer. Under the CDM[15] regulations, any roof work will at least require access from a scaffold, tower or cherrypicker.

At the eaves, an angled tilting fillet would be used to ensure the bottom row of slates kicked up to align with the rest of the slate roof.

Lofts

Lofts are rarely well insulated and often contain past debris of old slates, flashings, tiles etc. and are often used as storage for unwanted materials. This can make them difficult to insulate. Collars and ties may be loose and evidence of previous water penetration can usually be seen on the underside of the sarking boards and near the party and gable walls.

If the loft is not well ventilated, then mould growth can appear on the underside of sarking due to condensation. Nails which protruded through the sarking will often appear wet from condensation. The problem is exacerbated when loft insulation is increased, as the loft becomes cooler. Slate and ridge vents may be installed to prevent this, although it is best to have a vapour permeable membrane over the sarking before the roof is reslated or retiled.

Cast-iron access hatches are installed in most cases, although in earlier Georgian tenements timber hatches which are lead covered are the norm.

Slated roofs

Early medieval and Renaissance roofs that were slated would often use thick slates from Easdale, fixed with timber pegs into inch thick sarking timbers, although this practice was mainly for more prestigious tenements.

Traditional blue/black slates were quarried in various parts of Scotland and around 1700 would have been mostly used on castles and mansions. When slates were used in tenements around 1800, they were mostly undersized Scottish slates that were too thick to be sold in the English markets. However, around 1800 some Welsh slate was imported by sea to Dumfries. Selkirk and the Borders might have imported Welsh or Cumbrian slates. Welsh slates only really came into Scotland when the Scottish quarries were producing less and less slate after 1910. Welsh slate was not undersized but usually 12 × 8in., 13 × 7in. or 14 × 8in. Welsh Moss slates[16] were also popular in Scotland between 1860 and 1900 due to their thickness which

15 CDM Construction Design & Management Regulations.

16 Moss slate from Wales was considered 'spoiled rock' given its thickness which was not suitable for the English market.

[far left] Recycled slates combined

[left] Easdale slate

[right] Ballachulish slate

mirrored the Scottish slate, but did have a slight heather coloured hue which eventually mossed down to a colour more recognisable as a dark grey.

Most Scottish slates came from Easdale and Ellanbeich Island[17] or Ballachulish[18] or other smaller quarries in the West Highlands such as Aberfoyle and Craiglea. There was also a slate quarry at Birnam. They were commonly about 16in. long by 10in broad and about ⅜in. thick. Because they came in varying sizes, it was best to sark the whole roof rather than apply battens to rafters which was a more common practice in England where Welsh slates[19] were used.

Slates at the eaves are formed of two rows: firstly the under eaves course, usually set over a tilting fillet to ensure the slate angle matches the rest, then the eaves course. The under eaves course is laid with its bed uppermost and the eaves course on top staggered at joints. As the slates rise to the ridge they usually reduce in size and so are laid in diminishing courses. This is traditional Scottish practice as it makes most use of the variety of slate sizes, so ensuring less wastage. The smallest top row of slates would then be nailed and mortared into place before a ridge tile or ridge capping was installed.

The slates nearest the eaves (or gutter) will receive most of the rainfall that lands on a roof, whilst the ones near the ridge will receive less rainfall, so it makes sense to have the larger slates near the eaves.

Although bye-laws of the time asked for double nailing of each slate, this was rarely done. Good practice is to head nail each slate and double nail each slate every third course which still allows any defective slating to be easily removed and replaced. Double nailing on larger slates may be required, particularly on lower pitches, when large slates are required.

17 Ellanbeich quarry was 200feet deep; the entire quarry flooded in 1881 after a great storm.

18 At Easdale and Ballachulish, the slate was not only quarried but also split and trimmed to standard sizes by crews of five or six men. By the middle of the 19th century, Easdale was producing on average 7 million slates and Ballachulish 15 million. Quarrying at Easdale ceased in 1911 and Ballachulish closed in 1955.

19 Welsh slates began to be used when the railways allowed good transport links, although the thicker Scottish slates were available until the quarries closed in the 1950s.

Most tenement slated roofs have used ordinary galvanised slate nails; however when roofs are now reslated, copper or stainless slate nails are preferred.[20] Side nailing between each slate is sometimes used, particularly when repairing slated roofs and near the ridge.

Originally, in good practice, a type of breathable felt was laid over the timber sarking. A bitumen based felt material came into use around 1930 and from about 2000, breathable reinforced felts came into use under slates and tiles. The advantage of the waterproof, yet vapour permeable felts (introduced in 1983), is that they allow the attic space to be vented and prevent condensation, and thus mould growth, occurring on the underside of the sarking.

Concrete tiled roofs

After the great storm of 1968, so many Glasgow roofs, along with chimneys, were damaged, that many had to be renewed. Repairs were undertaken over two years, often at speed to prevent water damage. After stripping the slates[21], affected roofs were covered with bitumen felt laid over the sarking boards then secured with timber battens. At this time there were no conservation areas.[22] The storm led to a massive programme of retiling tenement roofs in Glasgow which carried on well into the 1970s and 1980s with Housing Action Areas and Tenement Improvement Plans. Slates were simply placed into skips and carted off to landfill.

Sarked roofs were felted with rolls of bitumen felt (up to 2000, breathable roofs and non-breathable in terms of liquid plastics, followed) and battens fixed over them at spacings to suit the concrete tiles. The concrete tiles were expected to last about 30 years and mostly they achieved that, but this was far shorter than the lifespan of a well slated roof which with good maintenance could last for well over 100 years. Certain problems arose with these early concrete tiled roofs: a lack of ventilation under the tiles led to water collecting at battens. In

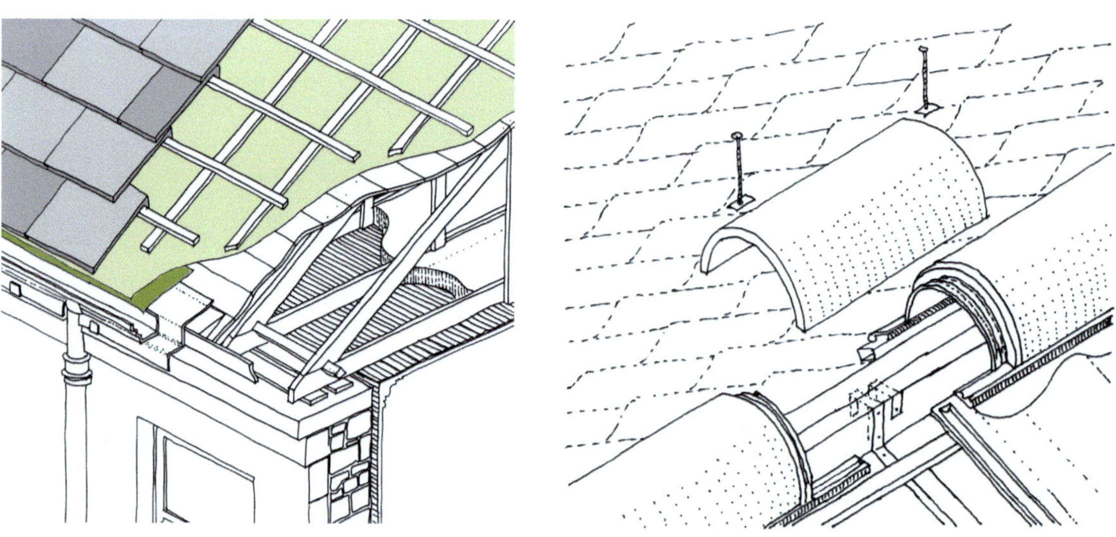

[left] The battens and outer battens for a tiled roof

[right] Note the under cloak under the gutter, dry ridge system

20 When a hammer strikes a galvanised nail it can damage the galvanising and lead to rusting of the nail.

21 Glaswegians believe that most of their slates ended up repairing Edinburgh's slated roofs.

22 The first conservation area in Scotland did not come until 1974 although there were some small local areas before that.

the 1980s the specification was changed to having counter battens and battens. By 2014[23] there was a requirement to clip and nail the tiles as they were sometimes laid with limited nailing. Ventilation into the loft spaces was required as the bitumen felt prevented it and mould tended to grow on the underside of sarking.

Concrete ridges which were bedded in cement mortar were also a problem as the mortar bedding often cracked up and ridge tiles blew off. Now dry ridges are available.

Flat platform roofs

As noted above, these are mostly found in Edinburgh. These would likely have been covered in bays of lead, but the use of asphalt and then bitumen felt became common, certainly by 1910. Hot applied asphalt was initially used on concrete and some timber roofs, commonly found in rear shop extensions where a 'high back' also acted as a drying green for residents. Rock asphalt[24] was usually laid in three coats over timber flat roofs and had the ability to resist wear, so it was used when foot traffic was expected.

Lead (or possibly zinc) would have been used originally on most flat timber roofs with raised roll joints, but over time a majority have been replaced with bitumen felt.

When lead is used it should be at least code 6[25] and laid at about a 1:60 fall with 1½–2/¼in. (55–60mm) stepped joints. Bays would be formed over wood rolls about 1¾in. (45mm) high secured to the roof sarking at about 26in. centres (675mm crs).

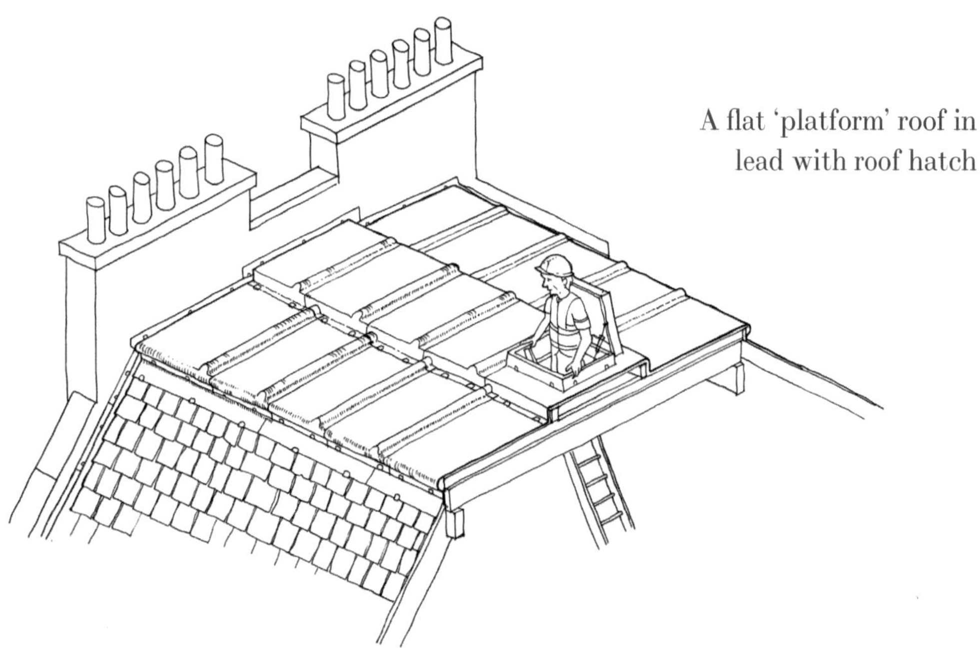

A flat 'platform' roof in lead with roof hatch

23 Roofs: Repairing Scottish Roofs (2006); Short Guide 11 Climate Change for Traditional Building (2016); Slaterwork, Glasgow West Conservation Trust (1999): NFRC Technical Bulletin No 43 describing Scottish Slating Practice.

24 Surprisingly, Walter Raleigh discovered a black sticky substance in a lake in Trinidad, back in 1595 and used the substance to prevent leakage in his ships. Barrels were shipped back to Britain and mastic asphalt soon became popular. In 1902, Edgar Hooley, a surveyor, patented 'Tar Macadam' after a lucky accident. Hooley's company was then taken over and Tarmac Ltd was formed, producing a cold laid asphalt in 1938. In 1968 Tarmac merged with William Briggs from Deeside, and also acquired Amasco, the asphalt part of the company being called 'Briggs Amasco'.

25 Lead is sold by the thickness of lead and lead codes denote the thickness.

Bitumen felt roofs

Composition roofs, originally made from tarred sheathing paper, were originally used in Greenock in 1792. Sheathing paper was dipped in tar heated to boiling point, dried for two days, then dipped in tar at a lower temperature. When the membrane was laid, it was then covered with a coat of tar and ashes spread over it to prevent the tar melting. This sort of roofing, which was cheap, began to replace thatching, then in 1818 and 1849 two cotton mills in Glasgow which had very low-pitched roofs were coated in tarred paper. William Briggs, who was tired of recoating the roofs with tar, developed the tarred paper in Dundee in 1865, called bituminised felt, which has been in existence since then.

In the early 20th century the felt would have been laid in three layers. The first layer being laid over molten bitumen poured onto the timber deck, then subsequent layers bonded onto the first layer, again with molten bitumen.

As felts developed they became modified, so that the bitumen was impregnated to the underside of the felt and, when heated, melted and formed adhesion with the surface below. This 'torch-on' technique of laying speeded up the process. The finished layer might also have a reflective coating to reduce thermal expansion from solar gain.

However, some roofing companies use a bitumen felt which is first nailed to the sarking around the edges, then further layers might be added with a self-adhesive edge or an applied adhesive. This approach is really only suitable for shed roofs.

In recent storms in Edinburgh, some of these poorly applied felt roofs have been entirely blown off, exposing the timber sarking boards. Wind creates a difference of pressure inside the attic space and above it, so there can be a huge force which acts to 'suck' material off any base.[26]

Bituminous felt[27] became cheaper and easier to lay (even in three layers), and only requires local torching. The upper surface may be finished with a small stone aggregate to help reflect the heat which can crack the felt. It was in common use around 1900, although it has a limited lifespan. The felt, like the asphalt, has to be flashed to a wall (the skew) or chimney, so it has to be dressed up and protected by a lead cover flashing which should be raggled into the stonework. Bituminous felt roofs have improved greatly in quality over the last 20 years; provided they are laid correctly, they should be able to last 30 years.

Most felted roofs on a timber base will not have any insulation directly below the layer, so it is important to ensure the attic space is well ventilated to prevent condensation building up on the underside of the membrane. This also applies to any lead covering.

Concrete roofs

Concrete (in situ) roofs at high backs and rear extensions were usually coated in rock asphalt. The concrete to make these Victorian roofs was often specified 'to be one part of Portland cement, one part of sand and four parts of clean broken brick.'[28] These concrete roofs might

26 High winds create different pressures under felt and above it, tending to create suction which lifts roof membranes off their base.

27 Commonly termed 'Roofing Felt'. Read *Roofing Felt* (1998) by Jens Lindhe and the Danish Roofing Advisory Board.

28 *Building Construction and Architectural Drawing* by John Reid, who taught at the Glasgow School of Art 1911, page 15.

be set between wrought iron T sections, but rarely did they have reinforcement, as this didn't come into common use until after 1898. The bits of broken brick make these roofs problematic and, in some cases, additional strengthening or replacement is required.

They were also found on outhouses in backcourts, but the exposed concrete was not asphalted.

Leaded bay and oriel roofs

Earlier bay roofs were usually formed as small slated roofs similar to dormer roofs. These can be found in Pollokshields and Edinburgh. However, the majority of late Victorian oriel roofs are formed as flat lead roofs which drain to the main front gutters. Some are formed behind small parapets so they are similar to parapet gutters. Such bay roofs should have the lead laid in bays over wood rolls. In Glasgow most lead roofs were replaced with a bitumen material in the 1970s and 1980s, possibly because of the risk of theft. Lead is now commonly used as it has proved how long it can last.

Leaded bay roof with roll joints

The flat bay roofs still had a large timber bressumer beam which carried the ceiling joists and smaller trimmers extended from this main beam to rest on a wall plate formed above the bay window lintels. The lintels were often joined with iron (or ideally wrought iron) staples, which in time can corrode and cause a diagonal crack at the lintel joint.

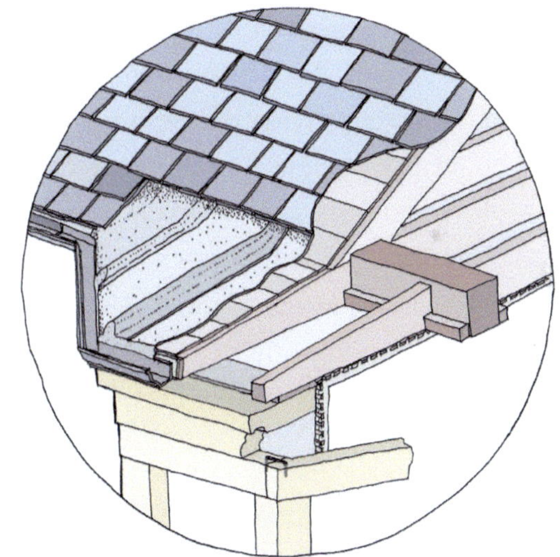

Leaded bay roof showing construction under it and also iron staple which joins the stone lintels

Slated bay and oriel roofs

These are often found in Glasgow, Edinburgh and other Scottish cities and commonly consist of five slated flanks with the ridges covered with a lead flashing. The valley between the slated bay roof and main roof would be in lead and drain into the front gutter.

Flat high back roofs

These are often found in tenements from 1890, where, because of the limitation of a site, the tenement drying green had to be located on the flat roof. Initially these would have been timber decked roofs with thick floorboards then covered in two or three coats of asphalt[29], taken up a parapet wall and flashed with a lead cover flashing. When concrete flat roofs

29 Each asphalt coat was about 20mm thick.

became more common around 1900, they were sometimes built above extended shop units or pubs and then covered in two coats of asphalt. The concrete was often rather poor, containing brick rubble and set between steel T sections, as reinforced concrete had not become the norm until about 1910–20, although it had been invented, using wrought iron bars, in 1854. In the earlier concrete roofs, the steel T sections often rusted and the concrete mix was often of low quality containing brick fragments. Rusting of the T bars caused cracking and movement of these decks. Wrought iron hooks would be bedded into perimeter walls and wrought iron clothes poles formed as part of the metal railings, although by 1910 they were often using unprotected steel as well as wrought iron, so corrosion became a problem.

Cupolas

When the close stairs are in the centre of a tenement, as occurs when tenements turn a corner, for most Edinburgh tenements it is necessary to light the stairs with a glazed cupola or a skylight. A hanging stair with an open well allows the light to penetrate down.

Cupolas, mostly found in Edinburgh, can be elliptical, round, square or hexagonal. They are formed out of timber and the inside walls are plastered to reflect the light. Externally any upright surrounds are usually leaded. The glazing bars support the single glazing and a small draining channel may be provided to cope with any condensation.[30] There is usually an air vent at the head which should remain open, but is often sealed. Cupolas need to be ventilated at the ridge and at the base,[31] to combat condensation and if the flat platform roof is raised (to insulate it), then the upstage of the cupola would also need to be raised which usually meant also having to rebuild the cupola.

Cupolas have often been 'modernised' with barrel roofed plastic cupolas and other modern interventions. UPVC cupolas can become brittle with UV light and the glazing bar and glass joints are prone to decay.

Sometimes Victorians created a top floor attic room (possible in a maisonette flat) which has a cupola to light the internal hallway. Here there may be a laylight below the cupola to prevent dust and heat loss to the attic space. These laylights often had decorative and coloured glass in them.

Skylights

Large close skylights are often found in corner tenements, to top light the close stairs, and are usually formed at the internal valley corner. Glazing bars were often planted directly onto rafters. The single glazing was originally bedded onto putty in the glazing bar, but not puttied on the outside. When it is puttied on the outside, moisture collects and the glazing bars can rot. Sometimes lead flashings are placed over the exposed joints. Originally the glass sheets would have been laid in short panes and had to overlap at each pane.

From about 1900, galvanised metal glazing bars with lead cover flashings allowed glass to be bedded on an asbestos rope and the joint protected with lead. Similar products are still available today, but with exclusion of the asbestos rope!

30 Sometimes the drain was formed as a lead tray with a weephole to drain away any moisture.

31 This can be a minimal gap, the thickness of lead clips.

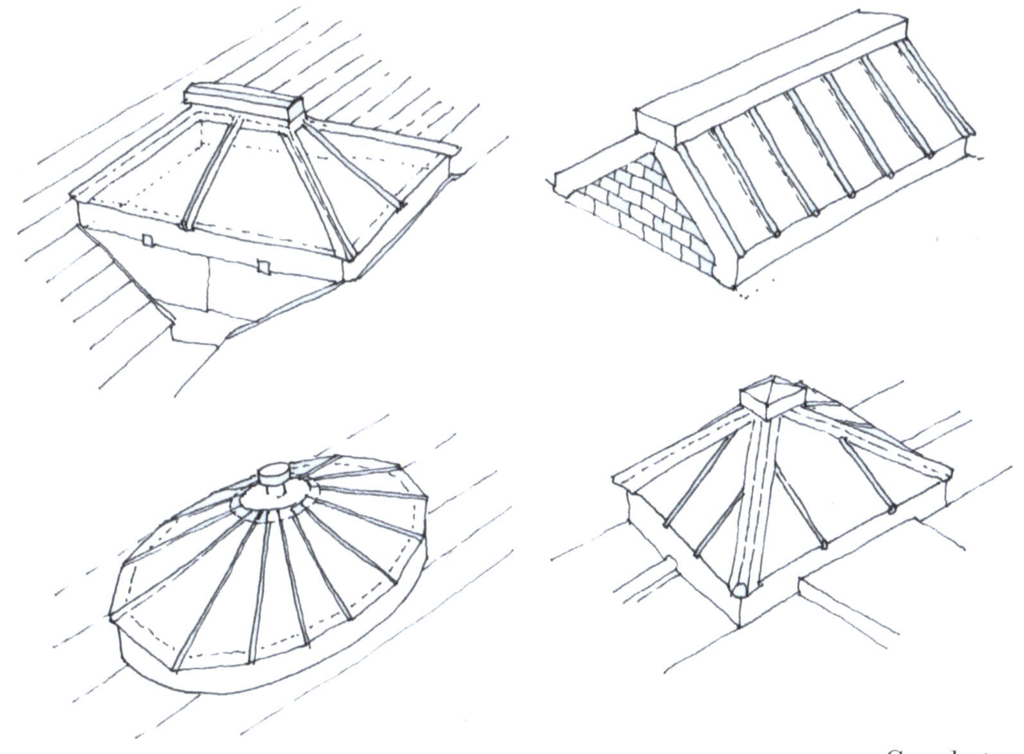

Cupola types

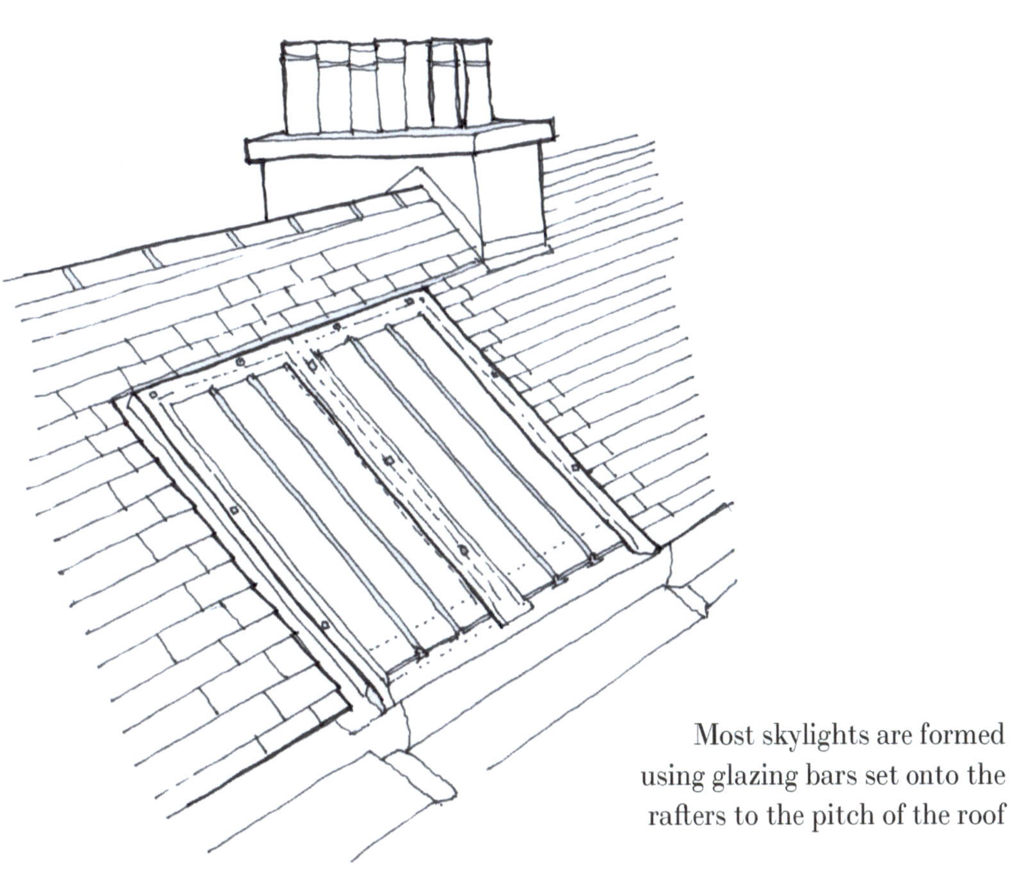

Most skylights are formed using glazing bars set onto the rafters to the pitch of the roof

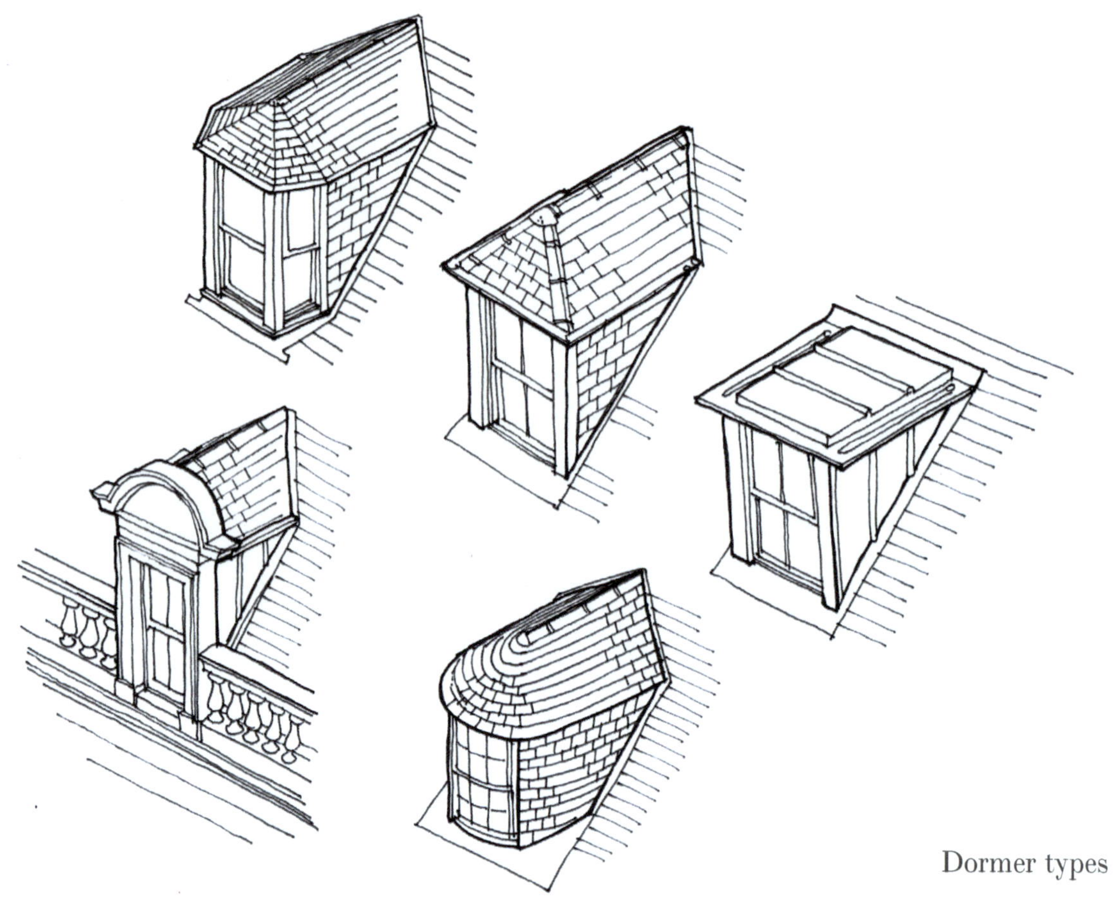

Dormer types

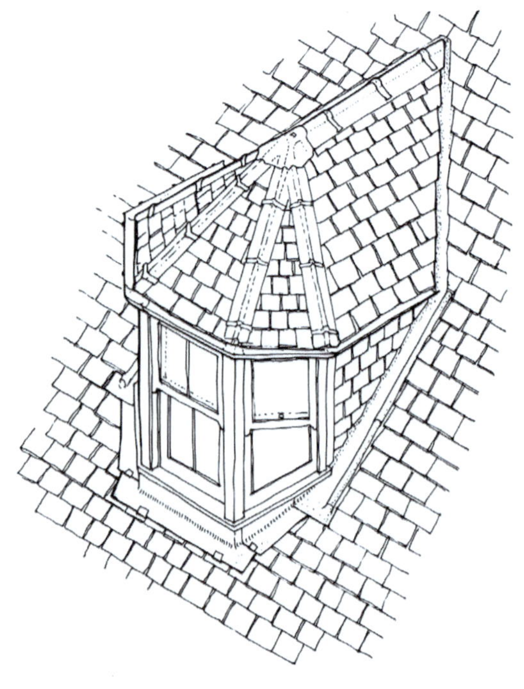

[left] Dormer in Aberdeen tenement

[right] The five-sided dormer is most common

In the 1970s and 80s there was a tendency to replace the glass with Georgian wired glass for safety reasons. From the 1980s aluminium patent glazing bars came into common use. However, lead would still be used at the sides, head and apron flashings. There was also a tendency in this period to replace the skylights with two large Velux skylights, which came with their own flashings. However, they did reduce the amount of light reaching the stairs.

Dormers

In the 18th century, developers would use the attic space in the steeply sloping roofs that were initially formed. The windows were called Stormont windows[32] and were sometimes not allowed under local bye-laws. These developed into catslide windows[33] which were vertical windows housed under a slated or tiled roof set at a slightly lower pitch than the steep pitch of the main roof. Wallhead dormers were formed by raising the stonework up and forming a small pediment, with the dormer cutting into the roof space. Wallhead dormers were still being built in the 19th century but sparingly; they were eventually replaced by bay fronted dormers that we know today.

Where a front chimney head rises up and a window is inserted between flues, this was called a nepus gable,[34] some of which can still be found.

Dormers were often added later when people decided to develop the void roof space, although tenements with slated mansard roofs usually had simple rectangular or bow roofed dormers, as found in early Georgian tenements, and common in Aberdeen. The three-window bay fronted dormers were common in cottages and manses, and eventually moved into tenements, becoming common between 1840 and 1920, with curved bow fronted dormers first appearing in 1830 and barrel roofed dormers appearing around 1900.

In order to insert a new dormer in an existing roof, rafters have to be cut and this can weaken the roof structure, so care needs to be taken to ensure that additional support purlins are added to stiffen the structure. When a dormer has been installed retrospectively, sometimes the builder neglects sufficient ties, trimmers or purlins to strengthen the roof structure with the result that the roof splays out and the ridge line sags.

Flashings, valleys, leadwork, ridges, skews and gutters

Most of the causes of damp from a roof occur not from the slating, but from the gutters, as they leak, overflow or become blocked with leaves and pigeon guano. Also, the flashings, which provide a joint with the skews, chimney heads and ridges, can be blown off, exposing the top row of slates. This section covers these important areas.

Leadwork

Lead is a very durable material and can last for hundreds of years. It is soft and pliable which makes it ideal when it has to be formed into complex junctions to keep a building watertight. However, as a metal it can expand and contract with changes of temperature, so fixing and

32 'Storming' is another word for a skew, and refers to the slope of the roof, a Stormont window is set to the slope of the roof, like a modern Velux window.

33 Catslide windows simply suggest that a cat sitting on the roof will slide down it.

34 The term 'Nepus gable' is also used where small gables containing a window rise above the wallhead.

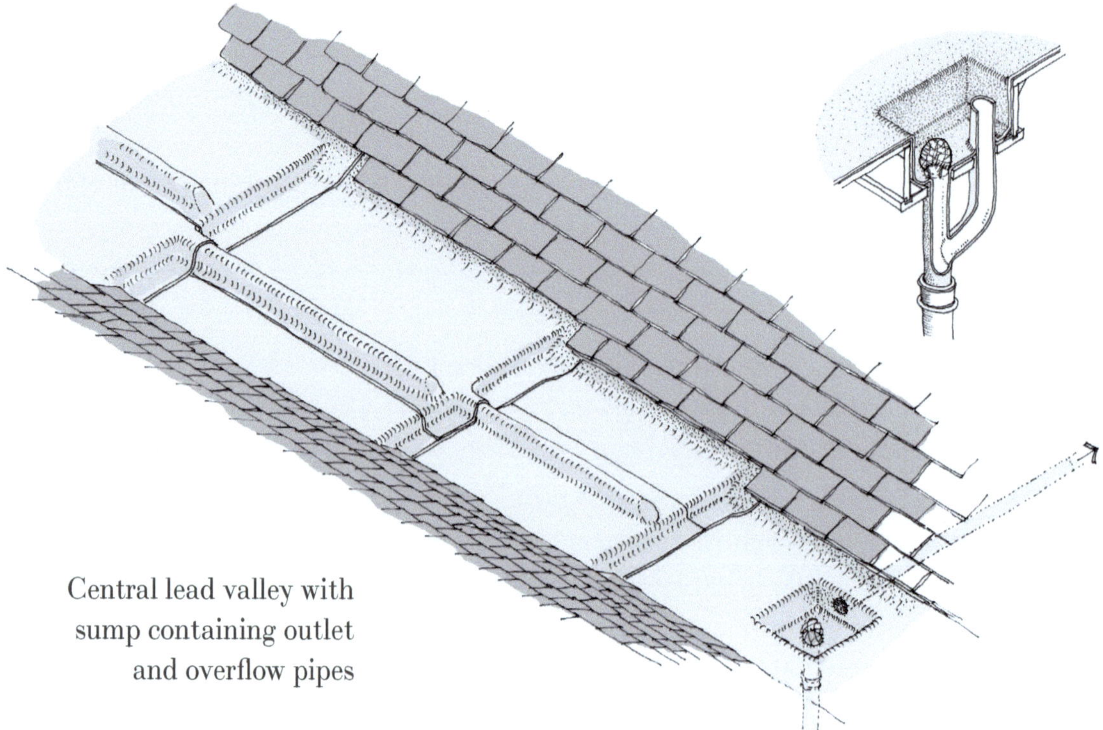

Central lead valley with sump containing outlet and overflow pipes

detailing leadwork is important to ensure the lead can move with temperature changes. Also, because it is so flexible, it is sometimes laid by inexperienced roofers not trained in leadwork. This can result in the wrong weight of lead being selected, often because of cost cutting, or where lead is over dressed with a hammer it can thin to a point that it cracks. Also fixing of lead into raggles[35] and clipping of lead is often poorly done, unless formed by a plumber trained in lead work.

Leadwork should be laid on flat timbers with a building paper below it. Gaps, called penny gaps, between the sarking[36] will allow the lead to be vented from below. If the lead is not vented, then there is a risk of condensation forming on the underside of lead which can cause it to corrode.

Lead comes in different weights referred to as codes of lead, and each code will have a different thickness as well as a different maximum expansion that has to be allowed for when forming steps and rolled joints in lead.[37] (*Code 4 lead is used for soakers; code 5 lead for valleys & cover flashings; code 6 lead for parapet gutters; code 7 & 8 lead for flat roofs and parapet gutters*). As a rule of thumb, in a gutter lining, the maximum length between any stepped joint for the codes of lead may be as follows: Code 5 & 6 lead – 2 metres; code 7 lead – 2.5 metres; code 8 lead – 3 metres.

Some joints may need to be site welded, a skill that should only be carried out by a trained plumber and a hot work permit should be obtained as there is always a danger that the heat could ignite the timbers under any lead.

35 Raggles are chases or grooves cut into the stone to allow lead to be secured to a chimney, skew or stone upstand.

36 Usually called a 'penny gap'.

37 The Lead Sheet Association publishes a very well written guide on using lead; unfortunately it is not free or downloadable. The weights of lead and thicknesses are: code 3-1.32mm; code 4-1.8mm; code 5-2.24mm; code 6-2.65mm; code 7-3.15mm; code 8-3.55mm.
See NFRC Leadsheet.co.uk.

Lead is normally fixed at the highest point – on timber, using copper flat head nails about 50mm apart. When fixed to stone the lead may be secured into raggles with lead wedges or stainless-steel screws.

Lead platform roofs should be laid to a fall of about 1:60.

Where there is a likelihood of lead theft, then the lead may need to be protected. In a listed building, this may be through some additional stainless mesh secured over the area. However, if the building is not listed then a substitute material may be used which combines a bitumen felt material with an embedded aluminium mesh. This looks similar to lead and is formed in the same way.

Types of sheet lead

Sheet lead can be formed by:

- Sand casting involves pouring molten lead over a prepared bed of sand and levelling it to the required thickness.
- Machine casting: invented around 1950, but only in use in the 1980s. The sheet is formed on a rotating drum set in a bath of molten lead. Once cooled, the lead is peeled off the drum. Thickness of lead is determined by the speed of the drum.
- Milled lead sheet rolls a block of lead between two cylinder rollers, with the rollers being brought together progressively until the desired thickness is obtained. This method started in the 18th century and is still the most common method used today.

Alternatives to lead

Because of the risk of theft, particularly if lead is located at lower levels, substitutes may be used provided planning consent is not needed. In the 1980s, a bitumen-based material called 'Nuralite' or 'Aqualite' was used, which was cheaper and not prone to theft. It was commonly used to undercloak gutters and at other flashings where lead should have been used. Generally it has not lasted.

Modern alternatives today include material which is a laminated polyester with embedded aluminium mesh which has a much longer life (but not as long as lead!).

Ridges

At the apex of a pitched roof, the rafters are fixed to a ridge board. The rafter ends are trimmed and butt into the ridge board and are then nail fixed. The timber sarking covers the rafters up to the ridge board.

As the top rows of slates are usually the smallest, they are sometimes just single head nailed and the slates are bedded in a hair lime mortar which prevents them from being blown away at this vulnerable area. Sometimes the slates at the ridge are also side nailed. In the past they used special T-shaped nails[38] which were driven between each slate and secured them. Galvanised nails can still be used but copper or stainless-steel nails should now be used to ensure a long life for any slate roof.

38 Also known as 'check nails'.

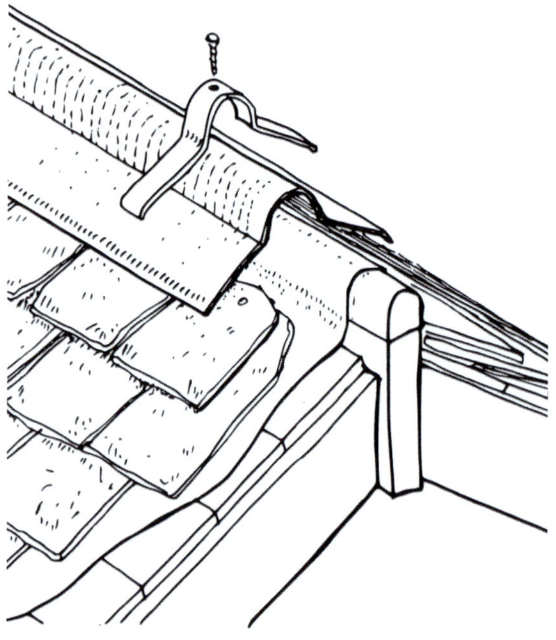

[above] Zinc ridging to a slated roof (and young seagull)

[right] Each ridge should be clipped

The metal ridge (usually zinc) is then fixed to the ridge board with galvanised metal clips at about every 600mm. In listed buildings, a lead ridge[39] may be installed which would be fixed to copper clips secured to the ridge board and formed under and over the edges of lead.

Some slated roofs are finished with terracotta ridge tiles which are bedded in a lime mortar. Depending on the age of the building, the ridge tiles may be plain or crested in some way and sometimes mixed. If re-bedding displaced terracotta ridge tiles it may be prudent to include a stainless-steel mesh secured to the ridge board first to provide a good fixing for any lime mortar, and to mechanically fix any end ridge tiles that are exposed.

When slated roofs have been replaced with concrete tiles, the tiles are laid on battens and counter-battens and in this case the ridge tiles are usually concrete ridge tiles bedded in cement mortar, although, because of the problem of the mortar cracking and tiles falling off, a dry ridge tile system is more commonly used (see illustration 8.25). Here the ridge tiles are clipped into a plastic carrier which is secured to the ridge board and then the ridge tiles are mechanically fixed. They also provide ventilation at the ridge. We are unsure how long such plastic fittings will last in UV light, but probably as long as the concrete tiles (30-60 years).

Ridges on steep square plan faceted turrets may be formed in lead and dressed under the slate with a small secret gutter. The slates are butted close to the ridge roll. If the lead is dressed over the slating, individual lead soakers may be installed under each slate to provide a weathertight joint.

Removing rainwater: lead soakers, weatherings, aprons and finials

Sometimes slates are joined together at ridges with lead soakers. These are small pieces of lead which are set under each slate course and nailed at the head, then slated over. It can give a neat joint but the two rows of slates need to butt tightly to one another.

39 Lead ridges are usually code 6 lead. Copper clips would be secured with copper nails.

Lead weatherings are sometimes formed over string courses and mouldings of stone that project from the building and tend to weather badly because of exposure. Here code 5 is used and the lead may be clipped on the underside with copper tags as well as being dressed into the mortar joint above the string course. Wider projections like stone hoods above windows may require lead cover flashings as well as fixings at every welded joint. As we have increasingly wetter weather, this is becoming more necessary, although disliked by some conservation officers.

Apron flashings are formed under skylights and at the head of vertical slating on dormer cheeks. They need to be well secured with copper nails. It is best to treat lead ridges and apron flashings with patination oil as new lead can produce a white carbonate stain onto slates. Patination oil prevents this.

Finials come in many different sizes and are usually found on top of turrets[40] and spires. They may be entirely formed out of lead which is dressed over a timber formwork. The lead will need to be site welded. Some finials are formed from wrought iron or cast iron which has a long spindle which penetrates the turret and is secured to a timber framework below. A lead flashing provides weatherproofing around the finial, often shaped with decorative edges if Victorian. New leadwork finials should be treated with patination oil.

Skews and skew flashings

Skews are formed by the top of gable and party walls which rise up through the roof and act as a firebreak between tenements. They are normally about one foot higher than the slating.[41] Although this was formalised in 1850, party walls rose above the roofs before that to protect

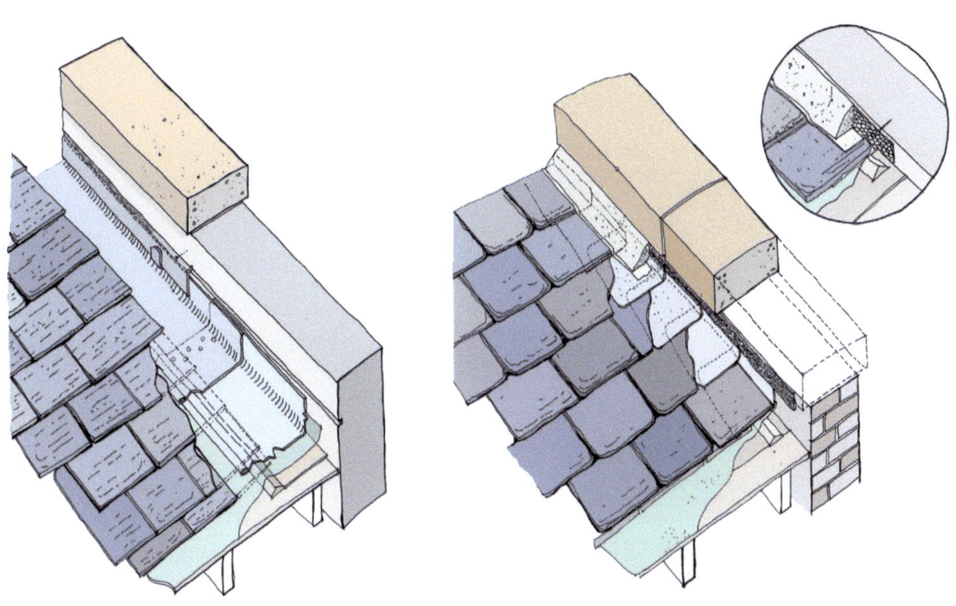

[left] Lead skew flashing

[right] Lead soakers with mortar fillet at skew

40 See Historic Environment Scotland (2008) *Inform Guide: Finials and Terminals*.

41 The 1850 Police and Improvement Act (Scotland) stipulated that in improvement and construction of all buildings, in all Scottish burghs, the party wall had to be carried through the roof to form a parapet no less than 12 inches above the slated roof (measured at right angles). The 1879 Edinburgh Municipal and Police Act required the skew to be carried 24 inches above any roof timbers. Page 60 para 163.

against fire spread.[42] The joints between slating and skew were initially made by having a small tilting fillet to raise the edge of the slate near the skew and to apply a lime mortar fillet between slates and skew. This worked although, as slates moved, the mortar cracked and would often need to be redone. When this is done today, a paper tape is applied over the slates and the lime mortar flaunched over the tape and onto the wall. A better way is to insert lead soakers under each slate and dress them partly up the skew, fixing them at the head, then applying a lime mortar fillet which, in exposed areas, should be an hydraulic (NHL5) lime mortar.

Lime mortared skews are more common on the east coast of Scotland, whereas lead gutters at skew flashings are more common on the wetter west coast. In most tenements the lead valley would be formed by placing a length of lead against the skew and dressing it over a timber fillet to form a watergate. The leadwork is copper nailed at the head only and lapped by as much as 150mm. A lead cover flashing is then set into a raggle (which may be formed into the skew stone, or under it) and secured with lead wedges. In good practice the lead would also be secured with copper or lead clips. Although the raggle joint of the lead cover flashing would originally have been lime mortared in place, it is now better to use a lead mastic compound which caters for some movement in the lead.

The slates are fixed to dress over the raised watergate. Sometimes the lead valley is clearly visible and sometimes the slates are dressed tight to the skew.

Club skews[43] are placed at the lowest point of the skew to prevent the skew stones from slipping down. There are many different types of club skews (see Chapter 6), although in many tenements the skew copings may be fixed into the party or gable wall with metal dowels.

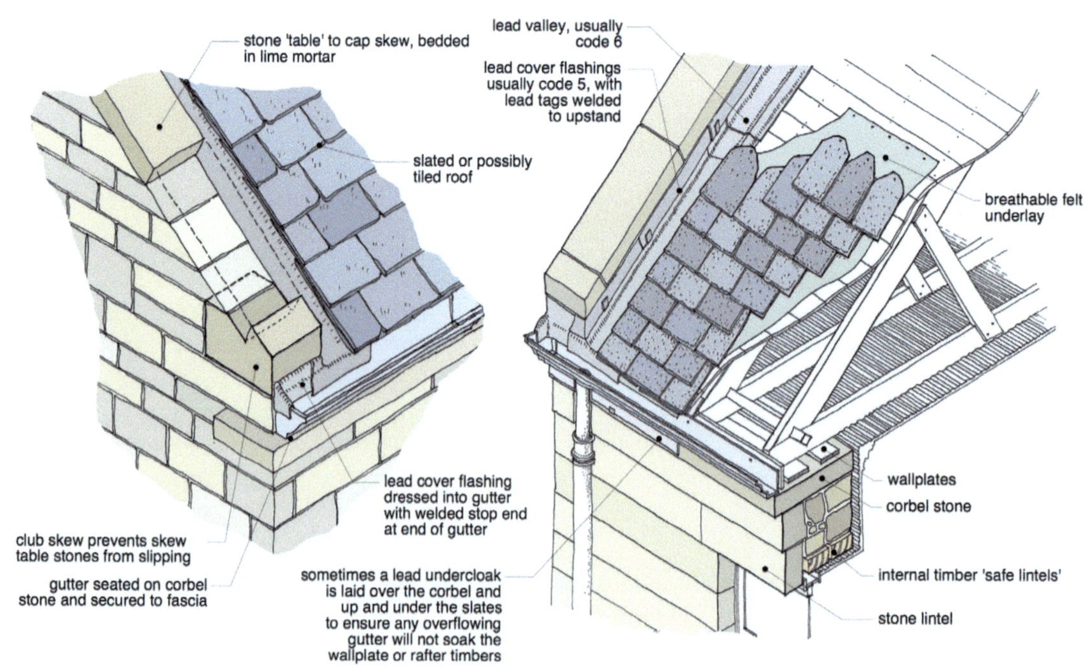

Club skew and lead skew flashing

42 Also formalised in the Glasgow Building Regulation Act of 1892.

43 Club skews are stones built into the external gable structure, so they can support the downward sliding force from the skew copes.

As the stones are exposed to the weather, moisture can seep into the skews at joints. When this occurs, the skew joints may need to be repointed, or the skew leaded over, or partially leaded over.

Flashings to chimneys

At the mutual chimneys, the stone can become badly worn, or poorly maintained, as here the stone is fully exposed to the elements and over time salts from the flues may have led to internal stone decay. So the chimney may need more work before any new flashings are installed. Where a lead flashing is formed around a chimney, similar to the skew flashing, the lead cover flashing is set into the stone in a diagonally cut raggle which cuts across the stones and bedding joints. When the chimney stonework is found to absorb moisture, larger areas of the stone chimney head may require to be dressed in lead.

If brick chimneys are flashed, then individual stepped lead cover flashings may be secured into brick courses. Many chimneys have been rebuilt in brick, usually with a facing brick or engineering brick. A lead tray which acts as a moisture barrier can be built into the new brickwork. Individual cover flashings can be dressed into each brick bed to follow the pitch of the roof. Note that dense engineering bricks can act as a damp-proof course, but a lead tray is still advisable. At wallhead chimneys, in Edinburgh, these are formed with a small slated roof to the rear and lead flashings between roof and chimney. These roofs are called 'saddles'.

In Glasgow, the wallhead chimneys are flashed with a simple lead flat at the rear and sides. This chimney would also originally have had an iron ladder stay for strength and access for the sweep. In both cases it is best to have a lead flashing at the front, between the chimney and into the gutter (assuming there is one).

Side flashings and storm rolls

At certain critical areas, a 'storm roll' might be formed in lead to prevent a deluge of water from washing up and under the slates.

Storm roll at side flashing to dormer

Secret valleys

In secret valleys at M-type roofs, the width of the valley is likely to be wider and so the lead has to be laid in sections and stepped at certain points to allow it to move. Jointing of the lead when on the flat is formed over wooden rolls. The valley will fall to the main outlet which is usually in the centre of the tenement, and there should also be an overflow outlet in case there is a blockage in the main outlet. This outlet (or warning pipe) should go through the roof and ideally discharge over the front door as it is a warning to tell owners that something needs to be sorted. Originally these overflow pipes were

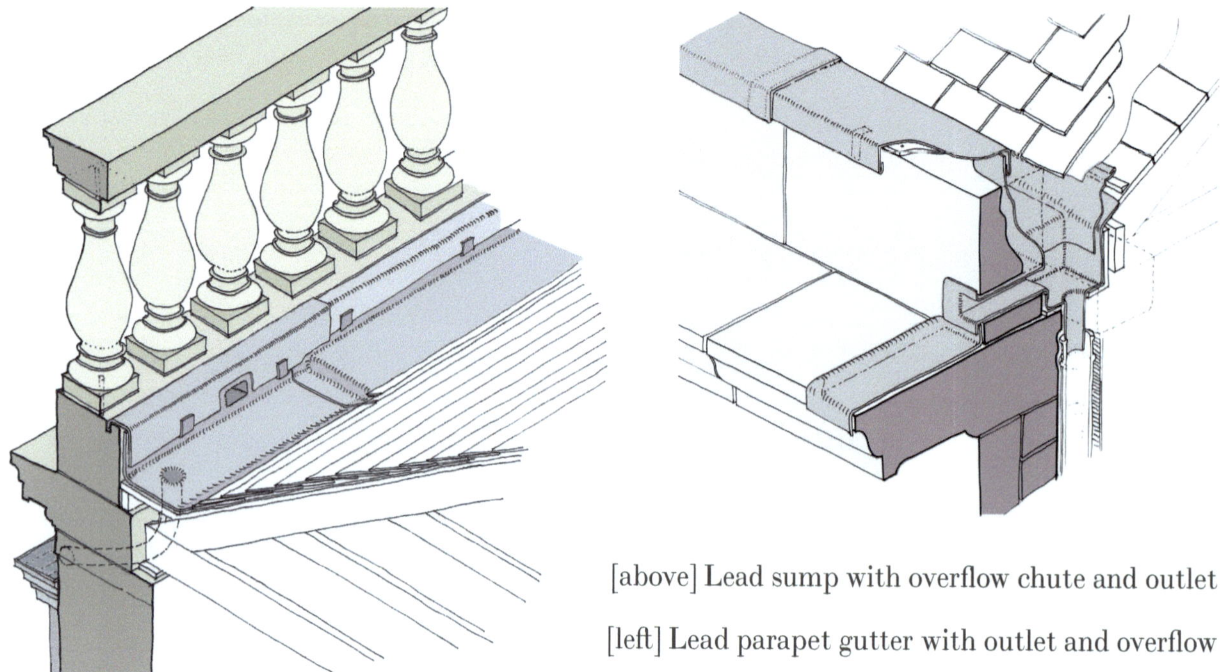

[above] Lead sump with overflow chute and outlet

[left] Lead parapet gutter with outlet and overflow

formed in lead and through time they have sagged as they may not have had a support board under them.

The outlets may be formed as a sump and there are many ways to detail it.

Parapet valley gutters

These are formed behind parapets and blocking courses that cap cornices. They are also prone to blockages and overflow problems. Parapet valleys usually run the full length of the frontage, so the leadwork has to be stepped to allow for expansion. However, the space behind the parapet may not be sufficient to form appropriate steps in the lead and so cracks develop in it. This then gets coated with a bitumen compound which may repair the crack temporarily, but rarely lasts.

The correct weight of lead also needs to be specified to satisfy the length of the lead valley.[44] It will also need lead cover flashings into (or over) the parapet stonework. A lead or copper outlet usually drains through the parapet under the cornice into a cast-iron hopper and then into a downpipe. Alternatively, a lead chute drains the water through a rectangular hole in the parapet wall and then into a hopper. In this case, a separate overflow is not required. Overflow outlets, where required, should be formed in lead, through an outlet large enough to take a deluge of water.

Stone cornice gutters

Sometimes the topmost stone cornice has an incised gutter formed in its length. These were popular in Georgian times and each stone would be jointed to the adjacent stone with hot asphalt. This was not very effective so the cornice itself may have been coated with asphalt at some stage.

44 In some cases it may be appropriate to use a T-Pren rubber joint to weld lengths of lead together and allow for expansion of the lead. These are effective but have less of a life expectancy than stepped lead joints.

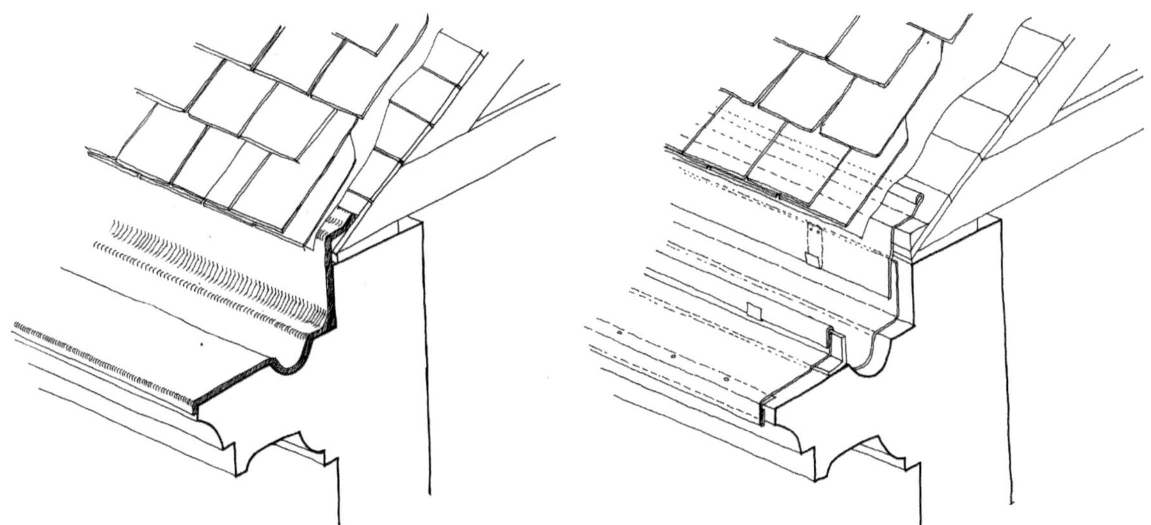

[left] Stone cornice gutter with asphalt [right] with raised angle under new lead to increase capacity of gutter

The main problem with such gutters is that they are both shallow and have very limited falls. Lead can be formed over these gutters provided T-Pren expansion joints are used. Also, the depth of the gutter can be raised by installing a stainless-steel L section or a timber roll, which the lead is dressed over. Sometimes, as a shortcut, roofers install an added aluminium or cast-iron gutter over the cornice and extend the eaves to drain into the new raised gutter.

Gutters

Front moulded gutters[45]

Traditionally these would be cast-iron gutters[46], or rhones, in six-foot lengths joined with red lead putty and bolted together with a galvanised nut and bolt at the base of the fauceted joint. They would be seated on a projecting stone corbel and secured to a rear timber fascia with brass screws. They would either drain directly into a cast-iron downpipe or into a cast-iron hopper at the head of the downpipe. 7in. × 4½in. gutters were commonly used although these have often been replaced with smaller gutters over the years.

Front moulded gutters are usually laid level, with water naturally draining to the outlet. Sometimes there are insufficient outlets to take the amount of rainfall falling on a roof slope, particularly with climate change.

When curved bays are formed, the cast-iron gutter has to be made to the same curvature. Cast-iron gutters are still being made today, both in Scotland and in England, and curved moulded gutters can be found for most curvatures. At regular bay roofs, offset angles join the gutter at 45°.

Cast iron is a very appropriate material for gutters, although it should still be painted. If the rear wall of the gutter rusts, water in the gutter can overflow into the roof and soak the

45 Also known as 'ogee' gutters, which refers to the profile shape.

46 In Scotland these are called 'rhones' (possibly derived from the river Rhone in France).

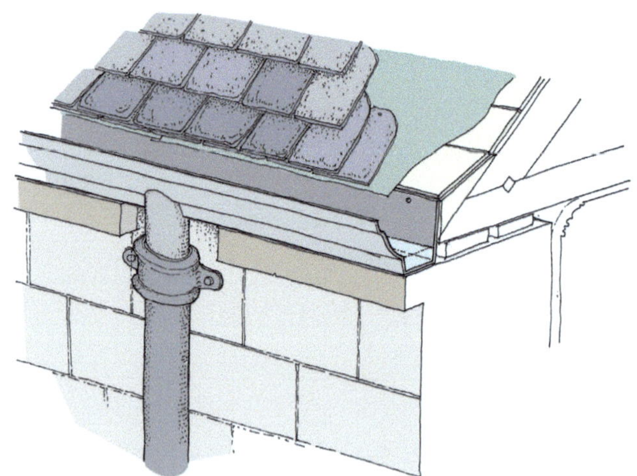

[left] Cast-iron gutter on stone corbel

[right] Replacement angle in cast iron at bay

wall plate and rafter ends, leading to rot. This will also happen if the gutter is left to overflow because weeds gather in the gutter and cause a blockage. In the 1980s, the solution to this was to form a bitumen undercook under the gutter and up the back of the gutter to dress under the slates. This is still good practice, but is now done in lead as it ensures that any overflow will not soak the roof timbers.

Alternative materials for gutters have been used in the past such as cast aluminium and cheaper sheet aluminium. The latter is often used to form gutters on a curve because small sections can be welded together in a workshop. It works but is not attractive. Other materials like glass fibre have been used but they have not stood up to snow loading on the roof, even when they had top fixings. I do not recommend plastic gutters on tenements because of UV degradation and strength. Deep-flow plastic gutters with the correctly spaced fixings may be used, but avoid them if you are repairing a listed (or conservation area) tenement.

Rear gutters

These were originally beaded cast-iron half round gutters, again in 6ft lengths joined with red leaded putty. They are usually supported on galvanised (or wrought iron) support brackets secured to the timber sarking, three fixings for each 6ft length. The back gutter is usually 5in.×3½in. Rear gutters are often replaced with plastic gutters; whilst their capacity may be greater, their fixing to the roof is usually wanting as brackets are not spaced close enough to prevent damage from snow loading. Cast-iron gutters are still preferred and larger sections are recommended.

Rear half round cast-iron gutter secured with galvanised rafter brackets

Snow guards

These were traditionally installed as timber boards secured to wrought iron upstands fixed into the slates near the eaves. They prevented a sudden fall of snow, particularly on steeper pitches, which could harm the public below. Snow guards are often installed in areas prone to snowfall, such as Fort William, Aberdeen and the Highlands. Modern galvanised mesh snow guards can be fitted near the eaves.

Downpipes and drainage systems

Downpipes (conductors are rhones) are essential in tenements to safely remove the amount of rainfall landing on a roof. Clearly the size of the gutter is also important, given increasingly wet weather, but it must also discharge into the drainage system and some gutters may require more than one downpipe to adequately remove the water in a period of heavy rain. If the downpipe is undersized, the gutters can overflow and cause damage to the stone. Water landing on the roof is called surface drainage and is similar to water landing on the ground which might get soaked up if the ground is permeable or if it needs to be connected via a gulley to the drainage system if a hard surface.

Downpipes

In Georgian times, downpipes on frontages were sometimes recessed into a *chase*[47] in the wall so that they were less intrusive. Also, front downpipes were sometimes completely hidden as they were placed inside; if taking rainwater from a secret valley, they would descend internally, sometimes in the close or possibly in a flat covered over with lath and plaster finishes. This made it difficult to repair internal downpipes and when recessed in a chase externally, difficult to paint. External wall mounted downpipes secured with cast-iron *holder-bats*[48] were favoured

[left] Georgian downpipe set into a chase

[centre] Victorian downpipe with cast-iron ears

[right] Internal downpipes can cause damage

47 Chase: a vertical slot or gap in the front wall which houses the downpipe.

48 Holder-bats are pipe fixings to secure a pipe to a wall, also called 'Haudfasts'. When cast in with the pipe they were called 'ear wings'. Galvanised fixings are now commonly used.

in most tenements, although over time decorative holder-bats have become simple galvanised iron hooks driven into timber *dooks* in the wall.

Downpipes would originally have been cast iron, in 6ft lengths with *fauceted* joints. Joints would have been *caulked* with hemp rope then lead filled. Rectangular section downpipes were sometimes used for more prestigious tenements. Downpipes should have an access eye at the base to allow for any blockage to be cleaned by rodding.

Cast iron should be painted at regular intervals, ideally every five years. The back of downpipes are often unpainted and this is where corrosion often starts, cracks then develop in the cast iron and water can then soak into the adjoining stonework, leading to rot in any embedded timbers.

As Georgian and Victorian tenements are connected to the main street drainage system through a single-pipe system, the rainwater downpipe at the front should either drain into open galleys at the base of the pipe, or pass through a *Disconnecting Trap*, often called a *Buchan Trap*[49], before connecting to the tenement drainage pipes. The Buchan trap was used in many tenements, front and back, from 1875 onwards, although rear downpipes were meant to drain directly into the main drain and be vented by a rainwater pipe to have a branch to vent above gutter level.

Rainwater downpipes

Front and rear gutters connect into drainpipes, sometimes via a hopper, and are usually three inches in diameter and made of cast iron. Some downpipes are overloaded because a neighbour's gutter has been made to drain into it, making it a *'mutual downpipe'*. The gutters should drain into vented gulleys before they reach the main drainage system,

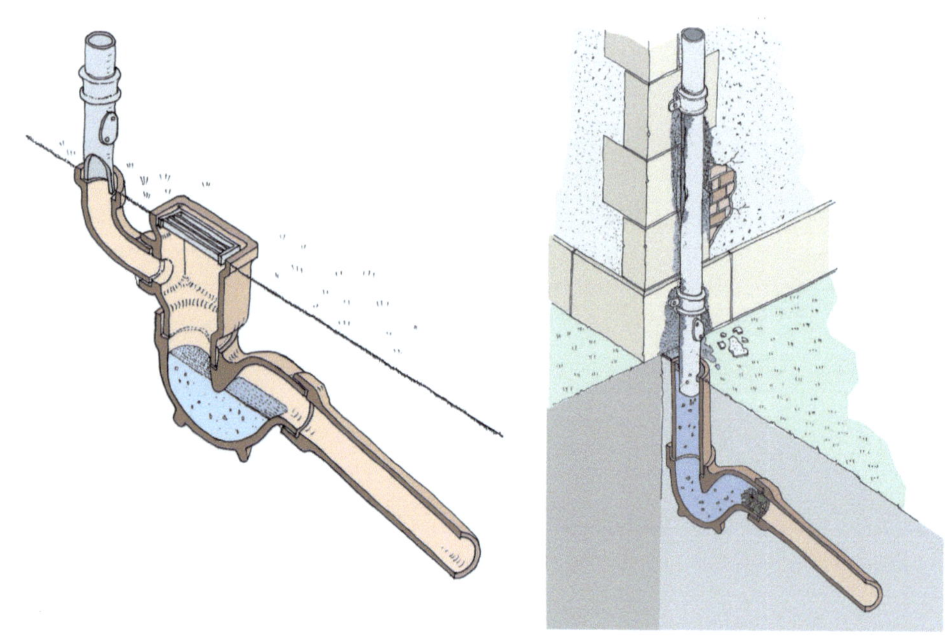

Downpipe into back inlet gulley, downpipe into sealed gulley trap

49 William Paton Buchan patented the trap in 1875. He got the idea from Mr Wallace, a Glasgow plumber who resolved a smell at a house in Garnkirk by cutting a hole on the house side of the trap and venting it to the roof.

although some will drain into an unvented gulley which can be harder to clean. There should be a removable access eye at the base of any cast-iron downpipe which allows for cleaning[50]. Cast-iron downpipes are robust but they do require to be painted regularly. Painters often miss painting the rear of the downpipe, leading to rust and cracking at the rear.

Tenement drainage system

In early tenements there was no drainage system apart from street surface drains[51]. Dungsteads were placed in street locations where the night soil would be collected and then sold on. In the 1750s at 10 o'clock in the evening, people were at liberty to throw out their waste to the street below.

In 1840, Robert Cowan, a Professor of Medicine at Glasgow University, identified the cause of epidemics which, at that time, were frequent. He referred to: '*the accumulation, for weeks or months together, of filth of every description in our public and private dunghills*'.[52]

By the end of the 1860s, the Medical Officer for Health was trying to ensure water was available on the stairhead at least on every second floor of tenements and that one water tap would be provided for ten families, at a minimum. Glasgow got its new water supply in 1859 which was piped from Loch Katrine, 45 miles to the west of the city.

In 1850 the Burgh Police Act was passed which, amongst other things, required that

Added brick-built toilet

Disused toilet on half landing; toilets built using timber housing off brick stairwell

50 The base access eye should have two bolts which can be unscrewed to gain access.

51 Unlike today, when there is a dual pipe drainage system which takes soil and waste to the drains and rainwater to a different system which may be incorporated with a SUDS system (Sustainable Urban Drainage).

52 As Burt describes in his letter about lodging off the High Street in Edinburgh around 1750: 'Being in my retreat to pass through a long narrow 'wynde' or alley, to go to my new lodgings, a guide was assigned to me, who went before me to prevent my disgrace, crying out all the way, with a loud voice 'Hud your hand". The throwing up of a sash, or otherwise opening a window, made me tremble, while behind and before me, at some little distance, fell a terrible shower' and '...for the smell of the filth, thrown out by the neighbours on the backside of the house, came pouring into my room to such a degree, I was almost poisoned by the stench.'

all houses had a water supply and if lived in by a family, a water closet to be installed. It also required every house which did not have a WC to have one constructed apart from where a WC was shared by two or more families[53]. The 1862 Burgh Police Act required every owner of a house to introduce a water supply to the house (*plus crane*[54] *sink and waste pipe*) as well as provide each house occupied by a family with a WC. In Glasgow, this was followed by the City Improvement Act of 1870 which required two WCs[55] on each half landing. These were often added on as timber additions although most were built in brick on the back elevation.

Initially, up to the 1850s, water largely came either from water butts which collected rainwater, or from natural springs routed through lead pipes to water pumps in the street. Private houses in better off areas were supplied with water, although the quality varied. In Edinburgh, 'Water Carriers' (or *Water Caddies* in Dundee) made a living by carrying water up closes to certain houses. Around 1800 the main lead supply pipes from reservoirs and springs were replaced with cast-iron pipes to increase the flow.

The spring water was often contaminated with drainage and in 1830 half the patients admitted to hospital had either typhus or cholera, both water-borne diseases[56].

In 1842 Edwin Chadwick published his *Report on the Sanitary Conditions of the Labouring Population of Great Britain*: '*narrow wynds that got no sun, no water or sewers, animals and people crowded together*'[57]. It showed that cholera and typhus were caused by poor conditions and recommended the provision of clean drinking water and better drainage. It was not until 1849 that it was understood that cholera was a water-borne disease.

In 1862 Dr Henry Littlejohn was appointed Edinburgh's Medical Officer in Health, following the collapse of a tenement on the High Street in 1861, resulting in a great many fatalities. He carried out a detailed survey of the closes and wynds in Edinburgh and used statistics to illustrate the problem. The *Report of the Sanitary Condition of the City of Edinburgh* was a landmark in public health. In a census taken in 1841 in Edinburgh, few if any houses had an internal water supply.

This led to further changes in many burghs. The Burgh Police Act of 1862 required houses in Glasgow to have a water supply and, if a family, a water closet. However, not all burghs adopted the 1862 Act.

Littlejohn fought to have Police Acts to require better drainage, sanitation and water supply, but it was not until 1897 the Public Health Scotland Act was published[58] (also the 1892 Burgh Police Scotland Act) which prevented contamination of a water source and improved drainage.

Sometimes the water supply would simply be a piped supply up the close, draining into a 'jawbox' mounted under a window of the close stair, so it could be communal.

By 1865 most tenements had some kind of water supply. Tenements for the middle class would have had a water supply before this, as well as toilets and washing facilities. Most

53 It also said that where a house was used as a school or workplace, then a WC should be provided for each sex.

54 Crane, Cran or faucet simply means a tap, from the Dutch Kraan

55 Glasgow Building Regulations Act para 53 & 54.

56 The first outbreak of cholera occurred in 1832 and again in 1848-49; 1853-54 and 1866

57 Edwin Chadwick: Report on the Sanitary Conditions of the Labouring Populations of Great Britain July 1842.

58 1897 Public Health Scotland Act.

[above] Lead-lined communal water tanks

[upper right] 'Belfast sink' at window; upmarket twinned ceramic sink to wash clothes

[right] Lead water pipe in loft, poorly insulated with hair felt

Victorian tenements had an incoming water supply via *a rising main*, which was in two legs, rising through each house at the rear kitchen, feeding the kitchen tap and entering the loft to supply a large communal water storage tank. This would be constructed from a large timber box with dovetailed joints which was then lead lined.

The water supply would fill the cistern until full and a floating ball valve would control the supply. There would also have been a lead overflow pipe to the rear elevation. There would be one communal cistern on either side of the close, assuming a tenement with eight flats. Originally they would have had lids and be vented to the roof via lead pipes, although in later times the hot water expansion pipes might drain into these tanks rather than pass through the original lead 'gooseneck'[59] pipes found on old roofs.

The rising main would be fed by the new street supply, with a Tee off branch which would pass through a 'toby'[60] in the pavement. The maintenance of the supply is the owner's responsibility from the toby to each house.

From the communal cistern, cold water feeds would descend to supply each hot water cylinder in each kitchen, usually with a 'stop cock'[61] before connection. The copper cylinders would have a flow and return to the kitchen stove which would have a back boiler. Another hot water supply would go to the bath and wash basins. Such pipes would be taken from the

59 Also sometimes referred to as 'swan-necks'.

60 The water toby requires a water key to switch it off. It simply requires a quarter turn of the key, once the cast-iron lid is hinged back. The name toby is thought to derive from Shanti Irish word 'tobar' meaning a street.

61 A valve which shuts off the cold-water supply to the flat. There would also be a valve at the hot water cylinder to terminate the hot water.

top of the cylinder and insulated with hair felt wrapped with copper wire. Each cylinder would be vented to the roof through a gooseneck vent pipe.

A secondary cold feed would also be taken to the bath and basins where they existed.

After 1892, any new WC had to be located on and next to an external wall, with an openable window which was at least 6 square feet. Tenement drainage systems up until 1920 were 'combined' systems, which simply means that the toilet, bathroom and kitchen wastes also drain into the same sewer pipe in the street. The main issue was to prevent smells and back siphonage, so that soil pipes and waste pipes had to rise 2ft above the rear gutter level. After 1920, surface water was treated separately from the sewerage system as the water does not need treatment.

The main drain would be formed in *vitrified fireclay* or stoneware pipes with *fauceted* joints which were fully *caulked* with mortar. The drain would be laid on a bed of stone gravel (hardcore) and carried from the back of the tenement to the front through the close, being set at least below the level of the lowest floor, and usually a lot lower. After 1900 the pipes were also bedded in concrete, although when they had to pass through and under a wall, a space of 3in. had to be allowed to cater for settlement. At the rear, soil pipes which are usually 4in. diameter, along with waste pipes from baths, basins and sinks (usually 3in. diameter), feed into a rear manhole, to connect to the main drain under the close. Any rainwater pipe that has a waste connection into it should pass through a trap to ensure sewage smells don't enter the house.

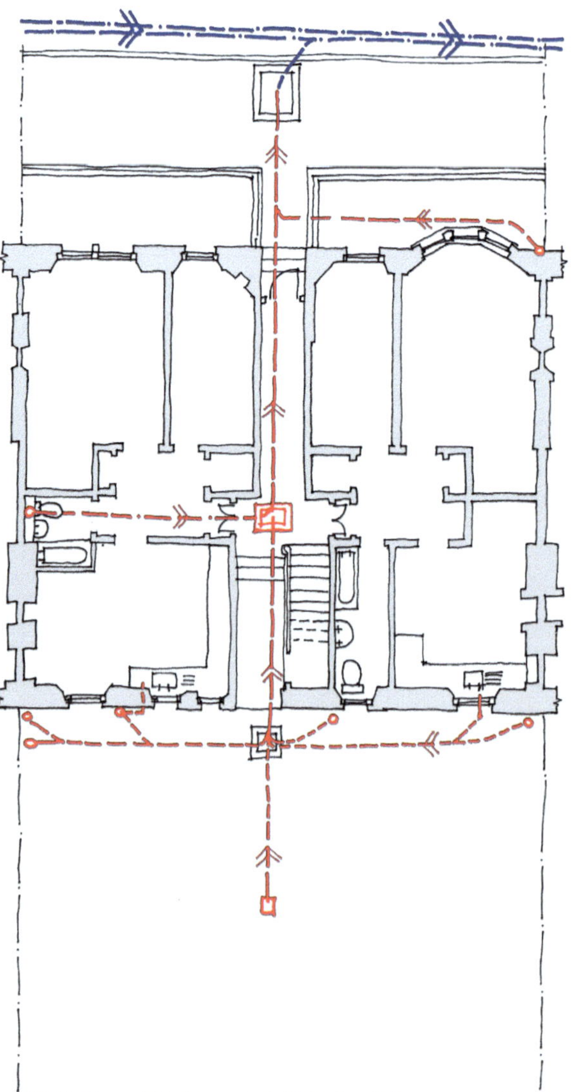

[above] Buchan trap with rodding eye, old sealed gulley traps

[right] Typical tenement drainage plan

Any front rainwater pipes will connect to the drain but should pass through a trap, which could be a Buchan trap or a gulley trap.

The drain should then pass through a manhole or 'disconnecting trap' before it connects to the main street sewer. Responsibility for the tenement drainage system is with the owners as far as the front street disconnecting trap.

Originally, most Victorian tenements would have had a copper tub and sometimes a boiler, along with sinks in a small communal wash house placed in the backcourt. This also needed a water supply and drainage, which would be connected to the rear drain and would also have syphon traps. However, most of the rear drainage from toilets, roof and backcourt is fed into a single drain which then travels under the close to a manhole (*intercepting trap*) in the street and then to the main drain.

Tenement drainage after improvement

Where a tenement has undergone modernisation under a Housing Action Area provision and has had an internal bathroom installed in the bed recess or the kitchen altered, then the drainage and water supply system will have been renewed. The soil pipes may have been renewed in cast iron or plastic UPVC. As these soil pipes pass through intermediate floors they need to be fire stopped at each floor. For plastic pipes this means fitting an *intumescent collar*[62] around the pipe which expands in any fire and closes the hole left by the pipe. For cast-iron pipes, the area around the pipe was often filled with mortar or *vermiculite* although now there is a requirement that it is also fitted with an intumescent collar.

Water tobies in the street

However, the surface water will still discharge into the combined drainage system.

The waste from the bath and basins will connect into the soil pipe which is vented above the roof (or fitted with an *air admittance valve*). The communal stack will go under the ground floor and may go directly to the outside or more often to a manhole that is built within the close to allow connection to the drain under the close. (See Chapter 11 on tenement upgrade 1970–1990s).

Internal fittings connected to the soil and waste were originally vented through vent pipes, although in modern plumbing special devices called air admittance valves can be installed which allow air in without letting smells out. These prevent the back siphonage which can occur when other fittings use the same soil or waste stack.

Although the UK knew that lead was poisoning the water back in 1919, it was not until 1969 that it was banned and was gradually replaced with polyethylene rising mains. Lead lined water tanks have also been replaced with plastic or glass fibre tanks. although there is much less need for them as modern gas boilers take water directly from the mains supply these days, so many communal tanks are only supplying the few remaining hot water cylinders[63].

62 Intumescent material wraps around plastic pipes as a collar and, in the outbreak of a fire, the material expands and closes the void left by the plastic pipe.

63 Which work with modern electric immersion heaters.

Chapter 6

The close and close walls and ceilings

Closes, pends or wynds were needed to get through to the backlands when the frontage was all tenemental. Forestairs at the front allowed access to the upstairs dwelling when there was a workplace below. But pends developed when rear access stairs became common and were accessed from courtyards. When the turnpike stair, which was initially at the front of tenements was moved to the rear of buildings, it was essential to have a pend or close to provide access as the courtyard might be entirely enclosed.

In Edinburgh, access to flats is usually referred to as 'the stair', whilst in the West, the term 'close' is used. Edinburgh's stairs in the New Town are either in the centre, at the front or rear, whereas Glasgow's are invariably at the rear. In this chapter I commonly use the term 'close' to refer to the common access stair.

The turnpike stair

Turnpike stair typically has a handrail along the outer periphery only, and on the inner side may have just a central pole. There are not many left. They used to go all the way to the topmost floor. Some of these are positioned at the front of the building, and some at the rear. You can see what Georgian builders got to in the forming of helical stairs (see Chapter 2).

Open well stairs

Top lit skylights or cupola, were common in the Georgian New Town where the depth of the tenement was wider, but, even in working class Edinburgh tenements where there were four flats to a landing, two flats facing to the front and two to the rear, top lit skylights or cupolas were in use as stairs were in the centre of tenements.

This was achieved by using individual stone treads that were penchecked[1] together to act as a unified whole. One end of each tread set into the close walls allowing the load of the staircase to be transferred onto each landing. Stair flights with no central spine wall rely on the pencheck[1] joints to transfer forces between each step.

1 Pen(d)check stairs. A pencheck is simply a splayed rebate which allows the stairs to interlock with one another.

Where past movements have occurred, these joints can open up and gaps should be infilled with a suitable grout otherwise they can become prone to sudden and progressive failure, which can affect the whole stair construction. The curvature of the stair plan and the installation of wrought iron or cast-iron balusters with hardwood handrails helped to stiffen the structure. The open well allowed more light into the close and was extensively used in Edinburgh where the stairs were lit from above. Some turnpike stairs were open well stairs, and open well skale stone stairs were also adopted using the same principle, although, after 1900, steel T beams were used to provide added support to Glasgow's square stairs.

Edinburgh tenements developed in a similar way up until the late 18th century, although they were often higher due to topography and the density of housing at that time. One tenement had fourteen floors; at one level you rose seven floors and could also descend seven floors, but tenements of eight, ten and twelve storeys were not uncommon in that period (16th century). The New Town development tended to set the standard for later tenements with a centrally located stair accessed by a close which was top lit by a cupola. The stair was formed in a circular, ellipsoid or D-shaped plan as a hanging penchecked stair[2] which allowed light through the open well. At corner tenements and in the Edinburgh New Town, these internal stairs were often D shaped in plan, or elliptical, allowing a very open stairwell that allowed light to penetrate down the stairwell from a skylight above. This developed into a square plan with corner landings. Glasgow tenements also had square shapes; the latter would have had one well on the baluster side.

The Burgh Police Scotland Act of 1892 limited the number of dwellings to each common stair to 12; however, if there was an external access deck, the developer could build up to 24 houses to a common stair. Although Dundee had built tenements[3] before 1890 (one in 1870 was nine storeys high), the tenement boom slowed down in 1880, but because of the demand for workers in the jute industry many more *platties* were built, encouraged by the new legislation.[4] Many platties have unfortunately been demolished but a few still remain in Dundee, Glasgow and Edinburgh.

Circular stairs found in Pollokshields

2 The obtuse angle of a pencheck (similar to the angle a quill pen is cut to), is perfectly aligned to resist forces tending each individual step to rotate.

3 Dundee used a local stone which was of variable quality (or perhaps wrongly bedded).

4 The plattie form of housing with rear access balconies was already a traditional form in Dundee before 1890.

[above] Close with turned steps

[upper right] Hanging stairs

[right] A skail stair with 'wally' close tiles

Underside of a plattie stair

Today there are few left and, even now, more of these unique tenements are being compulsorily purchased and demolished, often against residents' wishes.

Close stairs (skale)

Early tenements from 1500 to about 1820 would have been formed as square edged steps set on the courses of stone and found externally, either at the front or back, to allow access to the first floor. Any higher would likely have been via internal timber stairs. As the staircase came into the tenement, two straight flights would rise up and be turned in a curve, with treads spanning between the close walls and a central spine wall. At the half landing, a base of sarking timber would be formed between the joists and a bed of ash and lime formed onto which individual Caithness stone pavement stones would be set.

This developed into two straight flights, each of about ten treads and a rectangular half-landing which was easier to build, with a close window sited on the half landing. In early tenements the spine wall remained, but in later middle-class tenements there was an open well. In Glasgow, there was a concern that the stairs could restrict the light into the stairwell so in 1892 the Glasgow Building Regulation Act set down the minimum size of such a stairwell. Stone stairs were built into these half-brick walls although in better class tenements the close walls would have been a full brick thick.

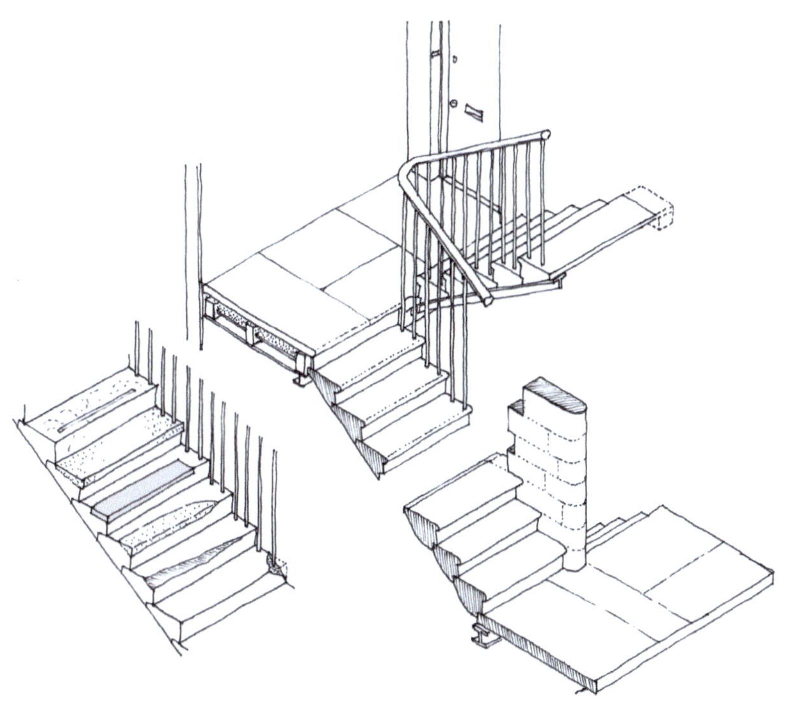

[left] Central spine wall or dividing wall

[right] Square 'skale' stair with a half landing

[upper left] Stone treads become worn, sometimes repacked with concrete

[upper right] Added strengthening to a hanging stair (poorly done)

[left] Strengthening to underside of an open well turnpike stair

The sandstone for the treads was selected for its wearing ability, which in the Edinburgh area was usually stone from the Hailes[5] quarry. In Victorian times the stair nosing developed to form a sort of bullnose edge. The width of the stairs was to be at least 4in[6]. Although stairs were largely stone, granolithic concrete was invented around 1880 and began to be used in stairs, landings and closes. It was a concrete mix which contained hard wearing but fine aggregate, usually granite, which could be formed over a concrete base and allowed steps to be formed 'in situ' using shutters to give a bullnose *edge*. By about 1890, open well tenement stairs were built with additional T-section stringers under the open edge and connected to larger steel joists on the underside of each landing.

By 1910, concrete stairs, using a granolithic concrete mix were fashionable, combined with timber handrails and newel posts and usually enhanced by a *wally close*.

When a hanging stair collapsed in Woodlands Road in Glasgow in the 1970s, the Council decided to strengthen all the stairs that had treads that were only supported by the close walls. Although the railings and interlocking nature of these pencheck stairs ensured that the load was transferred to each landing, they decided to install steel T bars and beams under the open treads of the stairs as well as at landings. Thankfully, this has not been done in other cities as it can be quite unsightly. Some open edged turnpike stairs have had cracks which are often repaired locally with steel straps, secured to the underside of the stone tread, rarely well done.

Timber stairs

Aberdeen is known for its timber stairs in closes. This may have developed because granite was more difficult to shape into stair treads, or may not have had sufficient tensile strength. However, they also exist in smaller towns like Forfar, although this was usually only for a two-storey tenement, so the builder may have found it simpler to make a timber stair.

Close entrances

Most close entrances in Glasgow were open onto the street, although middle-class tenements often had close doors. In the early 1970s, metal close gates began to be installed, but new glazed close doors only started to be installed in Glasgow in the mid 1970s, probably due to door intercoms becoming available. Close doors have been traditionally installed at close stair entrances in Edinburgh tenements[7] since Georgian times,[8] as well as having main door houses which had an additional door to the close to allow access to the back green.

As noted previously, close doors were only fitted in more middle-class tenements. Although some were fitted in Glasgow, most are found in Edinburgh where the close door tradition continued when sandstone tenements were built for the working class. In Edinburgh, the close doors are usually six panelled doors with a top rail, intermediate rail, lock rail and a bottom rail as well as styles and a central muntin; the panels are flat but raised *bolection* mouldings

5 Hailes Quarry tended to produce a whiter stone than Craigleith which was yellower.

6 As set by the 1892 Burgh Police (Scotland) Act and earlier acts.

7 Close doors were fitted with offset bottom hinges so they would close under gravity. They were controlled by a wire system where a brass pull knob alarmed the flat dweller who then went into the close to raise a lever which had a wire which released the door latch.

8 Complete with a bell system operated by wires and levers located on each landing (see Doors section).

[upper left] Timber close stairs in Aberdeen, with a timber dado

[lower left] Two flat entrances

[upper right] Also timber stairs in Forfar and [right] Aberdeen

Original bell pulls on an Edinburgh front close door.

Offset hinge to ensure close door closes

Glasgow closes were originally open to the street, then metal gates were installed

In time, door intercoms were introduced, and close doors were installed

secure the panels externally. A top glazed fanlight is common. The door is usually hinged with a bottom mounted offset pin hinge[7] which ensured the door closes.

In Edinburgh tenements, a bell in each flat would be activated by a pull knob at the close entrance and the latch would also be controlled by pull wires which allowed the door latch to be opened by pulling a lever in a slot with a lever placed at each landing level[8]. Few of these still work having been replaced by electronic door entry systems which allow visitors to call and ask owners to allow access. Locks vary, but magnetic locks are now commonly used.

In the mid 1970s when metal gates started to be installed, the gates were not secured. When door intercoms were developed, glazed close doors giving more security were introduced where owners could afford it. As a result of the Tenement Improvement Programme in the 1970s and 1980s, most closes were fitted with close doors and door intercoms.

Close and pend openings

The close is usually about 4ft (1220mm) wide[9] although not actually stipulated by law until the 1866 Police Act. It was certainly a traditional width from 1690, although some 17th-century *wynds* were down to 2ft 6in. (762mm) and access into houses off a turnpike stair was often very narrow, well under 4ft. Where a rear turnpike stair was the main access, a wynd or 'close'[10] was taken through to the back. Some wider pends were made to ensure carts and horses could get through to the back where there might have been workshops or dung steads.

9 Many architects and developers preferred a width of 4ft 6in.

10 Close, pend or wynd, meaning a passageway leading into a courtyard; any narrow passage or alley.

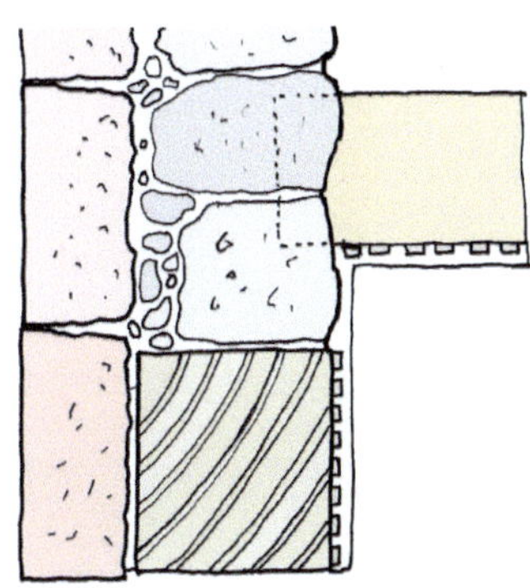

[upper left] Large section timber found at close entrances [upper right] and [left]

The main landing in a typical rectangular close would be supported on a large timber beam[11]. Smaller joists spanning between close walls would be toothed into the beam. At the close entrance, most tenements built around 1850–90 had external stone lintels with a timber baulk lintel inside to take the weight of the wall above. These timber baulks[12] were sometimes covered in lathes and plastered on the inside.

Large square section timber baulks were often used at close mouths and at shop window openings, although after 1910 they were usually replaced with steel or even cast-iron lintels

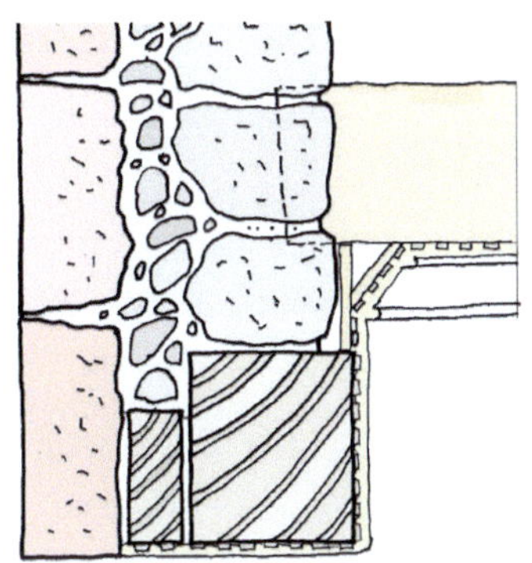

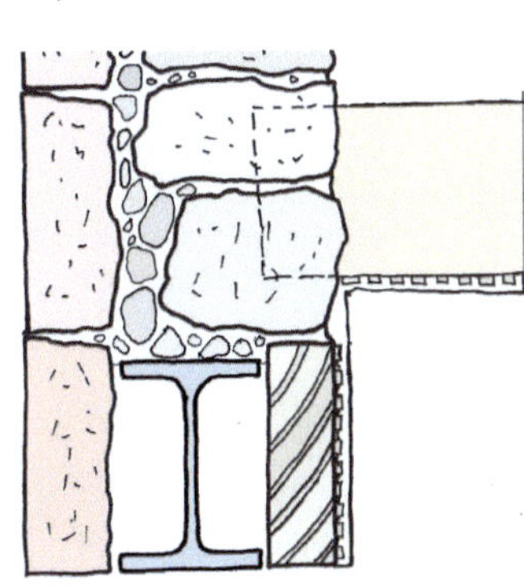

Timber beams at close entrances, sometimes timber baulks, sometimes steel or cast-iron beams

11 Termed a *Bressumer* beam, originally *'breast summer'* from the French *sommier*, a *'pack horse'*, meaning bearing great burden or weight.

12 Baulks were large square section timber beams about 12in. × 12in,, also referred to as bauks or *baulks although the term has also been used for collars in roofs.*

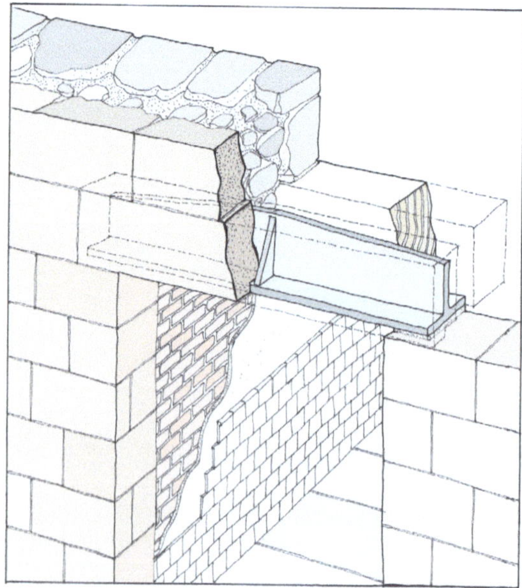

[above] Cast-iron beams with the stone checked into the plate. When the iron rusts, the stone spalls, exposing the edge of the original beam

[upper right] Cast-iron rusting and decay in the stone surround

[right] Typical of Edinburgh tenements although also found in more upmarket Glasgow tenements. Moulded cast-iron beam over a wider pend

(*see Chapter 8 on shops and shopfronts*). Cast-iron lintels are often found at pend openings and at shops, as they allowed a much greater span. After 1890 the use of concrete became more common, with each landing being supported on two steel joists.[13]

In Glasgow, around 1895 cast-iron lintels[14] were used, sometimes simply thick cast-iron plates, with the outer stone lintel checked and set into the web of the cast-iron lintel.

With increasing wet weather, corrosion can develop on the cast iron and this leads in time to damage to the outer stone lintel.

External exposed cast-iron lintels[15] are also found on close entrances and at ground floor window lintels in Edinburgh around 1890–1900, often cast to mimic any adjacent stone profile. Where the lintels are exposed, they show rust but have weathered well and many have been painted. They are also often found at larger spans, for shop windows and wider closes and pends.

13 4in. × 3in. RSJ joists.

14 Cast-iron beams and lintels were often used in shops. Although used from around 1810, the lintels were eventually replaced with RSJ (steel beams) after 1895 although cast-iron standards and columns were still in production up to 1930.

15 There are various suggestions as to why cast-iron lintels were used in Edinburgh rather than stone in this period, but the most plausible appears to be that it was done in areas where the ground was considered unstable because of a burn, or difficult ground conditions.

[above] Cast-iron lintel over a large bressumer, to form the entrance to the backlands

[right] Cast-iron stanchions supported the bressumer beam that carries the main load of this tenement

After 1900, rolled steel joists (RSJs) were used instead of cast-iron steel; internal timber baulk beams were still used to help support the wall thickness. Sometimes, when shops were built into the original design, the large timber baulks (beams) would carry through to also support the close opening, although cast-iron beams, then steel RSJs[16] were also used.

Bressumer beams are basically long baulks and are often used at bay windows, shop openings and above close entrances (*see section on Shopfronts*). They were also used in shops to support the load of floors and walls above, and as mid beam supports in Edinburgh tenements. When installed in shops they are often supported on cast-iron columns. Sometimes the column head does not spread the load enough, and the timber baulk can be crushed by the load above.

Close floors

These would be formed from Caithness stone slabs spanning between the close walls, sometimes with under-building of brick support walls. As the ground floors were timber, the joists had to be raised above ground level. Ventilation through the solum was designed to prevent moisture rising up the walls. Few tenement close walls have damp-proof courses, although if the Caithness slab rests on the full wall, it will act as a damp-proof course. In later construction the floors would be formed in concrete, usually set onto a permanent steel *shutter* or between steel T bars[17]; then the surface would be tiled, sometimes with tessellated tiles, sometimes with granolithic concrete or terrazzo flooring.

The close floor at the entrance was usually raised by a few steps, although some were level. The stones used in the floor would span about 1,200mm. Underneath was the solum

16 RSJ: rolled steel joist. Steel beams were being used from 1850, although they became much more common after 1880.

17 Embedded steel in this position is very prone to rusting.

No. ....

NOTICE TO TENANT

IT IS YOUR TURN TO SWEEP AND WASH THE COMMON

.... **THIS WEEK**

NOTE:— Please hand this card to the Householder next in rotation after having completed your turn.

Services for Communities .... 20 ......

[upper left] Caithness slabs in close floors.

[right] Spanning between close walls

[left] A sign that was passed to each neighbour advising them it was their time to sweep and wash the close

which, like the ground floor houses, would be ventilated to front and back. In some cases, these stone slabs have cracked and in one case opened a hole into the basement area.

Close floors could be constructed in different ways. If there was a basement below the close floor, the Caithness slabs might span the whole of the basement. Otherwise, the Caithness slabs would be seated on builder ashes, and sometimes, if there was a gap below the system, they might have used metal T-bars to support the slabs.

Drainage runs and other services often run beneath this close floor.

It was traditional to expect that every householder would clean the close and stairs about once a week, although in some areas, cleaners were paid to do this work. A pre-printed card would be left on your doorknob informing you when it was your turn to both sweep and wash (mop) the close.

The close would go to a rear door to access the back court, usually down at least as many steps as at the front. Sometimes there are small stores formed at the rear under the first flight of stairs and, depending on site levels, a basement, which may contain individual stores for proprietors or a basement for a shop above.

When the backlands were used by industry or workshops, a pend would be formed to allow carts through to the rear workshops. It also allowed access for the carts to remove the

[left] Close walls would have dados plastered with Keene's cement which was hard wearing

[right] Oil paints blistering from dampness in the spine wall

night soil. In such cases, a close entrance might be formed off the pend to access the stairs for the houses above.

Close walls

These are mostly formed in brick, usually half-brick thick and rendered on the close side and plastered on the house side. In some upmarket closes 9in. thick close walls are found in the initial flanking walls of the close and in the stair walls, but the rear close walls may reduce to half-brick thick. This would largely only be found in higher value middle-class tenements, as most Victorian tenements had close walls which were only half-brick thick.

In working class areas and early tenement close walls, a cement dado would be formed using Keene's cement[18] in three coats and three coats of lime plaster above it. The plaster would be lime washed and the dado would be gloss painted. In later Victorian and Edwardian middle-class tenements, the dado would be tiled, usually onto a cement mortar base coat using a variety of manufactured tiles. Each tile would be 'buttered' and set onto Portland cement, then the tile edges would be dipped in Keene's cement, before being set close together. In high-class closes these dado tiles would rise up the close stairs to the top forming a '*wally close*'.[19]

18 Keene's cement plaster was patented in 1838 by Richard Keene. It provides a durable finish which blends with lime putty to produce a smooth crack-free finish and is still produced.

19 A wally close is one that has been fully tiled up all the flights of stairs. Wally refers to a smooth material like teeth (mostly false teeth) or china tea sets and 'wally dogs'.

Stained and etched glass in close windows became more common after 1900

At the outer walls, the inner brick close walls connect to the outer stone walls and because brick coursing does not match the stone rubble coursing, brick stretchers were only tied into the outer wall every fourth course. This has not been enough to prevent separation cracks developing in the outer corners of such close walls.

Close windows

Early close windows were simple openings often with vented slats, but eventually sash and case windows were used. Sometimes they cross at a half landing, but most are quite simple with rendered dados and plastered ingoes to match the close. In later wally closes and more upmarket tenements, wider sashes would allow for margin panes with bright ruby or blue glass. Etched and painted glass was also used by the Victorians for a period. Stained glass panels were popular in some areas.

Railings and handrails

The decorative cast-iron railings were usually set into pockets in each stone tread and caulked into place with molten lead. The mason would form *dovetailed pockets* to bed the legs of each baluster into each tread using molten lead, though after 1900 this practice mostly died out. In earlier Georgian tenements of the New Town, they might be a simple square section wrought iron and offset at the base to make access up and down as wide as possible. Sometimes they would be shaped and pocketed into the side of each tread. Cast-iron balusters, as noted earlier, became more common after 1850.

Hardwood handrails were shaped and secured to a flat metal rail which was riveted to the top of each baluster (the malleable iron rivets are set into each cast-iron baluster). The hardwood handrail (mahogany or teak), which is chased to receive the steel rail, was then secured to this rail with screws. Sections of the handrail are bolted together with recessed bolts, but they are only accessible on removal of the handrail.

The handrail is normally finished in a *volute* at the base and to prevent people sliding down the handrail, the Victorians liked to fit brass nodules.

Hardwood newel posts and timber balusters became more common after 1900 in higher class tenements, often combined with granolithic, mosaic or tessellated tiled floors.

External cast-iron railings still exist, where they protect a stair down to a basement, or over a basement, or to prevent people falling into a dunny area. They were retained for safety in 1940 when most of the iron was removed for the war effort.[20]

20 And we understand some railings which were excess to requirements may have been dumped into the sea.

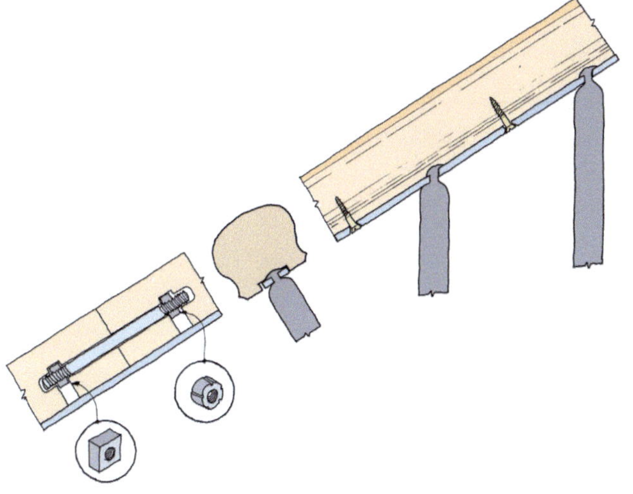

[above] The hardwood handrail is secured to a metal rail that is pinned to each cast-iron baluster. Handrail sections are jointed at a hidden bolt within the handrail

[upper right] A volute at the base and brass studs

[right] Raised buttons to stop youngsters sliding down the banister

[below] Hardwood newel post to link hardwood handrails

Cast iron is not good under tension, so if railings are not secured with handrails (metal or timber) they are at risk of cracking, as occurred one night in Glasgow when a young man fell through some broken cast-iron railings and cracked his skull (*see Chapter 9 front steps*).

Wear and tear: structure of stairs and landings

Stone stairs take a lot of foot traffic, particularly the lowest flights. In some cases the treads are now so worn that they have become unsafe. In the past the worn area was often repaired by cutting out the worn section and indenting new stone into the tread. Indeed this is still possible provided a good stonemason and match for the stone can be found. Since the 1980s, many stone treads in tenements have been resurfaced, usually by applying a granolithic mortar mix of mortar with an epoxy hardener, over the stone and nosing. It repairs the stone but loses the quality of the original surface. In some tenement refurbishment work, the treads and landings are completely resurfaced with Linotol which is a product made by the Veitchi company that consists of sawdust and resin compounds and allows the insertion of brass *nosing* and non-slip strips into the finished product.

Platties and galleries

After the 1892 Police Act, a tenement that allowed individual access via an external balcony or deck could be built with 24 houses, rather than 12. Platties (also referred to as balcony or gallery access tenements) developed in several cities with an external turnpike stair access. This form of housing had been quite traditional in Dundee for two-storey tenements which had a rear stair and deck access so it rapidly developed to house workers in Dundee's factories, where the jute industry was expanding and needing housing for workers.

Some much earlier tenements (Edinburgh) made use of the rear turnpike stair with decks to access each house from the outside

[upper left] Long gallery decks

[above right] Ramsay Place, Portobello

[left and lower image] High School Yards Edinburgh, was the first council housing in Edinburgh.

Brand Place Edinburgh

The Platties were supported on cast-iron columns and a steel frame structure supporting the concrete deck. They also existed in Glasgow, Edinburgh and Perth. In Edinburgh the access and decks were sometimes at the front (possibly '*ideal workers houses*' showing how well ventilated they were, after the Great Exhibition in 1851).

Terracotta balustrade reinforced with embedded steel

Gallery access flats still exist in Glasgow, at 3–11 Bain Street built in 1894 and in Portobello at 6–10 Ramsay Place, which has brick at the rear and stone at the front and a flat roof. In Edinburgh at 2–10 Brand Place (Abbeyhill) there is a small external deck, accessing two tenements with a landing and at 8–10 High School Yards (these are the first of Edinburgh's tenements), two tenements facing Holyrood Road with each deck accessing two houses and another two houses with doors in the close.

A much grander gallery access tenement is found at Caledonian Mansions[21] in Glasgow; it has terracotta balustrades and was built as an annex to the Central Hotel as it was built beside a railway line. The balcony parapets had to be repaired with replacement terracotta made by Shaws of Darwen. The photo above shows how they were tied together.

Dundee platties were the forerunner of deck access housing built abroad in the 1960s and in Scotland in the 1970s using 'system building' techniques.

Tragically the council is demolishing the few remaining platties although at least one in Inverness at Ardconnel Terrace is now listed. (*See Development of Tenement Plans for a plan of a typical Plattie.*)

21 Designed by James Miller and built in 1897.

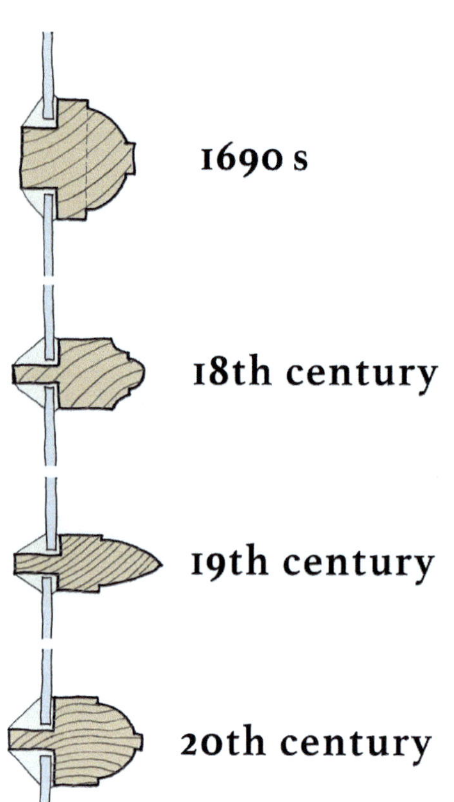

Astragal sections: early ones were chunky, Georgian ones were much finer

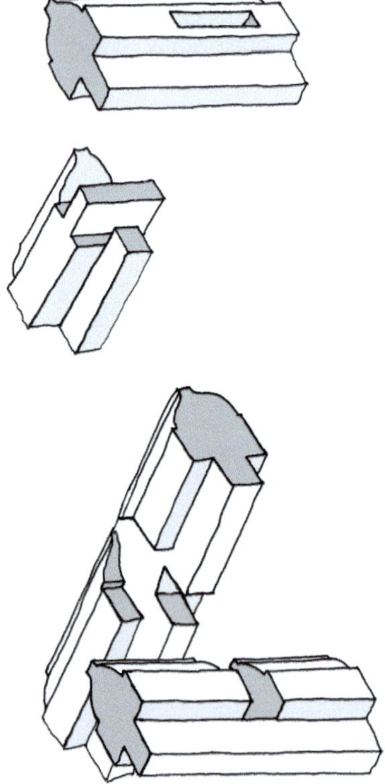

Joints used to join astragals

Chapter 7

Windows and doors

Windows

In medieval times, windows were rare, but openings would be formed in walls to let light in and for ventilation. Shutters would be used to cover the window openings on the outside.

In the 17th century, sash windows were used but sash and case windows only developed after about 1680 when pulley weights were incorporated as counterweights to each sash, to allow the windows to easily move up and down. These early sash windows were small and timbers quite chunky. As the only glass available then was crown glass, the panes had to be small and the glazing bars[1] quite thick with often eight panes in each sash.

By the 1750s the Georgian window had six panes to each sash and by 1830s the sash and case windows were big; although still with six pane sashes, the astragals (shaped or moulded glazing bars) had become much finer and more slender.

As glass technology improved in the Victorian age, cylinder glass allowed larger sheets of glass to be used, so a sash might be glazed in two panes with a central vertical astragal. Between these periods horizontal astragals were sometimes used as well. After 1850, larger sheets of glass were available. By 1860 windows in tenements might have one vertical astragal, eventually the sash would be fully glazed and horns were developed to strengthen the sash frame on the upper sash.

Around 1880, plate glass began to be used which was much heavier. This avoided the need for astragals although each sash frame joint was usually strengthened. Around 1900 the top sash was often made smaller and glazed with small panes, anything from six to nine panes, really for decorative effect.

The case[2] of the window has a timber sill at its base which is bedded onto the stone sill, in plaster or lime mortar. The case, or sides of the window frame, the top rail and sill are set behind stone scuntions which are checked to receive the case. The external stone lintel usually aligns with the top inside surface of the case. The internal lintel is formed from two (or three) red pine 'safe' lintels supporting the rubble stone above. There may be a relieving arch above these timber lintels as well. The window case is secured with timber wedges given in between the top of the case and the safe lintels. A modern installation will often use galvanised strapping to secure the case to the stonework.

1 Astragals, shaped glazing bars.

2 The case is the fixed frame or 'box' that the sashes fit into and slide up and down in.

Under Section 102 of the Glasgow Building Regulation Act of 1892, it was a requirement that: '*All window sashes in dwelling houses and in habitable rooms above ground floor shall…be so constructed to admit the outside of the window being cleaned from the inside the room.*' Edinburgh had a similar requirement.

Simplex[3] type fittings, to allow window cleaning, have been around for over 100 years, although they are now made as butterfly type hinges which allows the lower sash to be hinged in after the sash rope has been secured by a sash fastener. The Victorians[4] also made similar hinges that slid up and down on a brass rod, still allowing the lower sash to hinge in.

After 1910, some tenement designers used large timber inward opening casement windows, and these are still in use today in Glasgow's Hillhead Street tenements.

Close windows in the early 1700s would have been simply vented with timber slats, but they developed in a similar way as the main house windows, as sash windows, until in the 19th century, when margin panes were introduced around two central panes in each sash. The margin panes were often coloured and the central panes, at least in middle-class areas, might have been etched. Painted glass in these central panes was also popular, but few now survive.

Fanlights above front doors appeared after 1750 and are commonly found in Georgian era tenements, such as Edinburgh's New Town.

Sash and case construction

The window case[5] is formed from a timber lining, or pulley style, which is usually about an inch thick (25mm) with a central slot to house the parting beads that separate the two sashes. Rosalind Marshall's study of the third Duke of Hamilton shows that the restoration of Hamilton Place (1760–1800) would have featured continental type sash windows, perhaps the forerunner of the sash and case window.

The outer and inner lining, or facings, are grooved and the internal lining formed into a tongue and set into the groove of the inner and outer linings, forming a box, the *case*, which can house the cast-iron *counterweights*[6].

The outer sash is kept in place by the outer lining and the parting bead; the inner sash is kept in place by the same parting bead and the inside batten rod. The linings at the head of the case are usually smaller and stiffened with small pieces of angled strengthening blocks.

As the *parting bead* is removed, when the sash cords are replaced, it is not nailed in but held in place by friction. A pocket hole is formed in both pulley styles about seven inches (178mm) up from the sill and for a six-foot window about eighteen inches (457mm) high. The pocket hole is normally covered by the lower sash and is V pointed and set into a slot in the style, held in by the parting bead. It is important to keep the void inside the sash box clear of any debris that may fall down behind it, otherwise dampness can be retained which rots the adjacent timbers.

3 Someone called Brown seems to have invented 'Brown's Simplex window opener' as it was approved by the Glasgow City Improvement Trust in 1895 for use at 74 Kirk St in Glasgow.

4 Called the 'meeting rails'.

5 The frame that the two sash windows slide up and down in

6 Cast-iron counterweights, each stamped with their weight, were most commonly used, but lead counterweights are also used.

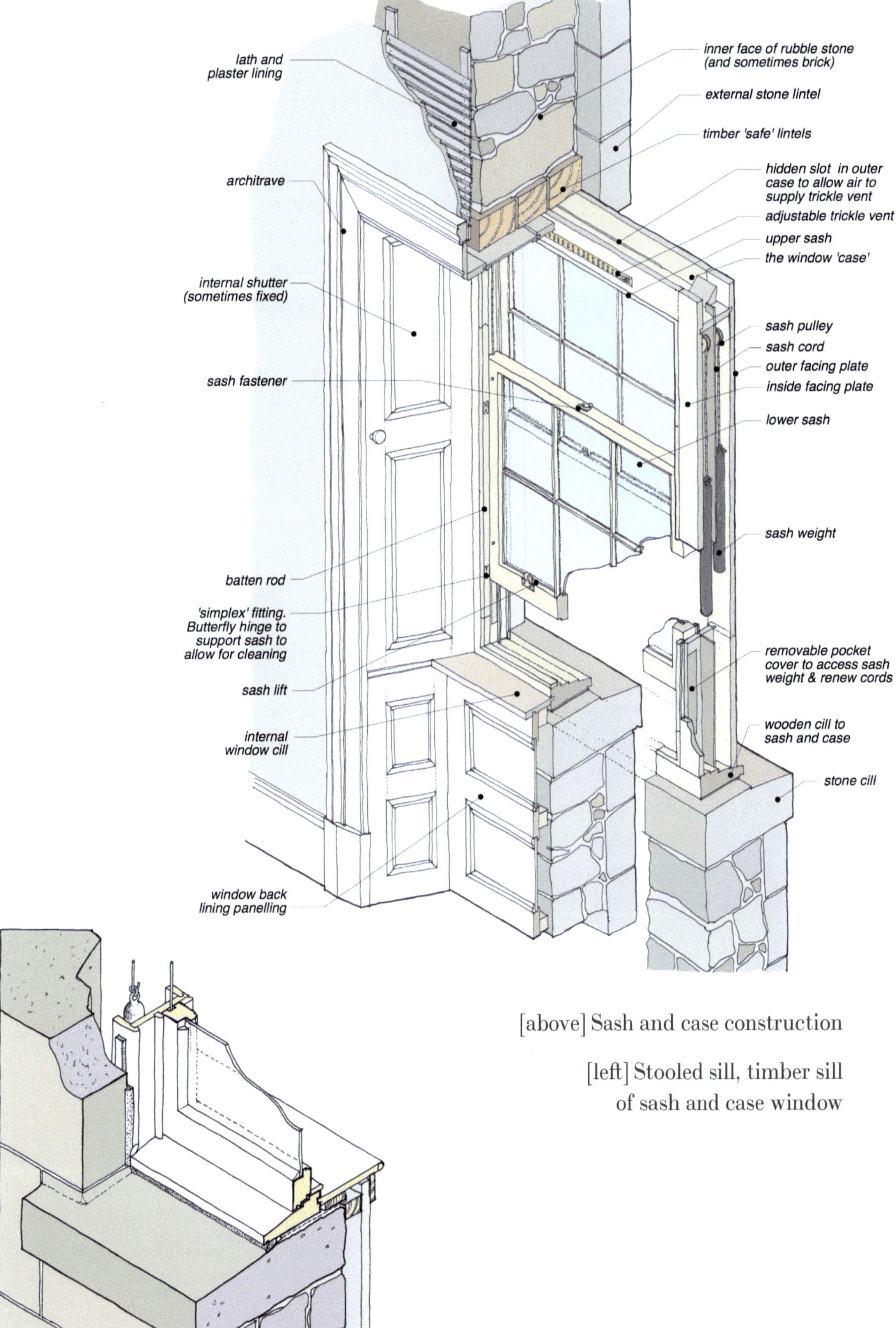

[above] Sash and case construction

[left] Stooled sill, timber sill of sash and case window

At the head of the style, brass pulleys are checked into the lining to receive the pulley cords which are secured into grooves in the sides of each sash and the other end connected to the counterweight. Each pulley is secured with a brass screw, top and bottom; care must be taken that long screws are not used as these can fray the sash rope. The sash ropes are usually made from hemp and secured to the sash with tacks, although when simplex fittings or heavy windows are in use, a recessed cord hole[7] or removable clip fixing may be used which clips onto a screw in the groove.

The sill, which is best made in Douglas fir or good quality pine, is checked to receive the internal lining and secured to the inner and outer linings. The rear of the case would sometimes be boxed in, to ensure the weights did not catch against any protruding stone. But this is usually only found in more expensive buildings. Twinned windows would be centred on a stone mullion.

Once the case is installed, a small gap of about ⅜in.(10mm) is left between the stonework and the case. The gap is sealed, initially with a flexible backing rod inset to the case, then with burnt linseed oil mastic, trowel-applied into the gap. The pre-prepared trowel mastic, often used, does not last.

Glazing

The sash window was introduced in the 17th century as a result of the introduction of crown glass. However, because crown glass was so expensive to produce, the most popular type of window remained a casement with leaded glazing. In the 18th century, sash design evolved, glazing bars (astragals) became thinner, and window sizes became more standardised, with the 'six over six'[8] being the most common arrangement for glazing a sash.

Before 1600, glass used in windows was often imported from Danzig (now Gdansk), but a glassworks was started in Scotland in 1619.

The modern glass industry only started to truly develop in 1845,[9] with the withdrawal on tax duty reducing the price of glass by 75 per cent, increasing demand. A cylinder sheet process was introduced, and sheet glass was then manufactured and used well into the 20th century. After 1960, Pilkington managed to make float glass commercial; this was glass formed on a layer of molten tin.Cylinder sheet glass would be about 2.3mm thick and bedded onto linseed oil putty set into the glazing bead and held in place initially by small springs. It would then be puttied to form a slender edge to the astragal.

Stained and painted glass was often found in tenements from 1880 to 1920.

Window tax

A window tax was introduced in Scotland between 1748 and 1851. It was more onerous than the window tax in England. If any house had seven or more windows, or a rent exceeding £5 a year, each additional window would be taxed. In some cases this led to window openings being built up to avoid the tax, although the opening was always behind the stonework to allow a window to be inserted when the tax was repealed.

7 The pulley cord is attached to a ring which is inserted into the hole which is recessed into the sash frame.

8 'Six over six' simply refers to a sash window divided into six smaller panes formed by astragals or glazing bars (usually in two rows of three panes).

9 Alloa was a major centre of glass making at this period.

However, quite a few window openings are blind windows, in that they were never designed to be windows, but only formed because of a Georgian need for balance and order. Classical architecture liked such repetition. Some blind windows have flues or staircases behind them so they could never be opened up.

Window ironmongery

Ironmongery is usually brass. The sash lock at the meeting rail can be a sliding type called a sash fastener, or a hinged nut which is secured to a small retainer called a Brighton pattern sash fastener. There is also a Fitch pattern sash fastener which rotates into a housing to lock.

The sash screw is a more modern invention which allows the two sashes to be locked together for better security. For children's safety a sash stop may be fitted, which also allows adjustment should the sash need to be raised more.

The sash lifts at the lower rail allow the lower sash to be raised and lowered, and the top sash may have a sash eye on the top rail which allows the top sash to be lowered with a sash hook. Sash lifts may be simple brass finger hooks, brass eyelets or brass handles (which are best for heavier sashes).

The pulley wheel may be brass, but now is often steel and a screw fixed into recessed pockets in the pulley styles.

The simplex fittings vary in style but usually consist of a pair of butterfly hinges which allow the lower sash to hook over two screw fixings in the case to allow the bottom sash to hinge inwards once the cord is released. The cord may be held in the bottom sash with a knot

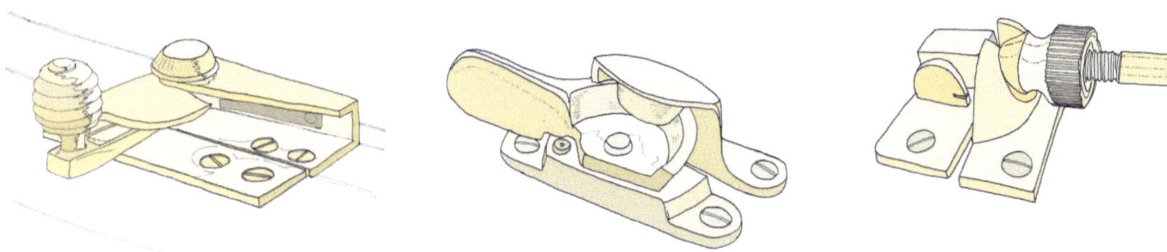

[left] Quadrant arm sash fastener

[centre] Fitch pattern

[right] Brighton pattern sash fastener

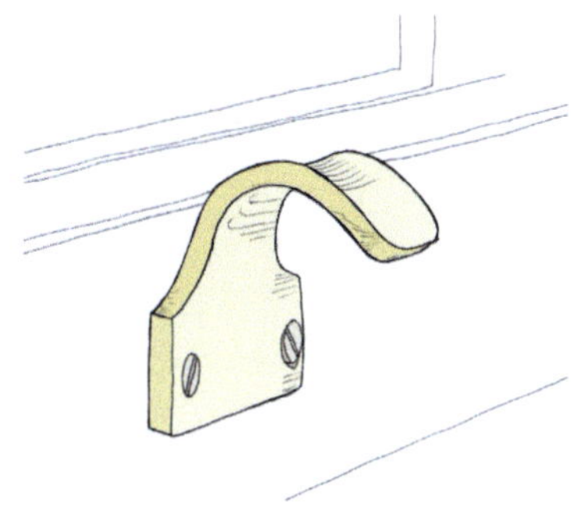

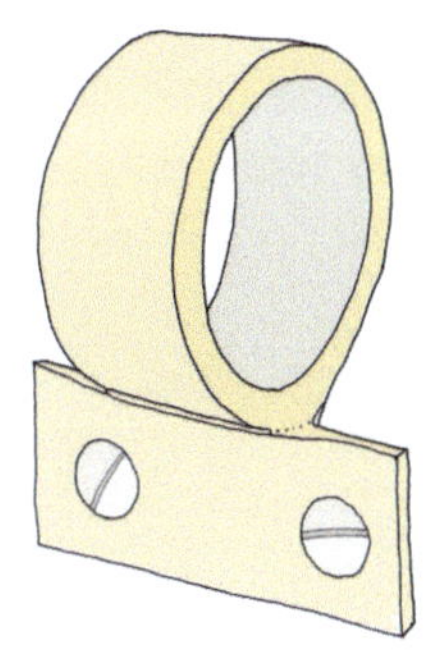

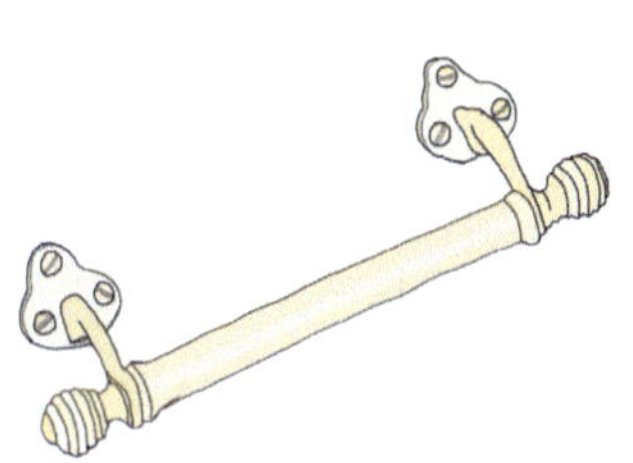

[left] Sash lift

[centre] Sash ring pull

[right] Brass bar handle sash lift

holder which fits into a recess in the side rail or a cord grip which latches to a screw in the sash slot. A cord clutch will temporally hold the sash cord when it is detached from the sash. A section of the batten rod at one side can be removed and is held in place by batten rod screws, although the rod is usually designed as friction held depending on design.

Side shutters, framing and skirtings

The stone face at each side inside (ingoes) and under the window sill was often given a single coat of plaster with the intention of helping the plaster absorb any moisture that might form on the inside, thus protecting the timber linings.

Older Georgian tenements and those in middle-class areas were more likely to have folding shutters which were hinged twice or three times and folded away into a recess behind an architrave. However most tenement ingoes or 'soffit linings' were simply panelled to gain the look of shutters, but were fixed in place.

Window repair

If you have a sash and case window which has not been painted for some time, the chances are that the timber sill has become rotten. The sill takes a lot of weathering, more than the glazing bars and sashes, but timber sills can be replaced provided you employ a skilled joiner who has the experience and tools to form new timber sills with the same profile as your existing window sill. We recommend that only good-quality red pine or Douglas fir is used for the sill.

The drawings below show the stages that will be required to replace a timber sill in a sash and case. The sill is usually cut out and a replacement sill to the same profile as the existing one is made. The bottom of the case is also usually replaced up to the pocket piece. The replacement case section is usually formed with a half lap joint to ensure a strong bond with the existing case. The repaired case is glued into the slot in the new sill. The outer lining is also usually repaired, often with a downward diagonal cut.

The sashes themselves can also be replaced. This is often needed if you are renewing glazing and the glazing bars are badly damaged. Slimlite glazing is available and has been used in some listed tenements. These glazed panels are double-glazed but have a small gap which is filled with a xenon gas. The advantage is that they can be set into a bed of putty to give a traditional appearance yet still use the same glazing rebate. As the glass may be heavier than the original glass, the sash weight may need to be replaced. Although they will improve energy efficiency, in the long term you might be better opting for secondary glazing units positioned inside. If you have money to spare, contact FINEO glazing as they make narrow glazing, despite it being triple glazed.

If you are retaining your sashes then they can be draught stripped by running a small groove into the sash rail and fitting a preformed draught strip.

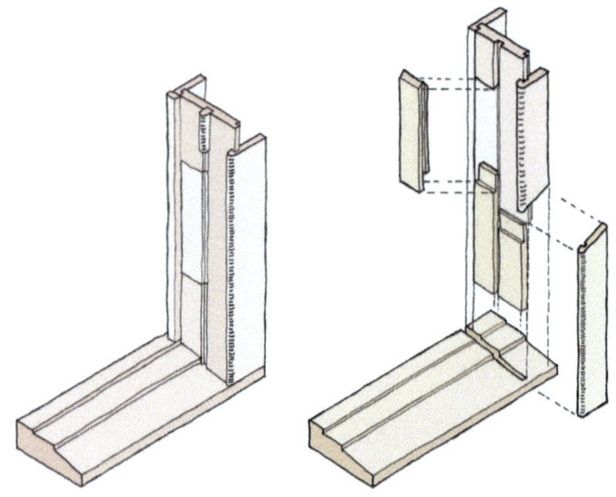

Sill repair in a sash and case

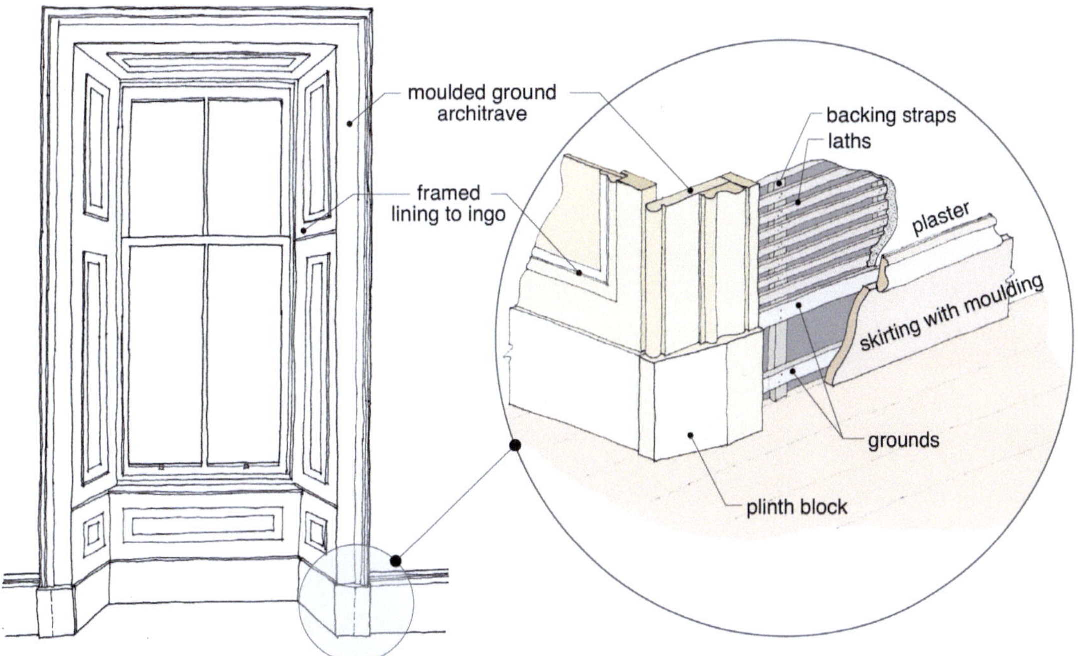

Window shutters, architrave and plinth block

[above] Fixed window shutters

[right] Opening shutters split between each sash

[left] Replacement sash

[right] Timber sash and case

Window replacement

New timber sash and case windows can be made to allow the use of thicker double-glazed units, although because of the width and weight of the units, planted timber glazing beads are required which are rarely approved if you are in a conservation area.

Householders often choose PVC windows because they appear to be cheaper and people think that they are 'maintenance free'.

This is rarely the case. Replacing sash and case windows in older tenements will eventually lower the value of the flat, and possibly the neighbours' flats. Installers of PVC windows usually remove the two sashes then plant the new PVC window into the old timber case. They may remove the old timber sill and fit a new PVC sill, although sometimes they simply face over the old sill in plastic. The result is that in a typical bay window, the glazed area can be reduced by as much as 10 square feet or a 28 per cent reduction in daylight.

PVC windows do come with ironmongery, locks, handles and catches, but you may find that in a few years' time you cannot find a replacement for that ironmongery and, if broken, the window sash can be impossible to repair.

Furthermore, if there was ever a fire in a flat, the PVC can burn and give off toxic fumes. If you must install UPVC then some makes are better than others. Sliding sash UPVC windows may be slightly more appropriate as the frame size is reduced. On all accounts, do not use UPVC if you are in a conservation area or a listed building, as you may be asked to remove it and replace it with a traditional sash and case window.

Why PVC windows are not the solution

- UPVC is often chosen because it is the cheapest solution and plastic window installers claim it will not require maintenance. UPVC windows may not require painting, but fittings can go out of production which makes the window impossible to repair, whereas a good softwood sash and case windows, if properly maintained, can last hundreds of years.
- PVC can release phthalate plasticiser additives which are beginning to be phased out under the EU Hazardous Substances Directive. However, UPVC (unplasticised polyvinyl chloride) windows and doors can be used as they don't emit these toxins.
- UPVC melts and decomposes in a fire. It may give off fumes but the fumes may not be toxic, although the smoke can be harmful. UPVC does add to the load of

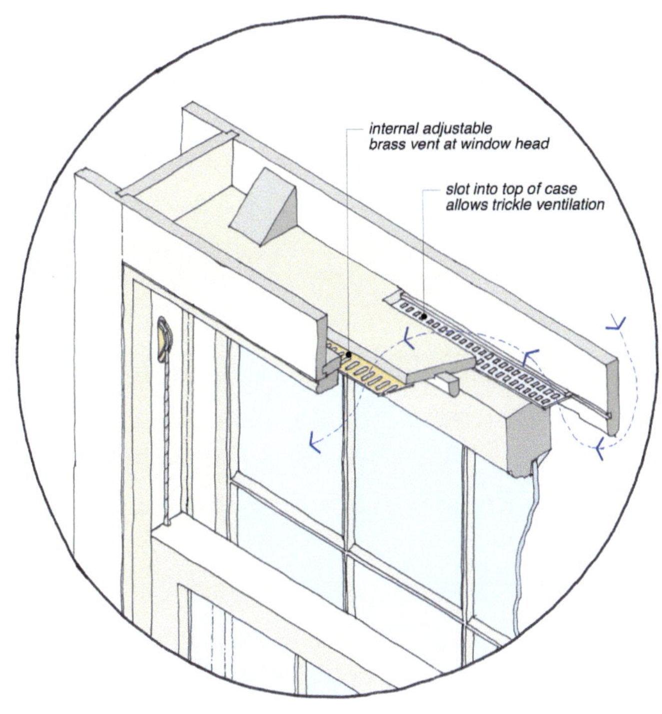

[above] Hidden and adjustable, trickle vent in a modern sash and case

[upper right] Replacement sash with slimline glazing

[right] Draught stripping a sash

a fire and is not as effective as a sold timber fire door. The Fire Brigades Union is concerned about UPVC use in buildings and fire officers in Scotland have warned against the use of UPVC fire doors in flats.

Draught-proofing an existing sash and case window

There are two main ways of doing this. The central parting bead can be removed, chased and fitted with a draught strip to provide a tighter fit between sashes and the case. However, the gasket can increase friction so it can be difficult to slide the sashes open. Also, when repainting, the gaskets may get painted and become difficult to remove.

The alternative approach is to remove both sashes and fit a chase in the unpainted sides of the frame that travel against the case. Then install draught strips into these chases. This is usually more effective and avoids any risk of overpainting the draught strips.

Doors and joinery

The humble door frame

Tenements were developed at quite a speed, usually with a small team of masons and labourers. The standard door frame helped the speed of erection, since, as a floor went in, the door frames were positioned, rising to full room height, and then other timber uprights were made which allowed the joists of the floor above to rest on the door frames and bed recess beams and so allow the next floor to be raised. Brickwork then filled the spaces between the outer sides of each door frame and other intermediate supports as well as tying[10] into the outer stone walls. These door frames are commonly termed H frames, so these 'humble' door frames are also supporting a load.

This allowed Glasgow tenements, and possibly others, to be built without the need for external scaffolding.

The door frame often included fanlights, although when there was a solid wall above the door transom the side frame would still rise to support the joists above and the thickness of the upright side of the door frame was reduced in thickness above the transom so that it could be plastered over. Sometimes additional laths were used over the reduced timber door frame to ensure the plasterwork would bond. As the timber transom was not strong enough to be a lintel, a brick relieving arch[11] was usually formed over the transom to help support the wall above.

The frames are commonly referred to as 'H-frames' and cracking is often common in the plaster which covers the hidden frame. The drawing on page 191 shows a typical tenement under construction with the floor joists supported on these H-frames.

Architraves and skirtings

The architrave would rest on a timber plinth block which formed a junction between the wall skirtings and the architrave at the door or window opening. The plinth block[12] would be secured to a timber plug set into the outer wall. The skirting may be scribed to ensure it meets the floor and appears level, although in good-quality work a chase or raggle is cut into the floorboard and the skirting inserted into the chase and nailed to the grounds in the wall.

Timber grounds would be fixed to the wall and the skirting would be fixed to these grounds. Skirtings vary in height but Victorian ones are usually about 10in. (254mm) high and formed out of two sections, a bolection moulding and the plane facing, although by 1900 single piece skirtings were in use.

Storm doors and vestibule doors at flat entrances

Front doors after 1900, called 'storm doors', were commonly fitted consisting of two doors which folded back into a space on each side of a small vestibule and with a fixed fanlight above. Each door would have two panels or three panels scribed into the rails and styles with simple mouldings around each panel. In addition to locks and knobs for the two leaves, iron stays might be installed to secure the half doors from inside.

---

10 Often difficult and internal brick walls are usually only tied in every fourth course of bricks.

11 Also called a 'discharging' arch.

12 Plinth blocks were also called 'base blocks'.

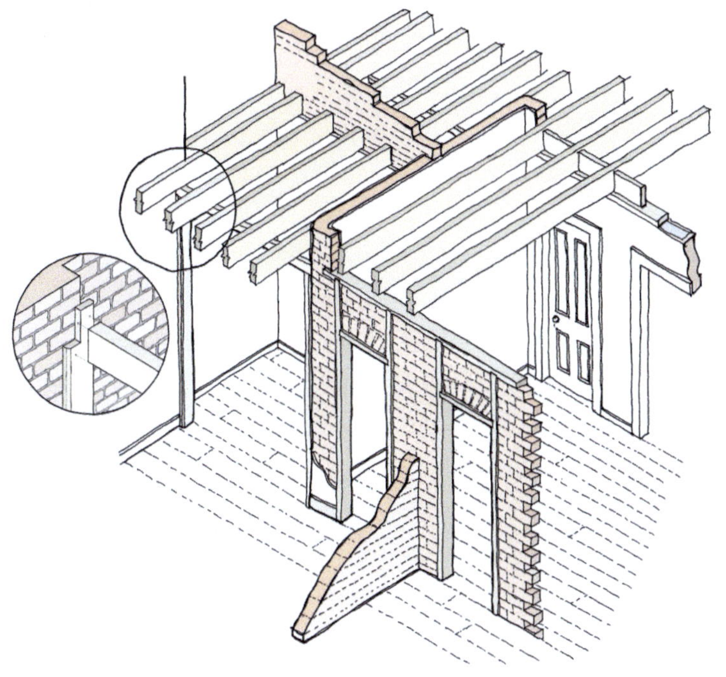

[above] H frames at doors and temporary beams at bed recesses allow for joists to be laid then bricks are infilled between timbers

[upper right] Doors which have an H frame, when improvements happen, require them to be removed and replaced by concrete

[right] H frame removed and precast lintel and brickwork placed above

A vestibule with 'storm door', leading to the internal glazed door

The storm doors provided security whereas the internal vestibule door allowed light into the hallway. The internal vestibule door would have normally been a glazed door with a solid base panel. The glazing might have been etched or stained glass, the door width was about 3in. (914mm) wide. Some flats had single-glazed panel doors into the close with a small stained-glass panel.

Over time, most of these original doors have been removed, particularly in housing which has been modernised as the main entrance door had to achieve a one-hour fire resistance[13] rating, which meant that fanlight glazing had to be replaced with Georgian wired glass[14] or PYRAN glass and doors would be replaced with solid core doors which also had intumescent strips in their edges and a self-closing device to ensure the fire door closed, thus preventing fire in one flat spreading into the one escape stair, as well as the adjoining flats.

Originally, most Victorian tenements would have had a borrowed light from the close into a cupboard or toilet (often protected with steel bars). However, these have long since been removed and bricked up to maintain fire resistance between each flat and the stair.

Internal doors

Internal doors were also of a panelled construction, with mid (lock), top and bottom rails tenoned into the styles with a central muntin[15], all openings filled with plain panels and trimmed with simple mouldings. The doors would be a bit smaller, around 2ft6in. (762mm) wide and some press doors even smaller. The mid rail or lock rail, would be where the door latch and knob was fitted, and most internal doors had a simple latch set into the rail with knobs to release. On bathroom, and sometimes kitchen doors, a surface rim lock might be fitted.

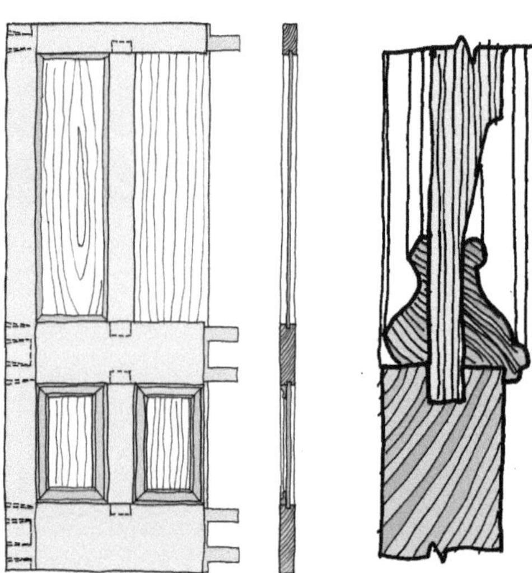

[left] Internal doors usually have simple inset mouldings on one side and a plain inset joint on the other

[right] Front doors, or important doors, usually have raised bolection mouldings on the outside

The door frames would have added checks; the joint with frame would be concealed by an architrave[16] and a plinth block. Architrave mouldings vary and sometimes a more prominent architrave moulding is used on certain doors on one room, while a simpler one was then used on the other side.

These old panelled doors have largely been replaced in modernisation work with flush

13 Fire resistance rating is set down in the Building (Scotland) Regulations.

14 Georgian wired glass fanlights could only give 30 minutes of fire protection so as flat doors required one hour's fire protection, newer Georgian wired glass which had a thicker embedded wire could be used and when this was used in frames with intumescent seals, achieved 60 minutes. The correct thickness of clear PYRAN glass can also achieve this.

15 The 'muntin' is the central vertical timber in a panelled door.

16 As architraves became simpler, they are often called 'facings' in Scotland and are now rarely fitted with plinth blocks.

panel doors and knob latches replaced with door handles. When this happens, the old H-frame is often removed and replaced with a new frame with rebuilt brickwork and concrete lintels.

Presses and press doors

Presses were formed into party walls adjacent to the outer walls, in part because the hearths and flues took up space which left a small space in each corner. Sometimes words can be heard from next door, as only brick (or Caithness slab) forms the separation between the two flats.

[left] Bricks on edge between tenements

[right] Wooden framing and rear panels with shelves inserted into slots in frame. Architrave and usually a press door

The kitchen press in the rear often housed a copper cylinder above the press, although it might be boxed out and covered with lining boards. The press door would be a simple four-panel door although six-panel doors do occur in more expensive areas. The rear brick, or stone wall of the press[17], would be given a single coat of plaster before half-inch lining boards were fitted to the back. The sides of the press would be single vertical planks about ⅞in. (65mm) which were slotted to receive a number of wooden shelves.

Borrowed lights in the close

These have nearly all been sealed up as they presented a fire risk. Originally small borrowed lights, either to light or vent a small press inside the flat, were formed through close walls. These small glazed fixed lights were set into wooden frames, at a level about as high as a fanlight, and for security high iron bars have also been fitted. Earlier borrowed lights that had an opening vent were much smaller and would likely have been installed to vent an internal toilet in Georgian times.

Back doors

These would have been ledged and braced doors which were sheeted with vertical tongued and grooved V-jointed or beaded panels, all set in a standard timber door frame, sometimes, but rarely, with a fanlight. A galvanised rim lock would secure the door. Similar doors would be found on basement store doors and the backcourt washhouse. New replacement doors will likely have been replaced with framed doors with a lock rail and also braced and sheeted externally with tongued and grooved boards. Some doors may have been partly glazed to allow light in if the back close was dark.

17 The back wall of a press is usually either 4.5in. brickwork or Caithness slabs laid upright. As the bricks were often added later and even on edge, it is thought that during construction the press openings allowed builders to move from one tenement to the next.

Where did all the timber come from?

In medieval Scotland, oak was mostly used, so it would have grown in Scotland. From 1450 the main structural timber supply switches from native Scottish oak to imported Scandinavian oak from Norway until around 1600 when most builders switched to using Scandinavian pine from Norway and then the Baltic.

Despite Royal prerogatives, most of the forests of England had been cut down to clear land for farming, also early ironmakers had used charcoal to make the iron. Scarcity in the 1600s led to Parliament seeking timber from Ireland (then a colony), but Irish forests had also largely been deforested. Iron workings which used charcoal also led to deforestation. Timber was imported from the Baltic but trade was affected by the Napoleonic Wars and in 1807 a levy was introduced on Baltic timber which made it uneconomic. Timber was then sought from the new colonies, firstly from Quebec in Canada, then Virginia (largely unsuccessful in 1600). After 1750 trade switches to eastern Baltic pine and then North American timber. Following the Napoleonic Wars (in 1819–20), British North America brought forestry products on clipper ships to the UK. In particular, Scotland had a leading part to play in timber imports in 1819. In Barbados, cedars were cleared to provide land for sugar cane (some 7,000 Scots left for America; 4,500 settled in the Caribbean Plantation). The timber from the Southern states and Barbados was largely facilitated by slave labour.[18]

Local homegrown firs were still available and sawmills had started to operate in Leith in 1695 and in Alloa in 1718 to cut large baulks of timber that arrived in the ports. However, homegrown timber was considered inferior to foreign wood, perhaps because it was unseasoned or foreign firs were slower grown, but the homegrown wood did not seem to last and sometimes more time was taken in preparing it, so its use was limited. Cost was also a factor, as labour was much cheaper in Russia and the Baltic states, so it was better to cut the timber into standard sizes before it was imported. However, the main difficulty in extracting homegrown timber was the difficulty of access because of the lack of roads. Transportation by ship from abroad was far easier.

Norwegian pine timber in 1754 averaged 12ft (3.6 metres) in length whereas timber from the Baltic countries could be 24ft (7.3 metres) long and even up to 36ft (11 metres) long. Whilst it was quicker to get timber from Norway than the Baltic, the length advantage won out.

There is information available on the timbers in the east coast, less about the west coast. After 1750 it switches to Eastern Baltic pine and then some signs of North American timber. Edinburgh's New Town was built with Baltic timber and it's likely that timber from Scandinavia and the Baltic continued to be imported to the east of Scotland in Victorian times.

In 1765 Greenock was the leading importer of fir timber, having surpassed Leith, and fir was coming from Poland and Russia, some distance away by ship. Timber was imported to Dundee, Leith, Perth, Montrose and Aberdeen.

The Canadians made timber ships such as brigantines and barques, often designed with flatter bottoms to take bigger loads, rather than for speed. With the advent of clipper ships which were much faster than the older and slower sailing brigs, timber was being imported from Quebec. The Atlantic timber trade developed from about 1868, restricted by the size of hatches

18 See *No Wood, No Kingdom* by Keith Pluymers, 2021, University of Pennsylvania Press.

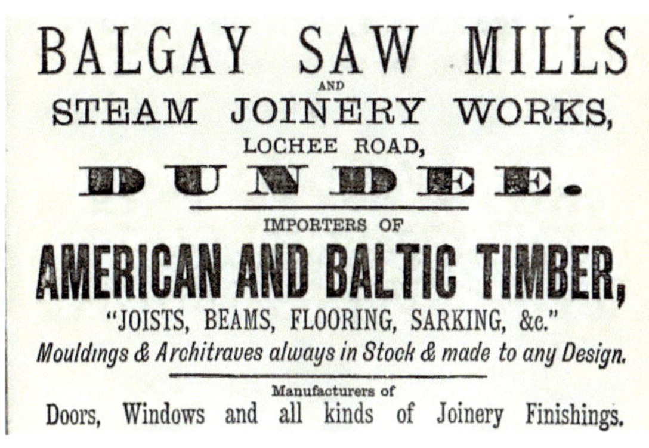

[above] Advert for saw mill in Dundee

[right] Timber balks as shipped

Fig. 43.—Pile of Oak Logs with Star-shakes. (From a photograph)

on clippers[19] until timbers could be stored on deck with the advent of larger metal ships.

Logs would be first lined and squared into baulks (balks) of timber to make them easier to freight.

It's likely that timbers for joists in tenements in the west of Scotland would have come from the Baltic or Quebec, the main provider of lumber to Britain from 1830 to 1880; after that America exported a greater tonnage. The timber was mainly pine, spruce or pitch pine from eastern North America and Oregon fir/Douglas fir, from Quebec in Canada.

Mahogany from Barbados and Honduras was imported to make handrails. Teak was imported from Burma and used for railway sleepers.

Regrettably no dendrochronology samples have been taken from timbers in Glasgow tenements built in the 19th century, so we do not have firm evidence of the likely source of the timber.

Timber imports in the past

As it was cheaper to have the timber cut abroad, as noted above, when cut it was also easier to transport as cargo. Timber section sizes varied a great deal depending on the species and place of export. The common terms are listed below. Baulks might have been used as bressumer beams in Glasgow or mid floor-beams in Edinburgh; the larger section planks or cut down balks would have been used for timber safe lintels.

Advert in *Lumberman*, a Canadian journal

19 The *Carrick* was a clipper once moored in Glasgow, originally called the *City of Adelaide* built in 1864 and is the last survivor of the timber trade between North America and the UK. Now being moved to Australia.

BALK

square section timber, easiest to transport size varies from 9"x9" up to 20" x 20" and Oregan Fir up to 30"x30"

hardwood

pitch pine

PLANK

in hardwood, about 9" wide x 1.75" thick, but Pitch Pine could be 3"-5" thick by 10"-18" wide

DEAL

over 2.25" thick and less than 10" wide

BATTEN

Less than 2" thick and less than 9" wide

BOARD

Less than 2" thick and over 6" wide

SCANTLING

Usually refers to miscellaneous timber sizes but typical scantling could be 2.75" thick and 10" wide

TIMBER.

		£	s	d		£	s	d
Teak, Burmah	per load	£13	0	0	to	£14	5	0
,, Bangkok	,, ...	9	5	0	,,	14	0	0
Quebec pine, yellow	,, ...	2	18	0	,,	4	18	0
,, Oak	,, ...	3	15	0	,,	5	0	0
,, Birch	,, ...	3	0	0	,,	4	13	6
,, Elm	,, ...	4	3	0	,,	5	3	0
,, Ash	,, ...	3	18	0	,,	5	3	0
Dantsic and Memel Oak	,, ...	2	0	0	,,	3	0	0
Fir	,, ...	2	13	0	,,	4	3	0
Wainscot, Riga p. log	,, ...	4	18	0	,,	5	18	0
Lath, Dantsic, p.f.	,, ...	4	10	0	,,	5	10	0
St. Petersburg	,, ...	5	0	0	,,	6	10	0
Greenheart	,, ...	8	0	0	,,	8	10	0
Box	,, ...	4	0	0	,,	15	0	0
Sequoia, U.S.A.	per cube foot	0	1	8	,,	0	1	10
Mahogany, Cuba, per super foot 1in. thick		0	0	5	,,	0	0	6¼
,, Honduras	,, ...	0	0	4½	,,	0	0	6¾
,, Mexican	,, ...	0	0	4	,,	0	0	5
Cedar, Cuba	,, ...	0	0	4	,,	0	0	4½
,, Honduras	,, ...	0	0	3¾	,,	0	0	4¾
Satinwood	,, ...	0	0	5	,,	0	1	0
Walnut, Italian	,, ...	0	0	3	,,	0	0	7
Deals, per St. Petersburg Standard, 120—12ft. by 1½in. by 11in. :—								
Quebec, Pine, 1st		£18	15	0	to	£21	15	0
,, 2nd		13	5	0	,,	15	15	0
,, 3rd		5	15	0	,,	9	5	0
Canada Spruce, 1st		10	15	0	,,	12	5	0
,, 2nd and 3rd		8	5	0	,,	9	5	0
New Brunswick		7	5	0	,,	8	5	0
Riga		7	5	0	,,	7	15	0
St. Petersburg		9	5	0	,,	13	5	0
Swedish		9	15	0	,,	16	5	0
Finland		8	15	0	,,	9	5	0
White Sea		9	15	0	,,	16	5	0
Battens, all sorts		5	0	0	,,	18	0	0
Flooring Boards, per square of 1in. :—								
1st prepared		£0	9	6	,,	£0	17	9
2nd ditto		0	8	0	,,	0	13	6
Other qualities		0	6	3	,,	0	7	0
Staves, per standard M :—								
Quebec pipe			—				—	
U.S. ditto		£35	0	0	,,	£42	10	0
Memel, cr. pipe		220	0	0	,,	230	0	0
Memel, brack		190	0	0	,,	200	0	0

[left] Timber section sizes used by Victorians

[right] List of timbers available in Scotland at the time

Softwood imports around 1900[20]

Northern Pine (also known as yellow deal or fir, red deal or fir, Norway pine and Scotch fir. Native of Northern Europe and Scotland.

The timber is imported as baulks,[21] deal, battens and wrought flooring from Russia (Archangel, Narva, Onega, Petersburgh, Riga, Wyborg), Prussia (Danzig (now Gdansk),[22] Konigsberg, Memel,[23] Stettin),[24] Norway (Christiania, Dram, Holmstrad) and Sweden (Gefle, Gottenburg, Holmsund, Stockholm, Soderham).

Largest balk timbers come from Stettin (average size 18–20in square, 35ft long). Best baulks from Memel (13in. square, 30–35ft long). Strongest balks for construction from Danzig (now Gdansk) (14–18in. square, 40–50ft long).

Best deals for joinery come from Russia, but also from Sweden. Timber from Norway is generally small, course and suitable for 'inferior' work. Some concern with Baltic supplies in 1900 after which forests in Siberia, Finland and Sweden were gradually taken on board.

20 See *Modern Practical Joinery* by George Ellis, 1902.

21 Balks, or baulks refer to large debarked and squared logs of timber which were often used as bressumer beams. In later Victorian periods balks sometimes referred to collars in roofs

22 Gdansk. In the 1750s was the most prosperous port in Europe but when incorporated into Prussia in 1793, trade diminished.

23 Klaipeda: now a Lithuanian seaport.

24 Now called the port of Szczecin in Poland.

Russian Redwood[25] was then imported via the Baltic ports in first and second qualities (although all qualities were imported).

Canadian Red Pine – comes from Canada, Nova Scotia (10–18in. square, 16–50ft long) and the US (where it is known as 'Norway pine').

White Pine (also called yellow pine in England) – came from Canada and North America. Used for interior work as it is soft and light.

Yellow Pine (often just called 'pine') – stronger and more durable than white pine but rarely imported to the UK. Yellow pine or white pine came mostly from North America and Quebec and was used in first-class joinery as it is free of knots, although concern was expressed in 1900 that the forests were being cut down too rapidly.

Pitch Pine – native of the southern states of America. This is the hardest, heaviest and strongest of the pine species but can be difficult to work with because of resins. Pitch pine is the name given to certain trees grown in the US. *Pinus rigida* (from northern American states) is a true pitch pine but was rarely imported, Instead yellow pine was imported and sold as pitch pine. It was good for heavy construction work. The pines often sold as pitch pine were Longleaf pine, Cuban pine, Shortleaf pine, Loblolly pine. *Pinus Australis* and *Pinus resinosa* are native trees of the southern states usually free of sap and shakes but full of resin which makes it difficult to work with, although strong and with an attractive grain so used in staircases and exposed ceilings. Tendency to shrink. Interestingly it was imported from Darien as well as other southern states.

Spruce Fir also called white Deal, native of Northern Europe. Norway spruce largely used for scaffold poles and boards, obtained from Russian ports of Narva, Omega and Riga.

American Spruce also called black spruce because of colour of bark. Native to cold regions of North America. Liable to twisting and shrinkage. White spruce that came from Canada was found to decay quickly.

Newfoundland Spruce from Nova Scotia. Turns reddish brown so also known as red spruce. Imported via Nova Scotia, Newfoundland.

Oregon Fir also known as Douglas pine or Douglas fir. Native of the North west coast of North America, having a straight grain, good for large carpentry and durable. Shipped in balk and plank from Puget Sound, Washington County. Some balks 30in. × 30in. square, 120ft long, planks 11–40in, on 12–40ft lengths.

Larch native of central Europe and Russia, straight grain but hard to work and tends to shrink. Now used for external cladding.

Cedar (*Havanah*), native tree of Cuba, also known as Red Cedar, soft, porous and brittle, reddish brown. Used for panelling and cigar boxes and fine joinery.

Cedar (*Australian*), native of Australia, used for internal joinery similar to red cedar.

Cedar (*New Zealand*) Native of North Island, light red colour, easy to work and durable, used in house joinery.

Hardwoods

Beech – native of England and central and southern Europe, mostly used in cabinet work or to make tools although Australian beech was used to make flooring in some cases.

25 Petersburg Redwood was highly prized.

Greenheart – from British Guiana and the West Indies – strong and durable but heavy and difficult to work with. Used for sills, warehouse floors, piles and jetties.

Mahogany – many different varieties of mahogany: Spanish mahogany from Cuba; Honduras mahogany from Honduras (Belize); African mahogany from Sierra Leone and Indian mahogany from Central Hindustan. All have different qualities but generally a hardwood favoured for internal joinery. Balks of Honduras mahogany were 2 to 3 feet square in 1900.

Oak – Most oaks had been largely difficult to find in Britain after 1450, as royal forests were protected for hunting and there was judicious felling of timber of which the English Oak was much prized and used extensively in shipping and the navy. Around 1600, the scarcity of timbers, particularly English oak, was recognised. Oak was gradually imported from many places:

Danzig (now Gdansk) Oak grown in Poland not as durable as English oak.

Riga Oak: From Southern Russia, often used in wainscots and sometimes called 'Wainscot Oak'; also used on dados and wall panelling.

American Oak was a white oak which grew from Mexico to Canada. Slightly redder than English oak but had a tendency to warp.

African Oak from West Africa imported from Sierra Leone. Hard and heavy and very durable in wet conditions but its weight limited its use in Britain.

Teak – A native of South and Central India, Burma (Myanmar), Ceylon (Sri Lanka). Lighter than oak but stronger and more durable in water. Burmese teak considered best for construction, used in window frames, sill, stairs and shopfronts. Around 1900 it was thought to be fireproof and sometimes specified because of that.

Chestnut – Spanish chestnut (not horse chestnut) was largely homegrown; it resembles coarse-grained oak, used in church benches and church fittings.

Timbers for different purposes

External doors on public buildings – Oak most frequently, then mahogany, teak, pitch pine.

Internal doors, for ordinary buildings – Yellow deal for framing, yellow pine for panels.

Floorboards – Older properties would have oak boards, then pitch pine, yellow deal then Swedish or Norwegian yellow or white deal.

Floor joists – Russian deals are best. Baltic fir was cheaper in 1900 and next best.

Half-timber framing – Pre Georgian would be oak, then teak and larch.

Pile foundations – Greenheart, oak, elm and sometimes memel fir.

Roof trusses – Pre Georgian would be oak, then chestnut, pitch pine, Baltic fir.

Shopfronts – Mahogany or teak.

Windows and Doors – Red pine.

Timber scarcity, then and now

Around 1600, concern was expressed in England that homegrown oak timber was becoming increasingly scarce. A pamphlet from 1610 advised how important trees were for England's security and that '*our mills of iron, and excess of building, have already turned our greater woods into pasture and champion, within these few years, neither the scattered forests of England nor the diminished groves of Ireland, will supply the defect of our navy.'*[26]

Indeed, the manufacture of iron required charcoal which came from trees, so much tree felling was to make charcoal for local iron foundries.

The English in 1600 thought they could rely on their colony in Ireland to supply the shortfall of timber, but, alas, much had already been deforested.

Initially they looked abroad. There was then a growing trade with the Baltic importing timber masts, pitch and tar and Norway became a main provider. However, the new colonies in America were increasingly looking for markets in Europe, although the shipping costs must have been problematic.

In America's southern states, slavery was used to clear land for products such as cotton, tobacco and, in the Caribbean, sugar. Some timber initially came from the Southern states but probably more came from the Caribbean:

'The sugar plantation of the English and Dutch Caribbean were uniquely manifestations of a colonial attitude towards the environment that blithely razed pre-existing landscapes to cultivate monoculture, maintained only through the brutal exploitation of enslaved people or they were one of the earliest examples of 'ecocide' in the Early Modern Period.'[27]

Colonists[28] would simply set fire to forests to clear land for farming. Indeed, much of Cuba had been deforested of mahogany by 1865.

Deforestation in the Amazon today

26 Plymers, K. (2021) *No Wood No Kingdom*, University of Pennsylvania Press, page 105.
27 Montano, J. (2011) *The Roots of English Colonialism in Ireland*, page 199.
28 Many colonists were Scottish.

Given that the same thing is happening now in the Amazon to create monocultures of soya and palm oil plantations, not a lot has changed.

Timber use in Scotland today

Approximately 80 per cent of our timber is imported into the UK. Canada and the USA are reducing their timber exports due to wildfires. Sweden, which supplies almost half of the structural wood used in the UK, has recorded its lowest stock levels for 20 years.

In 2010, forest cover in the UK was recorded as 11 per cent of land area; in Scotland it was 18 per cent. European countries averaged 38 per cent and Finland, Sweden and Slovenia were at 60 per cent of land area. In Germany, a highly industrialised country, 32.68 per cent of land is given to forests.

The main reason in Britain for the low forest coverage is that most of the country, apart from the Highlands, is flat and so forests are cleared for agriculture. However, land ownership has not encouraged good forestry practice with tax breaks for planting trees but no encouragement for making forests sustainable. The number of deer, cattle, sheep and livestock ensures that UK forests are not allowed to naturally regenerate like European forests, which are also helped by maintaining a forest culture.

Scotland's state-owned forests (Forestry and Land Scotland) comprise 92 per cent conifers and only 8 per cent broadleaves. Scotland's dependence on Sitka spruce which accounts for 58 per cent of the planted tree types is that it could be at risk from a pest or pathogen such as that which affected ash and elm. Sitka spruce is clear felled every 30 years, for short-term gain. In the past, trees would be planted and grown for future generations, only allowing selective felling to allow forests to regrow and protect the habitat, although there are changes in forestry management, such as those in Darnaway Forest on the Moray Estate.[29] I'd like to see Forestry Scotland with more power to plant some more trees and escape this process of having to clear cut forests.

29 Part of Darnaway has been forest for at least a thousand years and was formerly a royal hunting forest. The oldest tree in the forest is estimated to be approximately 750 years old.

Section through a tenement with a commercial unit (pub) on the ground floor and basement level. Extends to rear to form a 'high back' for the houses to use as a drying area. Steel RSJ beams support the openings with plated box steel standards to support the beams. Pavement lights help to light the basement

Chapter 8
Shops and shopfronts

Anyone interested in renovating a shopfront would be well advised to read *Scotland's Shops* by Lindsay Lennie, published by Historic Scotland[1] in 2010. This is an invaluable guide to Scotland's shops; the main issues are summarised below, but the book has a lot more detail. This section covers the structural elements of shops and issues concerning their restoration.

History of shops in Scotland

Market trading was the lifeblood of Scottish burghs in medieval times. Goods were displayed in the open air although burghs often provided special places to trade cattle (the Cowgate), Fish (Fish Market Close), otherwise markets were allowed in areas such as high streets, which might have been widened to cater for markets. Eventually a wider variety of produce appeared, displayed on tables or stalls. Some merchants built small extensions (called *Krames* in 1662) in front of their houses; they were mostly timber but provided a sheltered locked up space.

Piazzas, arcaded ground floor areas under tenements, and Luckenbooths, small locked booths or shops, appeared, but these additions often created a nuisance. Luckenbooths were removed in Edinburgh in 1812 and piazzas were banned in the 1800 Glasgow Police Act, although the form remained in towns in Scotland.

Shops as we know them only started to appear around 1780, usually as an alteration to the front of a merchant's house, or in a basement in Edinburgh's New Town. In smaller towns they appeared in properties between 1780 and 1790; some incorporated bow windows[2] as display windows (using small pieces of crown glass). Coffee houses were popular but shops in Edinburgh's Princes Street only started to appear after 1800. In 1785 a shop tax was introduced, although bakeries were excluded from it. Hanging signs were common, being pictorial rather than written as much of the population was illiterate; they were outlawed in 1772[3] after some pedestrians were killed in London by a falling sign, but reappeared in the 19th century.

1 Also see HES *Short Guide 12 Scottish Traditional Shopfronts*.

2 Bow windows were more common in England at this time, only a few appearing in Scotland.

3 Dates varied with Paris banning them in 1760, London in 1762 and Glasgow in 1772.

Shops with bow-fronted windows

By 1790 in Glasgow, shops would suit every taste; after 1790 there were also shops in every part of Scottish towns. Shopping tax returns paid anything from £5 to £10 per annum[4] (with a higher rental value of £25). Single bow-fronted shops had the bow separated from the door or it may be integral. The second type is a double bow where the shop is central and the third type is where the entire shopfront is projecting. Glass could certainly have been produced in larger sizes, but the glass tax meant only a few shopkeepers could afford it. After the repeal of the glass tax (1851), it became possible to have normal glass. Few survive. The panes are smaller (pane sizes vary in each city), but a smaller pane would attract the crown glass or hollow glass globe.

Pavement lights with prism glass set in a cast-iron frame let light into the basement; many have now been sealed or concreted over

When shops were built as part of a tenement the window frontage was quite small and glazed with sheets of cylinder glass about 2ft 3in. wide as, although *plate glass* had been invented in 1834 (by the Chance Brothers), the glass tax made it very expensive as it was taxed by the weight of glass. The use of glass in shops gradually increased, as did the production of cast iron to produce special castings.[5] Early shop signage was simply hand painted on the narrow fascia above the window and only increased in depth in Victorian times.

Renovating a shopfront: new tiling at entrance being installed

Shops added to existing tenements

Projecting shop frontages could occur when there was a small garden space or a basement in front of the tenement, so a lot of tenements after 1830 had conversions on the ground floor to create a shop.

Eventually shops were planned in tenement developments and the height of the ground floor increased as floor levels came close to street level, but rarely level with the street until Edwardian times. The plans of shopfronts vary, most having a recessed lobby area set back from the main shop window. Mid Victorian ones are square with storm doors; by late Victorian time around 1890, they have angled sides and in Edwardian times they could be curved.

Advertising the shop, use of mosaic tiles and tessellated tiles

4 Lennie, L. Scotland's Bow-fronted shops, *Architectural Heritage* 20, pages 75–91.

5 Montrose had three cast-iron foundries in 1861, although the main ones were in the Central Belt.

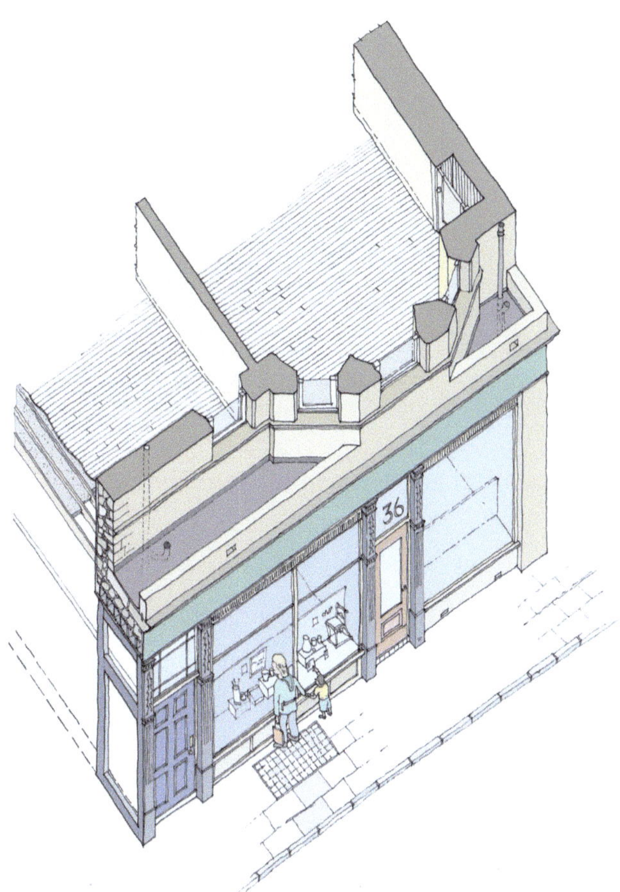

[left] Section of shop below a bay window, with small stone parapet and flat asphalted roof above cornice and shop fascia

[right] Victorian shop in Georgian tenement

Original shop sign made from solid section timbers, worth retaining, even as a base for new fascias

Most shopfronts have been changed over time but remnants of original shops are often there and worth conserving. Stall risers which are the low sections of wall under shop windows, were sometimes formed in timber and sometimes stone. In 1886,[6] polished granite stall risers were available from Aberdeen stone yards and they became quite popular.

In modern times, the glazing is putty fixed from inside rather than outside. In Victorian times it is likely that the glazing was puttied from the outside. Other features that developed from 1850 and are still seen in some shops today are as follows:

- Retractable roller blinds which provided shade from the sun and shelter for pedestrians

6 The 'Jenny Lind' machine used rotating discs and carborundum. It was patented in America and seen by John Henderson who managed to copy the idea when he was back in Aberdeen and used it, despite any patent laws, in 1886 to much effect.

- Retractable timber shutters, storm doors, metal and wooden gates to protect glass
- *Stall risers* in timber or sometimes cast iron to allow ventilation
- Decorative vent trims at the head of windows
- Cast-iron *pavement lights* with glass prisms to let light into basements from 1883. Pre-fabricated cast-iron frontages and columns from 1850
- Plate glass and curved plate glass for fashionable and expensive shops
- Shop fascias with side console brackets[7] 1880–90
- *Clerestory lights* in Edwardian times along with decorative tiles on walls and mosaic flooring, often advertising a brand Concrete and steel, used from 1910 onwards – new steel beams allowed greater spans and concrete decked roofs which were asphalt covered allowed shop areas to expand, sometimes creating 'high backcourts'
- Electric lighting in shops, which would have been gas originally, or whale oil before that. Although Edison had developed a working light bulb in 1878 it did not take off until tungsten light bulbs developed in 1910.

Structural aspects of shops

In many Georgian tenement areas, shops were added to the existing tenement. It might have been a main door house with a basement house or a ground floor house with a small garden. Shop units might then be added over the garden or basement area. These are called projecting shops. New lintels and columns would need to be installed (and sometimes foundations adjusted), and the original house floor level might have been lowered nearer to the street level, making the inside of the shop quite high.

Sometimes such changes were made on the building line, without encroaching forward.

It was probably only after about 1850 that shop units were designed as integrated with the tenement development, and standard cast-iron frames and columns as well as shaped stone mullions were then provided for building.

Whether the structural work was done during or after construction of the building, the outer wall still had to be supported. This might be done by a large timber bressumer beam (or two) and a cast-iron beam up to about 1900, after which steel beams were more commonly used, although mild steel beams were available and used from 1885. In fact, cast-iron beams were replaced by wrought iron beams and then steel beams. If a large span was required, a plate girder might be used – made from plates of steel and angles riveted together. These were eventually replaced with larger rolled steel joists.

These bressumer beams would be supported on cast-iron columns or cast-iron standards.[8] Cast-iron columns were sometimes used between standards with the shopfront on the outside. Cast-iron columns are also used inside the shop, sometimes down to a basement, supporting RSJ steel beams which carry the timber joists of the floors. The earlier cast-iron standards were quite wide, about 3ft (914mm), with a big web; later cast-iron standards would

7 In the 1880s and 1890s bookend consoles were used which are very large decorative features.
8 Standards, also called stanchions.

[left] Cast-iron beams

[right] Cast-iron standards at early Victorian shopfronts

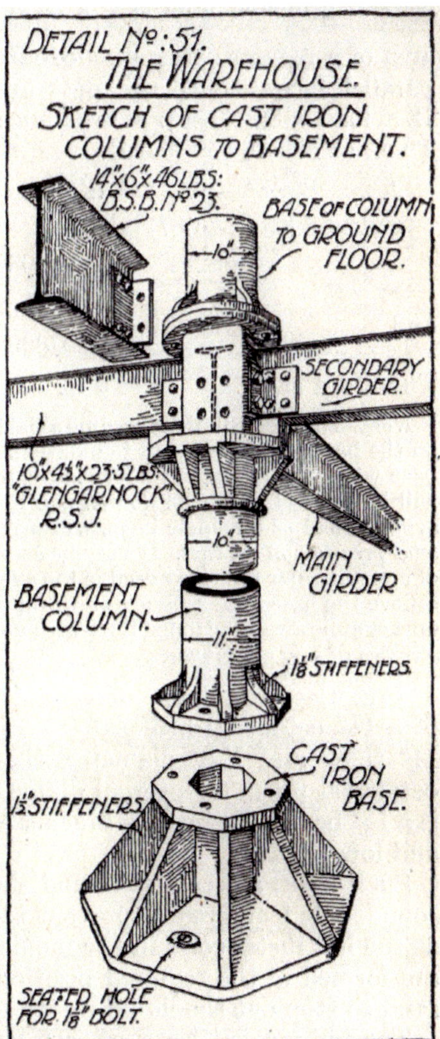

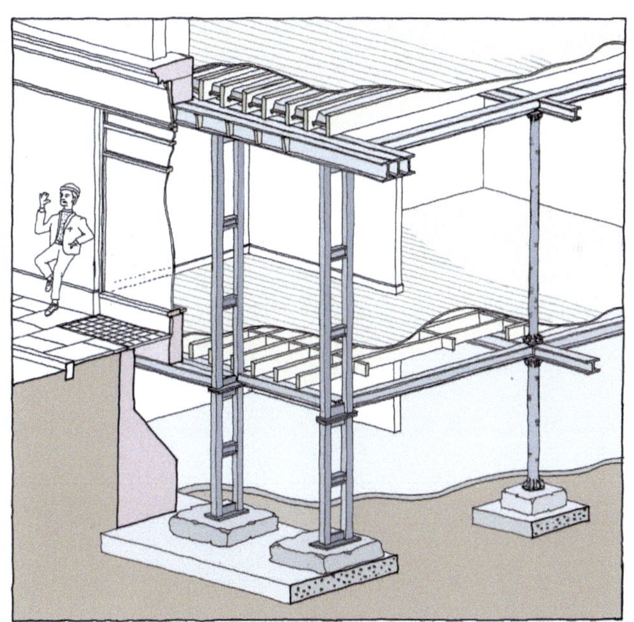

[left] Victorian detail of cast-iron column and connections

[right] Cut away drawing of a typical steel standard and steel RSJs to form a basement with internal support columns

John Dickson and Son shop at 21 Frederick St, Edinburgh around 1976. Formed in 1820, John Dickson and Co is today the oldest gunmaker in Scotland, although no longer in Frederick St.

[left] Cast-iron shopfronts were common around 1880–90, but few now remain. Sometimes also used at close entrances

[right] Curved glass was possible after 1900 but few examples survive, perhaps because it is expensive to replace

be narrower. These would be replaced after 1900 with steel I sections riveted together and eventually RSJs bolted together to form standards[9].

Inside shops, as there was a need to provide a large open space, internal columns would be formed to support timber beams which supported the internal walls of the houses above. Sometimes the load on these internal timber beams manages to crush the timber at the point it rests in a steel column which can lead to wall movement above.

Many shops were added later, so a ground floor apartment might be converted into a shop with new beams inserted. Some tenements were built to include shops, here 'the ground floor ceiling is abnormally high, to allow for storage space and adequate ventilation.'[10]

After 1900, when steel beams were commonly used, the steel had to be fireproofed which, if the steel was not fully bedded into the stone, entailed wrapping the steel beam with a metal mesh and coating it with cement mortar.

Some shops extended out the back with extensions which would be formed with concrete roofs. The structure would be formed in steel H beams (12in. × 5in. at 5ft centres) then have intermediate steel *T beams* spanning between the beams. Steel enforcement bars within the concrete did not start until after 1900 although patented around 1850 using wrought iron

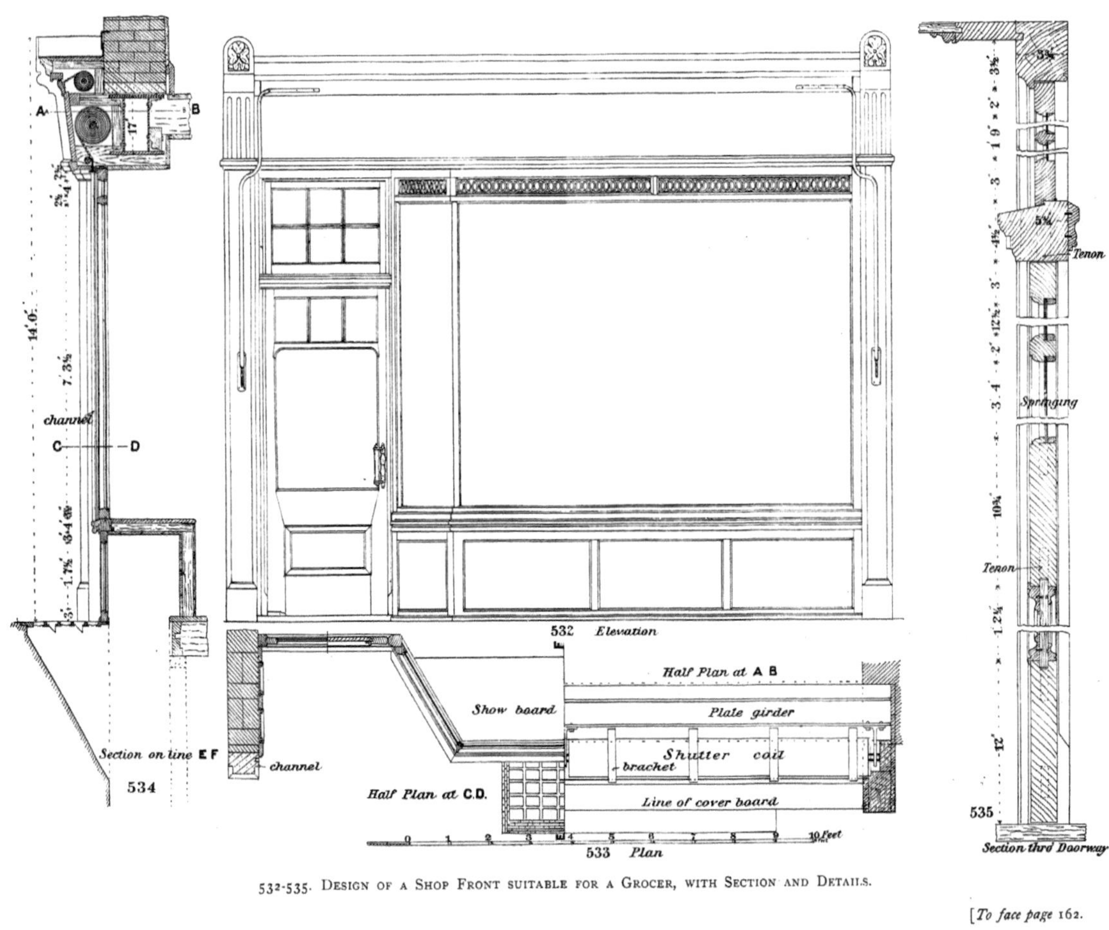

Original Victorian detail of a shopfront

9 Cast-iron standards were still used, both as columns and twinned U shaped columns and wider webbed columns, well after 1900.

10 Worsdall, F, *The Tenement: A Way of Life*, page 29.

bars. Concrete mixes[11] often contained old brick rubble and the embedded steel soon rusted, so these roofs can be problematic. They are usually covered in three coats of asphalt to provide a waterproof membrane for the roof and to act as a backcourt for the dwellers above.

When improving shop units

Each shop will be different and I can only advise any surveyor or architect to always inspect the void above any lowered ceiling. This should show whether any timber beams are rotten or crushed and what the actual beam construction is at the front. It is often the case that the original ceiling from an earlier shop has decayed or been dropped down, exposing the floor joists and deafening boards of the flat above. This will need to be addressed as it means that the fire protection between shop unit and house above is ineffective. Noise transfer may also be a problem, and even transfer of smells.

Look out for old signage and tiles. Wall tiles were used in food shops and were often covered over; with care, they can be restored. Also wall tiles at stall risers[12], often painted over. Floor tiles, mosaic or terrazzo is often used at entrances and over the years the entrance door location may have been changed.

Original photo of 110 Battlefield Road

11 Usually specified to have the concrete composed of: '1 part cement, 2 sharp sand and 5 hard clean broken brick passing a 1 inch ring.' Coke breeze was also specified (from *Building Construction* by Prof Adams, 1911).

12 Stall risers: the small section of wall or panelling under a shop window.

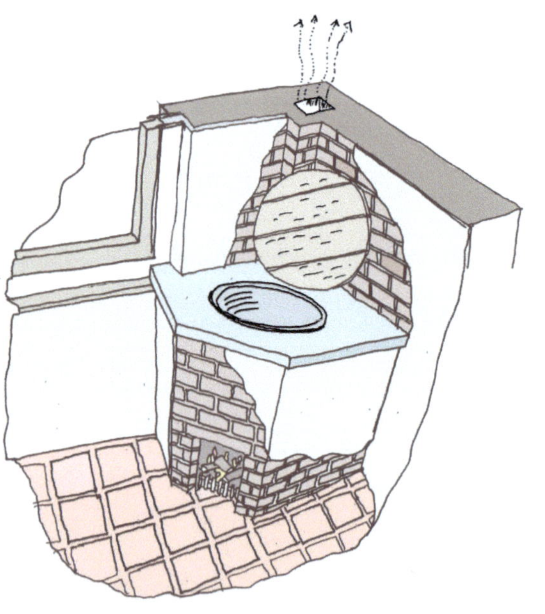

[upper left] Old wash houses in backcourts

[right] The copper was used to boil clothes.

[left] Wash houses in Pollokshields

Chapter 9
Backlands

In medieval times, the area behind houses, be they houses or tenements, were called 'backlands'. Under feudal law 'Liners',[1] who were burgh officials, had the task of ensuring that plot boundaries were respected and known, in some cases marking them out. The backlands then were busy places, used to grow crops, keep livestock, contain wells and house bread ovens as well as middens. The plots might be divided with primitive wattle fencing. Drainage was rudimentary.

The backlands developed to house workspaces, stables and manufacturing, increasing the density in urban areas. Access to the backlands might be through narrow wynds or pends to take the cart width, sometimes opening into a small rear court.

Tenements were initially accessed via fore-stairs at the front, although as they got higher, turnpike stairs were used and sited at the back (see Chapters 2 & 6).

Tenement backcourts or greens in Georgian & Victorian times

Georgian tenements would often have mews at the rear to house stables or servants' quarters, with a back lane to access the mews. As the cities grew, tenements were laid out following feuing arrangements which often dictated the plan form and what could be built in the backcourts. Typically this allowed a small wash house and bin stores to be built, usually in brick with slated roofs, although in later periods brick and concrete roofs. The backcourts that developed behind Georgian tenements were usually bounded by high stone walls with rounded stone copings. The wash house would have a copper bowl set into a brick surround and a small chimney to allow the smoke from the boiler to exhaust.

Most of these buildings were originally slated, but by the 1900s they were concrete roofed. A drain would be formed from the wash house to link in with the house drainage and water would be supplied by a lead pipe and *crane*.

Backcourt central spine walls in Victorian times would usually be in brick with glazed terracotta copings and metal railings dividing each backcourt.

Ground levels often required stone retaining walls to be built, along with steps up or down to backcourts. In poorer areas, workshops were often built in the backcourts, making the space for residents even tighter. Cast-iron clothes poles would be used to hang the washing on

1 Bailies and lesser officials monitored all aspects of town life. The 'Liners' were clerks for the burgh.

[left] Typical backcourt in the 1970s

[right] Timber chestnut pale fencing and derelict wash houses and store

Communal drying poles for individual clothes lines from each accessory deck in Portobello

lines, although each resident was usually told what day of the week they could dry clothes on. When there was no space below for drying clothes, rod washing poles would be hung outside scullery windows in order to dry clothes. In some cases, large communal poles were used with clothes lines on pulleys (still in existence in Portobello).

In 1940, the backcourt railings, along with any front railings, were removed for the war effort and never replaced. Most photos of tenement backcourts show them as 'slums' without railings or greenery, but before 1940 they were a place where children could play, as well as on the street.

Rubbish, storage and removal

Metal rubbish bins would be kept in backcourts and collected by the local council on a weekly basis. People had much less rubbish in the past and so one bin per family was sufficient. In the 1960s, the refuse department would also employ a 'sweeper upper' who would make sure each backcourt was tidy after the bin men had been. 'Midden-rakers' were usually unemployed men who searched the bins for anything worth selling.

In medieval times, people would defecate either into buckets or outside latrines and the dung would be collected and used as compost by the burgh. In Victorian times, the 1850 Burgh Police Act required commissioners to: '...*employ a sufficient number of scavengers, for sweeping, cleaning and watering the streets, and for the removing all dust, ashes, rubbish and filth therefrom, and from the houses and other premises therein, and for emptying privies and cesspools, in the manner by this Act directed*"[2].

High back drying area at Anderston Cross, Glasgow, circa 1966; note the stair windows are simply ventilated and open without windows

High backs

Some tenements were built so tightly together that there was no lane access into the backcourt. When space was tight, backcourts were built on rooftops above larger ground floor shops, complete with a drying area. Shops on corner tenements would often be built out at the back and a backcourt built on the roof at first, or second, floor level. These corner high backcourts were usually formed by asphalting over a concrete roof deck.

If there was a shop below, the roof might have a pyramidal glass light set into a cast-iron frame, to provide light to the shop below. These were similar to pavement lights in the pavement which often provide light to a basement area.

Backcourts on flat roofs of four-storey tenements were largely formed onto timber joists with a timber deck, which might be asphalted or felted. Concrete flat roofs after 1910 are found in Glasgow, but if the top floor was a flat roof, it

Flat roof in Govanhill, Glasgow

Loose railings

2 Police and Improvement (Scotland) Act 1850 page 48.

might only be a drying area, whereas lower high backs could house the bins. This would be done because of restrictions of space at the ground floor. Railings at this time might have been mild steel, although wrought iron was still being used for some items.

Railings

Cast-iron railings and metal railings in backcourts were mostly removed in the war. They were only left at stairs and sunny areas to ensure people didn't fall into a void. It's likely that the amount removed was surplus to requirements and some claim the excess iron was dumped in the sea; however, we have no proof of this. Unfortunately, after the war, they were not replaced. Backcourts became uncontrolled open areas, and fencing was often reinstated with timber palings. Cast iron that remains at the front of properties often needs repairing because it has very little tensile strength. Anyone leaning on a cast-iron railing at a landing can crack the railings. Cast iron should be repaired with cleats, mechanically fixed, although welding by a specialist is possible. Cast-iron sections are still being made by a Scottish firm in Bo'ness. Railings that crack and cause an injury can lead to claims against all the owners of a tenement, for not ensuring the common areas were safe for the public.

High back roof

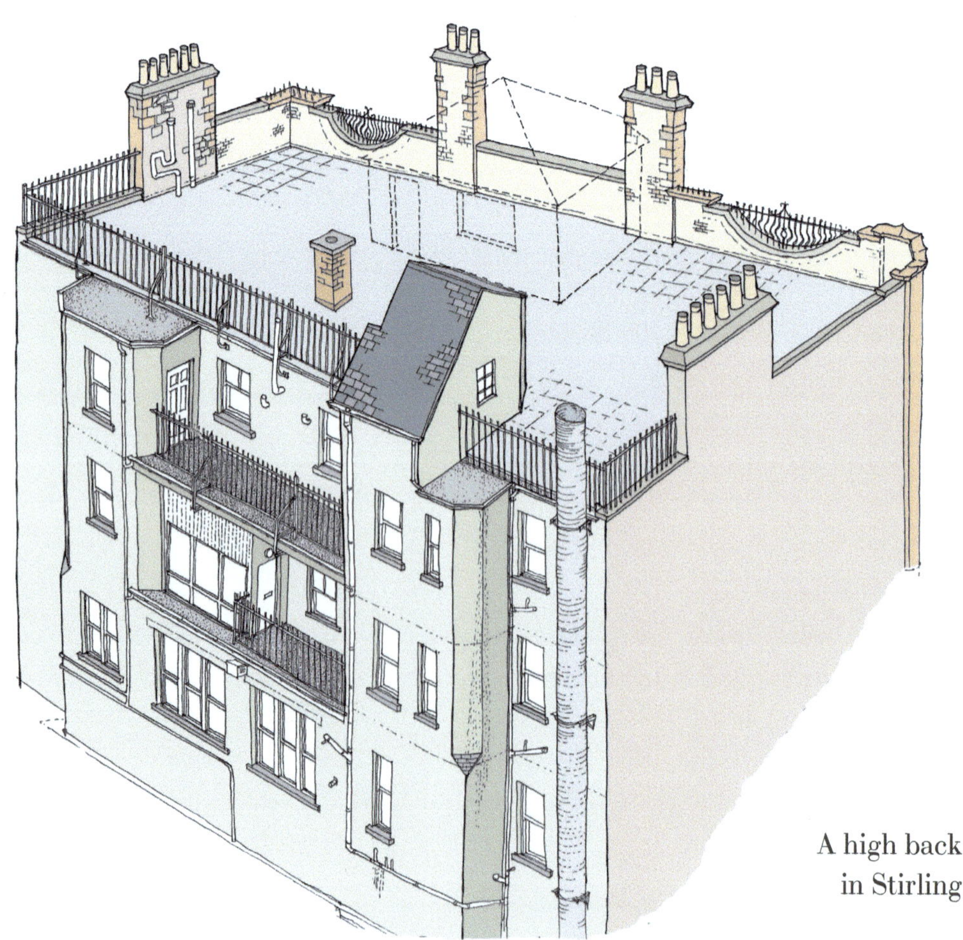

A high back in Stirling

[above] Front steps when a guest fell through

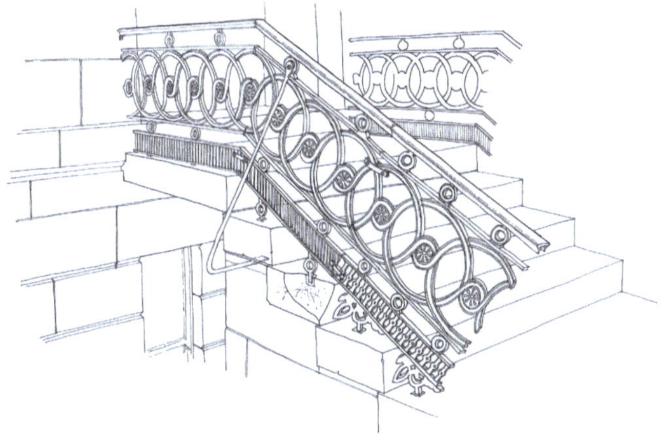

[right] Drawing of some front steps

Front steps

The advantage of front steps is that they were often allowed to remain. But this is the difficulty, as front steps can be very hard to repair. If you have a front step, make sure that it is in the common area as otherwise you could be threatened by legal action should someone fall down the hole.

It may be possible to fit a steel strengthening arm to the railing that would prevent any fall. Otherwise, a bolted plate may be necessary to resolve a defect in the cast-iron railing systems.

Street and close lighting

Gas lighting

The 1800 Glasgow Police Act distinguished between a 'public' and 'private' street and required the complete paving of the streets associated with any new tenemental development to be paved at the expense of the owners (or developers). A typical feuing contract for a Glasgow property in Bath Street in 1833 required an 8ft (2.5m) basement height and a 10ft (3m) wide pavement. Most domestic streets were built with 'coal holes' in the pavement to supply coal into the cellars under the pavement. Similar coal holes were to be found in smaller towns where basement light and openings can still be found.

The first streets to receive public street lighting were in Edinburgh in 1688, on Cowgate and High Street. By 1742 Edinburgh Town Council had 99 lamps which rose to 307; by 1786 there were 116 lamps in the New Town. The lamps needed Greenland whale oil[3].

In the 19th century Arctic whaling, conducted from ports right along the east coast[4] of the country as well as Greenock, was vital for the Scottish jute industry, especially for processing jute fibre in Dundee. It is estimated that between 1900 and 1999 nearly 3 million whales were killed, with over half a million killed[5] in the Atlantic.

3 Whale oil was also needed for the jute industry in Dundee.

4 The two main Scottish ports were Dundee and Peterhead.

5 Whaling ended in 1963 when Edinburgh-based Christian Salvesen ceased whaling, although, strangely enough, whale oil is still used by NASA.

In 1787 a resident of the New Town, John Hunter, installed a decorative lamp standard outside his house and copies of these wrought iron lamp posts can still be seen. They feature 'linkhorns' which were used to extinguish the link boy's torch[6].

Gas lighting was first introduced to Edinburgh in 1819 to light the North Bridge, following on from the start of the Edinburgh Gas Light Company in 1817. Houses would still be lit by candles or lanterns using whale oil. By 1820, 1,980 gas lamps lit Edinburgh's streets, although oil lamps were still in use.

Gas lamps were introduced to Glasgow in 1818; before that, oil lamps were in use since 1780, although only in some streets. Gas street lighting, using town gas, was the predominant form of lighting until the 1960s[7]; the last gas street light was extinguished in 1971 in Glasgow. Streets around Westminster in London are still lit by gas (2024 at time of writing).

Original lighting standard at close entrance in Edinburgh New Town

Gas lighting in closes became standard by 1860 and eventually Glasgow had 12,664 closes lit by gas. In 1885 incandescent gas mantles were invented and by 1895 they were cheap enough to have gas lighting in closes. Some houses in well-appointed tenements would have gas lighting plumbed in; this was done using 'block tin'[8] which can often be found behind plaster when tenements are restored. Gas lighting in tenement closes was still in use in 1976 in Springburn.[9]

Electric street lighting was initially installed in 1893 in Glasgow where there were 112 street arc lamps installed. Edinburgh had electric street lighting by 1898 and full conversion to electric lighting occurred around 1900 in Edinburgh. By 1914 Glasgow had 1,541 electric street lights. No doubt electric close lighting followed after the First World War.

Today, all communal areas should have provision for lighting; this normally includes the close and stairs, the main entrance, route to the backcourt and the backcourt itself. The burners are called 'scale burners', when upgraded, mostly to LED.[10] They take power from

6 Once the link boy had escorted someone to the house.

7 I lived in Northern Ireland; we had gas lighting outside our house which was lit every evening by the lamplighter, up to the late 1950s.

8 Block tin is much more valuable than lead.

9 At Mansel St and St Monance St. A lamplighter with a small ladder and a green hat came round late afternoon to light them and again early morning to turn them off.

10 Light Emitting Diode LED.

a separately metered landlord's supply located near the main entrance. Billing is usually managed either by the factor or property manager, although Edinburgh Council offer the service for free. The lights are designed to come on during the hours of darkness using a solar control unit.

Upgrading Scottish tenements

Chapter 10

The role of community-based housing associations in tenement renewal[1]

Setting the scene

There is nothing automatic or predictable about housing associations becoming involved in tenement renewal, still less becoming a major vehicle for renewal of thousands of homes in multiple neighbourhoods in Scottish towns and cities between 1974 and 1990. By the early 1970s, there were just a handful of associations in Scotland; some were set up by property professionals, but associations were mostly philanthropic or charitable societies. Associations established from 1974 onwards made a significant contribution to securing the built heritage of many neighbourhoods across the country, and tenements from different eras constitute a significant part of the stock of the housing association sector today.

Before we go any further, let's examine what a housing association is.[2] Within the array of associations there are some common factors. In a formal sense, associations typically started out as friendly societies with limited liability, incorporated under the Industrial and Provident Societies Act of 1965. They are neither private nor publicly/state-owned. An association is voluntary, in the sense that it is free to exist, unlike a local authority. It is not-for-profit, non-profit-distributing or non-profit-seeking. As a membership organisation, its members elect their Committee of Management, or Board, which is accountable for the decisions and actions of that association, though day to day operations are typically carried out by paid staff within a framework set by the Board/Committee. An association can acquire, own or let property, it can undertake whatever is required to develop or manage the property and it must comply with legal obligations such as maintaining the property so it is wind and

1 This chapter was authored by Mary Taylor drawing on written materials, personal experience as a protagonist in Glasgow in the 1980s and later in SFHA. The chapter also draws on conversations with colleagues involved in Dundee and Edinburgh. Comments gratefully received from activists of the period have helped to ensure it is relevant and accurate: the author is ultimately responsible.

2 The term housing association includes housing cooperatives, and is used in that sense generally here, except when referring specifically to cooperatives. A section towards the end of the chapter discusses various variations on the theme of resident control. The current terminology for a registered housing association is Registered Social Landlord (RSL).

watertight. Significantly from 1974, the law pertaining to registered association activities concern regulatory law and codes. Legislation also permits registered associations to levy rents and service charges, and enables public funding and private borrowing.

The year 1974 marked a watershed in renewal and especially in terms of the role of associations. A few associations set up before 1974 were intended to protect heritage buildings – such as in New Lanark or around Arbroath. Some early associations involved professionals channelling work to their firms – including lawyers working on conveyancing or factoring, or architects, surveyors and engineers. Their activities were limited and the associations soon withered or reformed once registration, comprehensive action, and funding came into play post 1974. Legislation and government funding priorities that year were critical in underpinning the new role of associations.

The early philanthropic associations worked in slums and with deprived urban communities. They were often church-based or inspired by Shelter.[3] They also drove and created new forms of accommodation and support to people whose needs had been neglected in mainstream housing provision – older people referred to as 'the elderly', and people with a physical or learning disability. But some associations focused on improving living conditions in deprived communities in older buildings, often pending demolition and replacement. Every association comes into existence voluntarily, although most go on to employ staff. Before 1974 the only public funding (for capital works) came very modestly from local councils. But the funding regime changed in favour of associations in 1974 when legislation was introduced to make it possible for the state to fund capital works by *registered* housing associations across the UK to create or redevelop housing. Such funding was made available on a very generous basis, and subsequently successive governments ploughed significant amounts of money across Great Britain into the work of associations, provided they were registered with an expanded quango – the Housing Corporation. Accepting funding meant accepting supervision by that body. This crucial registration, supervision and funding development was the culmination of many preceding strands of experimentation and pressure: one of these was tenement renewal in Glasgow which has dominated the national narrative about housing associations in Scotland and about tenement renewal, reflecting the severity of conditions in that industrial city in the 1960s.

This chapter considers the context for housing associations becoming the vehicle for renewal in Glasgow, Edinburgh and Dundee, the exploration of options and campaigns to secure funding, to build a movement and a network. These issues go to the heart of the people, organisational and money issues present in any housing initiative – in this case tenemental buildings. Many associations were established and registered for particular purposes in the years following 1974 and their number reached a peak of 150 in 2008 when the focus on registration and supervision gave way to regulation under a newly independent Scottish Housing Regulator (SHR[4]). Control of the capital funding system for associations was separated from regulation at that point and was/is retained by the Scottish Government.

3 For example, before 1974, the following associations had come into existence: Castle Rock, Hillcrest, Angus, New Lanark, Christian Action, Grampian, Bield, Albyn, Edinvar, Link. Some of these bodies still exist.

4 The earliest body responsible for regulation was the Housing Corporation governed from London, replaced in 1989 by an agency – Scottish Homes, replaced in due course by Communities Scotland – a department of the Scottish Executive until 2007. Devolution allows SHR to be accountable to the Scottish parliament.

Convergence and emergence

Tenement property ranges across a spectrum from gracious and spacious well-built mansions with highly decorative features inside and out, to basic and cheaply built flats consisting of one or two rooms in functional buildings with minimal amenity. It was usually the latter which became particularly run down and eventually attracted policymakers' attention towards renewal activity on the grounds of unfitness for habitation, where it could be economically renovated. Private rents had been controlled variously since 1915 following the mortgage interest crisis of the early First World War period. Ongoing rent restrictions over subsequent decades are commonly regarded as suppressing the appetite of landlord owners to maintain or upgrade property, with persistent lack of investment resulting in deep physical deterioration.

The buildings which came to be targeted for policy attention across the UK in that era had typically been built for industrial workers in the late 1800s, and usually within walking distance of workplaces. But declining employment in heavy and manufacturing industry, from the 1950s onwards, reduced the appetite for inner-city living. Car ownership was on the increase, and shopping was increasingly in shopping centres rather than corner shops.

Tenement life in Scottish towns and cities became less attractive in competition with better modern housing in the suburbs with gardens, for those who could afford to access mortgage finance or who had secured a council let in a new property. And various local authorities were building modern flats and houses for rent in the outskirts, purposely to attract people away from unhealthy, dense inner-city living. By the late 1960s tenements had come to be seen as rather old fashioned, even though tenement life was viewed with nostalgia and sentiment. Many sandstone tenements were demolished and replaced as a result of comprehensive redevelopment programmes to improve housing, to make way for new roads and other buildings such as universities.

However, the new high-rise buildings of the mid 50s were proving unpopular with families. They were distant from amenities and hard to heat, they were often badly maintained and revealed problems of dampness. And there was mounting economic evidence of the costs of rehabilitation of older housing providing good value for money, and at considerably lower cost than high-rise new housing.

By the early 1970s tenements were typically in multiple ownership, with individual flats occupied by their 'owners', sometimes following an informal system akin to hire purchase.[5] More typically, flats were rented out by private landlords, through a factor, if indeed there was still a factoring service in place.

The people who continued to live in inner-city tenement neighbourhoods in the late 1960s typically didn't have the resources either to improve or to leave. They were reliant on factoring services, often from lawyers, to address the increasingly complex problems of living conditions. Even if these problems were not intractable as such, they were beyond the ability of factors or owners to tackle – whether owners were resident or absent. A different approach was needed and pressure increased for public intervention.

The need for something to be done about tenement conditions arose from a number of strands randomly converging in the late 1960s and early 1970s. A central strand was a report

5 Mortgage finance was limited and it was at that time very difficult to secure a mortgage on a tenement flat.

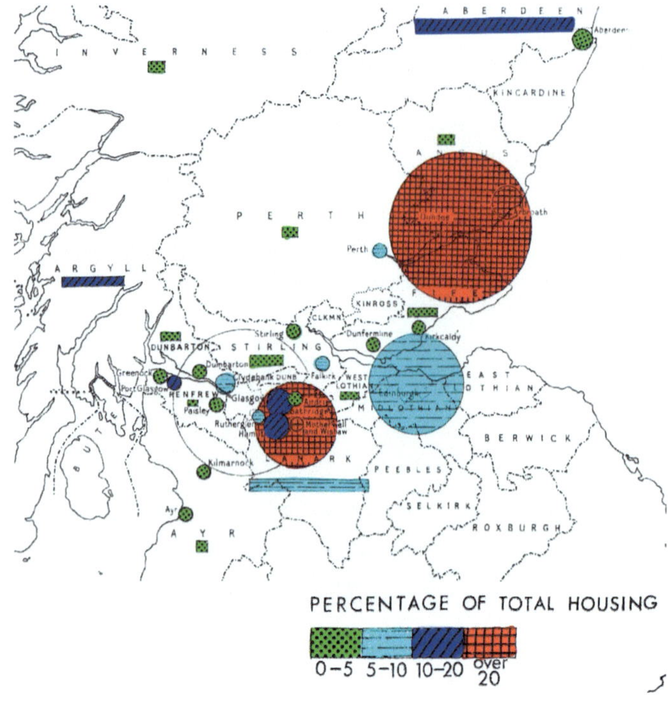

Detail map of affected places and percentage of housing

in 1967 from the Scottish Housing Advisory Committee under the chairmanship of J. B. Cullingworth, called *Scotland's Older Houses*. This coruscating report condemned the failure to address the dereliction of older buildings including, but not limited to, tenements.

Although some councils could and did estimate the scale of the problem, the extent of deficiencies was generally under-reported according to Cullingworth, whose report mapped the distribution of unfit properties reported by councils in 1964. Cullingworth estimated that 273,000 homes were 'unfit' – almost one in five of all homes in the country, and closer to treble the number estimated by local authorities at the time. Even so, as the map shows, the Dundee area had the highest percentage of local housing considered unfit, with over 20 per cent of all housing deemed unfit, a proportion comparable to Glasgow, but a higher number.[6]

Edinburgh, then smaller than Glasgow, also had a lower proportion of housing deemed unfit (5–10 per cent) and not as high as various smaller burghs across the country.

The Cullingworth report prompted the Scottish Development Department to propose legislation and improvement arrangements in 1969, incorporating some of the report's 58 recommendations. The most notable of the recommendations concerned the definition of a minimum 'tolerable standard', replacing the notion of unfitness, and a more aspirational 'satisfactory' standard.

Meanwhile a major storm in 1968 exposed to public gaze the shocking and unacceptable living conditions in many inner-city tenements in Glasgow. The storm had blown down chimneys and ripped off roofs on a huge scale, with repair costs beyond the means of owners. The council established a new department to inact emergency repairs. The damage provoked a groundswell of concern from residents clamouring for help from the council. The idea grew to repair and upgrade structurally sound tenements to provide at least a *temporary* solution.[7] This sense of a stopgap measure, rescuing tenements from dereliction, stemmed in part from the Cullingworth report, which identified a lifespan for improved properties of 10–15 years, pending demolition and replacement. Yet tenements are with us still and part of urban heritage.

6 In rural areas the highest proportion of unfit housing was in Ross and Cromarty (20 per cent), with other 'landward areas' coming in with 10–20 per cent of all housing deemed unfit and Aberdeenshire having the largest number. The unfit housing in rural areas was more likely to take the form of cottages than tenements.

7 Theo Crombie, former chief executive of Glasgow District Council, recalled that era in an interview saying 'Let the academics do an experiment in a quiet bit of Govan, and if it's a success, good. If it fails, we will just demolish them [tenements] anyway.'

Experimentation before establishing a framework from 1974

The extent of unfitness in various towns and cities prompted varied responses but there was no single tried and tested solution, or funding in place. In 1969 legislation was passed to introduce Housing Treatment Areas (HTA) in Scotland as a means of tackling physical disrepair in neighbourhoods based on a 'Tolerable Standard' of housing. Properties which didn't meet the tolerable standard were recorded as BTS – 'below the tolerable standard'. And each local authority was required to estimate and identify BTS properties across its jurisdiction. But progress in tackling disrepair was limited and slow, entailing considerable bureaucracy, and within five years Housing Action Areas (HAAs) with greater powers had replaced HTAs.[8]

Prior to 1974 every council in Scotland could take the opportunity to use its statutory powers to require action by owners of unfit or BTS properties – whether by compulsory repair or compulsory purchase. Some councils did use their powers but their experience of owner resistance to repairs in the form of time-consuming and expensive court action, made them less inclined to use area-based powers against owners. For example, Dundee Council embarked on issuing statutory notices but having been stung initially, simply avoided issuing any further. Nor did it pursue any further delineation of BTS properties even where the living conditions clearly demanded action to provide improved housing in blocks or areas, rather than individual dwellings. Arguably for other reasons, they demolished many of the city's tenements, perhaps not surprisingly as eye witnesses from the period refer to pockets of unimaginable squalor. With so much demolition and a limited improvement agenda from the local authority, local activists who were committed to improving conditions in Dundee needed to spend time and effort gathering and investigating options for run down tenement areas. Operating through embryonic housing associations, such activists had to engage and persuade owners without recourse to statutory powers.

Similarly in Edinburgh there were neighbourhoods on the south side, in Leith and around Tollcross or in Gorgie and Dalry requiring major attention. Some buildings only had cold running water facilities, and no electricity. Although the council might issue statutory notices to owners, they did not coordinate activity or resources ensuring action on the ground.

However, in Glasgow, the City Council declared 29 outline HTAs covering large parts of the inner city. The environmental health department had a tenement demolition programme based on 'worst first'. The demolition programme bore little or no relation to plans by the housing department, which was increasingly focused on housing standards and renewal in the private sector, as well as planning for construction of new housing and rehousing priorities arising from demolition programmes. The programme of HTAs proved very expensive, impossibly ambitious, slow and problematic to implement, and ill coordinated (though such criticisms were not limited to Glasgow). Poor coordination created delay and frustration: it took the City Council two years to rehouse residents from the Old Swan Treatment Area in Pollokshaws, before improving the whole block of about fifteen tenements. (Later the new housing association's approach to coordination, contract management and decanting was much more effective, allowing faster progress.) Bleak derelict sites strewn with rubble

8 For further information, see Chapter 8 by Keith Anderson in *Housing in Scotland*, published by CIH, 1996.

remained as eyesores, and communities which were the foundation of social networks and support were broken up. The City Council began to realise that this demolition and eventual new build would not solve Glasgow's housing problems. Arguably Glasgow was unique in this respect of council recognition of the extent of the problem and the need for a different approach. Having a council willing and able to take the lead on statutory action to address disrepair was essential to success. But not sufficient.

In *Annie's Loo*, author Raymond Young writes about the various experiments underway in Glasgow in the early 1970s when he was an architect activist interested primarily in participation in planning.[9] Many of these experiments involved senior officials of the council assiduously devising and testing innovative ideas in particular buildings, and trying out new approaches using lessons from previous experiments. Other chapters in this book account for the technical design challenge of introducing running water and flushing systems in tenements. But it's important to remember that it was PhD research at Strathclyde University Architecture School which demonstrated how to install a bathroom in a typical bed recess in a tenement flat with a plumbing stack installed down through four flats, one above another so that each owner could install an internal bathroom. In *Annie's Loo*, Raymond Young documents the significance not only of the technical problems but also the legal challenge, in terms of who had the right or authority to locate and access water supply and drainage through adjacent properties.

In 1971, the University of Strathclyde set up project ASSIST in Govan, with the aim of providing a locally based technical service to demonstrate to residents how voluntary improvements to tenements could be made. After an open day, many residents signed up with the help of the new Govan Housing Association. Govan HA had in turn been established backed by the New Govan Society and the Govan Area Resource Centre, with ideas developing into action. The Govan experiment became the foundation for forming associations which came to be known as community-based housing associations.

And the ASSIST practice grew, typically based in whatever community it was working for.

Meanwhile the years of experimentation and frustration in Glasgow prompted the creation of an informal group of officials, academics and technical advisers to develop a new system. It often took so long to find a way through a labyrinth of issues in an area that the individuals and groups trying different solutions ended up redirecting their efforts to more fruitful territory or approaches, sometimes devising greater powers. Eventually the relevant legislation came through Westminster and at a very slow pace relative to the need for change on the ground.

Meanwhile, where action was possible, council budgets allowed for fairly generous contributions to be made to owners' costs, sometimes greater than the grants eventually available to housing associations. But before 1974, council powers were fairly limited to intervene or to require or enforce action or cooperation, even when the council had the appetite to intervene. So a lot of work went on behind the scenes to prepare a scheme for legislation, which brought into existence a new strategic and comprehensive approach – Housing Action Areas (HAAs).

9 At the time consultation with residents and communities was limited or non-existent. The Skeffington Report on People and Planning in 1969 pointed a better way forward.

Legislation in 1974 took the form of two Housing Acts in the same year. The first of these was the Housing (Scotland) Act 1974 which:

- Established a tighter workable definition of a 'tolerable standard' in Scotland, and with the jargon BTS (Below the Tolerable Standard). Identifying a property as BTS permitted the responsible authorities (councils) to intervene to require the owner to bring the properties up to the required standard. An example of the required standard was a flushing toilet and running water inside the property.
- Permitted declaration of areas with a majority of property which failed to meet the tolerable standard as Housing Action Areas. Declared areas had been known as HTA but came to be known as HAAs with greater powers for councils to act to coordinate and fund activity, and enforce action.
- Gave powers to councils to fund grants to owner-occupiers and commercial owners for improvement purposes at 75–90 per cent of eligible costs.

The second piece of legislation was the Housing Act 1974 which applied across Great Britain. It updated and extended the role of the Housing Corporation (HC) first established in 1964. This Act:

- Extended the powers of the HC to act throughout GB to register housing associations, for the first time.
- Gave HC power to fund development for rent by registered housing associations: previously funding had only been available for development of new low-cost home ownership (including shared ownership).
- Gave HC power to support and supervise registered associations.
- Directed government investment funding to the HC in the form of grants, and loan facilities.

Within a very short period, funding started to flow from the budget of the Scottish Office[10] in Edinburgh not only to the Scottish office of the HC, already in Edinburgh promoting low-cost home ownership across the country, but to the newly established Glasgow office. The Glasgow office was directed from London, not Edinburgh, to undertake tenement renewal primarily in Glasgow. (The relevant budget and spending priorities in England came from the DoE (Department of the Environment).) There were national governing boards in each country accountable for priorities and programmes were overseen by the corporate management in London. The overall chair of the expanded HC took considerable interest in Scotland and tenement renewal in Glasgow in particular. The legacy of two Scottish offices of the HC with different priorities and cultures had repercussions for many years.

By the time the 1974 legislation was in place, Glasgow was already identifying and delineating/circling areas requiring improvement and investment. This defined the Glasgow

10 The Scottish Office was based in St Andrews House, Edinburgh, under a scheme of administrative devolution established between the wars, overseen by the Secretary of State for Scotland and relevant ministers. The Housing portfolio was one of the more important posts as housing was a key issue.

approach. The council would declare HAAs, in the knowledge that their partners in the HC would be proactive in creating local community capacity and moving towards registration and funding for improvement. Although a handful of associations already existed, they were not usually resident based, or focused on particular communities. So the HC worked to encourage local residents in the inner-city tenement areas targeted for HAA status to form steering committees to become formal registered housing associations. Such bodies were required by law to have a minimum of seven individuals each buying a share of £1 to be able to form an association (which would always be the minimum requirement). The members would register the association as an industrial and provident society (IPS) on the basis that it was non-profit-seeking. In principle, an association could also register as a charity but in practice that model was rarely promoted or chosen by most new associations forming in Glasgow. In cultural terms, charity was arguably the same as philanthropy, doing good to others; whereas mutuality – with and for neighbours – counted for more than philanthropy, certainly in (inner-city) Glasgow. And so almost all the new associations in Glasgow were first formed as IPSs.

Overall, agency was (and is) key to success in launching the transformation of renewal processes: that meant one or more competent bodies having: (1) powers to act; (2) resources to implement decisions; and (3) the (political) will to see things through and secure cooperation. These were the characteristics of the new framework for tenement renewal with housing associations in the vanguard as agents.

Majority ownership and coordination

We'll come back to Dundee and Edinburgh but let's focus for now on Glasgow.

From 1974 the council was defining target areas and purchasing properties for improvement. The bodies which acquired the purchased properties from Glasgow City Council were intentionally temporary. The HC promoted, cultivated and supported the local groups until they were ready to fledge and be registered, with capacity to own property. Associations would then seek registration with the HC typically on the basis of clear boundaries relative to adjacent associations. Once registered with HC, and only then, the association would be eligible to receive public subsidy from available funds (see funding below) and so start to thrive with seedcorn funding.

The Glasgow arrangements facilitated acquisition to the benefit of the local association. Where an HAA was declared, a property could be bought under compulsory purchase powers where the owner was not willing to cooperate, or was missing. This relied on the local authority, as in the case of Glasgow, exercising power of compulsory purchase. The properties could then be transferred to a housing association.

In the early days ownership would be very diverse with different owners participating in improvement schemes. Where the association responsible was able to acquire and control decision-making about all the units in a tenement or in a series of tenements, coordination was infinitely simpler. Conversely the more diverse the ownership pattern in a particular tenement or block, the greater the complexity of arrangements to plan and deliver an improvement scheme. (This issue of complexity from fragmented ownership comes around decades later with the impact of the right to buy.)

With or without an HAA, an association could also acquire property voluntarily from an owner, typically where a private landlord wanted to disinvest. An association would do that primarily to build majority ownership in a building scheduled for renovation. Where a landlord owned a portfolio or whole building,[11] the sale could be very quick and smooth. If there was a sitting tenant, they could remain in the same flat if it was habitable, or move to an alternative flat nearby, pending renovation work starting. Some tenement residents were owner-occupiers: they too had the option of selling to the association, moving away and avoiding the hassle of redevelopment themselves. However, the valuation and sale value achieved was not necessarily sufficient for the seller to purchase another property elsewhere. In this case the occupiers might sell and remain as tenants, or stay and go through redevelopment as owners, potentially with grants from the local authority. Some did stay on; some sold and moved away. In any case it was common to see the financial or tenure issues as less important to decisions about whether to sell up than community ties – people wanted to stay to be beside friends in familiar neighbourhoods.

Tenements on arterial routes often contained retail units on the ground floor. These might be contained within one tenement envelope, though some spanned two adjacent buildings which made apportioning share costs quite complicated. This was especially true where such matters were not defined in deeds or where custom and practice had shifted understanding of responsibilities in law.[12] Shop units had historically held quite a large share of the total repair liability of the tenement – say one quarter – whereas the business revenue might be precarious, informal, shrinking. This made for tricky negotiations for associations coordinating improvements in dealing with shop owners about their liabilities and payment: in some areas such potential complications led associations to avoid taking on property which involved commercial owners, whether small corner shops or corporate chain retail outlets. Debts could last for years and many associations faced carrying debts from unpaid bills for such work. And ultimately such debts could be written off.

Once the investment programme got going, the speed of acquisition of rented flats outpaced capacity to repair and improve. This had the effect of casting associations – established for the purposes of redevelopment and improvement – into the role of landlord, often having to manage tenancies in properties which were below the tolerable standard and so, at worst, of slum landlord. In the best case, this might just be for weeks or months, but in extreme cases, the disrepair lasted for years until the funding and coordination of improvement caught up with the prodigious speed of acquisition and programming.

Settled solution

The approach and the governance model developed in Govan and tested in a handful of other neighbourhoods (Govanhill, Queens Cross, Partick) became a template for people in

11 In some cases the ownership building remained in a single portfolio with one person or trust funding a pension or annuity from rents.

12 The author's first introduction to the long ribbons of tenements along arterial routes was through a short, early career stint with the Housing Corporation. I was tasked to use my inexperienced eye to assess on site and mark on a ground plan exactly where the dividing line was between tenements (the chimney stack line). The ultimate purpose was to allow office staff to verify accurately the apportionment of shares of renewal costs to shop owners. At the time the tenements seemed dingy, neglected, unloved and semi-derelict, difficult to imagine as the glowing cleaned sandstone tenements they became.

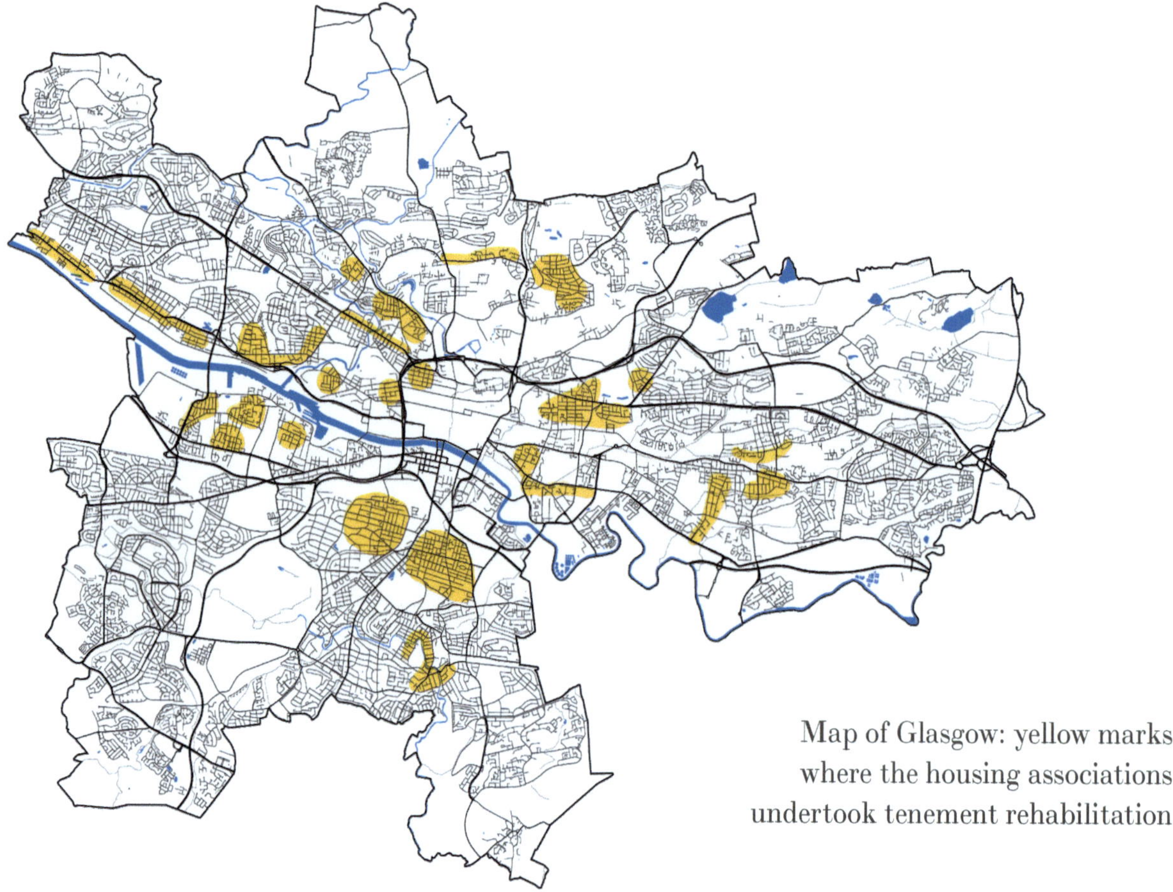

Map of Glasgow: yellow marks where the housing associations undertook tenement rehabilitation

other neighbourhoods in Glasgow. After Govan, came Govanhill[13] and then Reidvale Housing Association, with others soon following their example, as seen in the map.[14]

People in a few neighbourhoods in step with developments in Govan moved swiftly to create, establish and register a local association. In other neighbourhoods, while Glasgow City Council staff identified priority action areas, HC personnel would identify individuals, bring them together and galvanise residents to create associations. The associations were created and promoted by the HC under the provisional name Glasgow Fair (neighbourhood name) Association, pending any 'capacity building' required – development of a suitably skilled committee of local people, which would then take on the provisional mantle and assume formal responsibility for a particular neighbourhood. Within some five years there were about twenty associations in inner-city neighbourhoods of Glasgow, each taking on portfolios of between 500 and 1,000 properties.

The concept of a tenement renewal by a community-based association was not limited to Glasgow but radiated out to adjacent areas, with embryonic associations forming in many towns and neighbourhoods. In the greater Strathclyde area (defined by the boundaries of Strathclyde Regional Council) the Glasgow model inspired local people – for example in

13 A former ASSIST architect recalls working in the early days on an embryonic HAA in Govanhill, helping to secure community interests and fend off private sector predators. He talks of a clandestine meeting to alert the residents' association which moved quickly to form a housing association and get registered. The private sector cooperated with incentives to look elsewhere in more affluent areas.

14 The map highlights in yellow the various neighbourhoods in each sector of the City of Glasgow where housing associations were formed and undertook tenement rehabilitation. The map is based on the current boundaries of the city at 2022 and so does not include Rutherglen which is now in South Lanarkshire.

[upper left] Committee members of Charing Cross HA

[lower left] and [upper right] Residents queuing to see a recently improved flat

[lower right] At Langside Road, Govanhill, Glasgow. Back row left, close mouth and at windows: residents. Back row right: binmen, street sweepers. Front row left: GHA staff. Front row right: ASSIST staff including a student, Raymond Young (Housing Corporation), Councillor, and Ronnie Macdonald (local authority House Improvement team)

towns in Renfrewshire, Dunbartonshire, Argyll, Ayrshire, Lanarkshire. And at the same time the Glasgow office of the HC undertook to widen operations across Strathclyde rather than just in Glasgow, though in moving beyond Glasgow the housing association model operated in multiple neighbourhoods rather than just one. That could be due to smaller scale of tenement ownership outside the city, or a shift in funding and registration priorities.

There were comparable renewal activities in parts of Dundee and Edinburgh but very little in line with the concept of community-based associations as agents of tenement renewal.

The fate of tenements in Dundee varied widely. Even though tenements in the west end of the city were generally attractive and in reasonably good condition, many were demolished to make way for the expanding university sector. And the construction of a new road network linked to the Tay Bridge led to great swathes of demolition. Nevertheless many tenements remain, particularly in the west end of Dundee today, improved in the 1970s. Improvement came through market and private interventions in the very early period on a voluntary basis through HTA. Meanwhile tenements in more industrial neighbourhoods of Dundee (near mills and other manufacturing) were of poorer quality in the first place – smaller, denser, and in the 1970s more likely to be occupied by tenants than owners. Values were lower and scope to improve was occasionally in the balance because of the condition of buildings, cost of repair and potential impact.

As mentioned earlier, Dundee Council did not have the appetite evident in Glasgow to operate strategically and delineate areas for targeted investment, even though the scale of the challenge was at least as great as Glasgow. Various property professionals in Dundee had been quick off the mark in establishing associations in the late 1960s but without eventually registering them with HC. Lack of registration precluded access to finance for tenement renewal, except in the case of Hillcrest.

For its part, the Edinburgh office of the HC responsible for funding in the east of Scotland was considerably less committed to resident involvement, and kept a much tighter rein on spending on renewal. So staff in associations such as Hillcrest ended up creating property portfolios for the association, on a less strategic basis and sometimes through quite random connections. The combined effect of these various trends meant that a few associations in Dundee ended up spanning the city rather than confining their activities to neighbourhoods. In growing larger, they became more diverse, more independent and entrepreneurial – they would argue, of necessity.

Meanwhile by the early 1980s the council in Dundee was very focused on tenant participation and moved to set up very small resident-controlled cooperatives in peripheral estate council housing. Most of these have since been subsumed into larger bodies but such resident controlled bodies never developed the strengths to deal with the legal or financial complexities of tackling BTS housing acquired from the private sector. The contrast with Glasgow is quite marked.

Edinburgh tenements were also Victorian and occupied, but there were also some Georgian and older, semi-derelict and occupied only by pigeons. An important factor creating pockets of uninhabited tenement property was blight from road proposals (later aborted) on the south side near Edinburgh University. There were also pockets of disrepair in former industrial and working-class areas in Leith, Gorgie, Granton, Grassmarket, Tollcross, Holyrood and Fountainbridge. Establishing community-based associations involving local residents was difficult in Edinburgh. New associations were established in Gorgie and Dalry and in Leith with support from the HC. These operated on philanthropic principles, with limited resident involvement from the start. By contrast residents controlled new associations formed in the Old Town and in Fountainbridge, along Lauriston Place or off Leith Walk, buying services from previously established associations such as Edinvar, Link or Castle Rock.

Over the period 1974–8 Edinburgh City Council declared 40 HAAs across the city but stopped short of enabling or coordinating associations (or others) to act. Conservative-controlled Edinburgh was bedevilled by the various problems identified earlier in Labour-controlled Glasgow. These problems essentially boiled down to lack of coordination, with different departments acting seemingly independent of each other, promoting both improvement and demolition of the same buildings at the same time!

The consternation of local activists, both residents and professionals, prompted them to organise into a group, Housing Action Areas for Improvement, which compiled a hard-hitting 53-page report in 1977 criticising Edinburgh City Council. A collection of press cuttings in a scrapbook of the period showed the group's demands as fourfold:

- An agency responsible for implementation.
- A comprehensive programme of action for each HAA.
- Information for residents to enable proper consultation and control.
- Earlier planning and coordination by the council.

Scrapbook

The report caused a political furore with councillors, which rejected the criticisms and demands. They refused to engage with the group, ultimately prompting intervention by the Secretary of State in November 1978.

In Edinburgh overall, housing association involvement in tenement renewal ended up being undertaken primarily by larger, charitable bodies including those established before 1974 for other purposes (including shared ownership, housing for veterans or older and disabled people, temporary projects for homeless people, and so on). The culture of associations in Edinburgh was at that time less receptive to resident involvement on the same basis as in Glasgow. In many respects this was driven and certainly supported by senior HC personnel who did not recognise that tenants had capacity to take responsibility, very different from their HC counterparts in Glasgow.

The main exception to this inclination against mutuality in the case of tenement renewal in Edinburgh is Lister Housing Co-operative Ltd. which still owns, manages and upgrades four fine terraces of grade B-listed tenements on and near Lauriston Place, bought from the University of Edinburgh. The university had brought the buildings into single ownership from a trust and others, and transferred the entire portfolio to Lister. Since Lister became a fully mutual cooperative, it was not affected by the right to buy when it eventually applied to housing associations in Scotland and so remains in sole control of the tenement assets to this day.

Funding and finance

Having legal status and registration bringing access to funds was a start: enabling associations to acquire property provided the opportunity and imperative to engage in renewal. But without capital funding no progress could be made. Having a funding formula was thus the next achievement followed quickly by ensuring a regular predictable flow of funds to permit long-term planning. The last was the hardest to achieve and hold on to.

Public subsidy came to be known as a Housing Association Grant (HAG), which came in with the Housing Act 1974 across Great Britain. It was only available for creating or redeveloping housing for social rent. Owners participating in improvement schemes had access to other improvement or repair grant funding, administered by local councils.

Table 1.1 Worked example of HAG 1974–87: notional unit costs per tenement property

Type of cost	*Amount (£)*	*Notes and comments*
Acquisition	5,000	
Capital cost of renovation	44,000	
Fees	6,000	Related to the cost of works planned and covered arch, QS, engineering and legal
Environmental works	2,000	
Total costs	57,000	
Ineligible items	3,000	Could be above required standard or deemed as upgrading vs repair
Eligible for HAG	54,000	
Annualised rent	780	Based on weekly Fair Rent of £15 determined by Rent Registration Service
Less		
Allowable costs of management and maintenance	500	This amount was negotiated regularly with the Scottish Office, partly based on actual figures
Sinking Fund for major repairs	150	This was innovative provision calculated as a per cent of capital works and set aside for future repairs: it allowed reserves to be built up for longer term sustainability and independence
Balance of rental income available each year to repay loan	130	
Borrowing requirement	2,600	Based on formula estimating capacity to borrow discounted over 25 yearsNB not including ineligible items which might be funded separately or borrowed in addition
HAG requirement (HAG eligible amount less borrowing capacity)	51,400	Equates to over 95 per cent of the eligible costs

The delightfully named HAG was based on a formula which started with what could be afforded from the rental income. From a rent set as 'Fair', reasonable/permitted management costs could be deducted and the balance could be used to service borrowing over some twenty-five years. With rents set pretty low and permitted costs relatively high, the balance to be funded by borrowing was small: a notional worked example might help (see table).

This calculation produced the capital amount which required to be subsidised upfront by the state in the form of HAG. Typically this came out in excess of 95 per cent of the total cost, thus requiring a substantial public commitment. This system remained in place from 1974 until 1987 although there were adjustments from time to time around the levels at which various parameters were set, in light of changed circumstances, new evidence and as negotiations proved effective.

The only units which could be funded in this way were units to be rented by the association to tenants. Where other owners were involved, any grant assistance had to come from the local authority – at a rate of 75–90 per cent of *eligible* costs. Delays by or for an owner could hinder progress on a whole scheme/programme. So development staff from an association might inform an owner or help to complete application forms, but were not responsible for securing funds.

GHA in 1983. Marian Jacobs, student from Huddersfield; Beth Riley; Shirley Mullen; Betty; Andere Fyfe; John Borges; Dick Osprey; Lynsey looked after decants; Margaret McNeil (finance); Marjorie Cuthbert. Some temporary admin

Fair rents were set by the rent registration service (which also set rents for private landlords) and served as a mechanism for appeal by landlords and tenants in relation to periodic rent increases. The rent registration service sat within the office of the district valuer. While the register itself was open to public examination, the basis of decision-making was opaque and, it came to be believed, their determinations were susceptible to political influence, pushing rents ever higher. The higher the rent set, and the tighter the permitted management cost allowed, the greater the borrowing capacity. These factors produced a lower subsidy requirement so rent levels and what constituted 'eligible costs' became contentious from time to time.

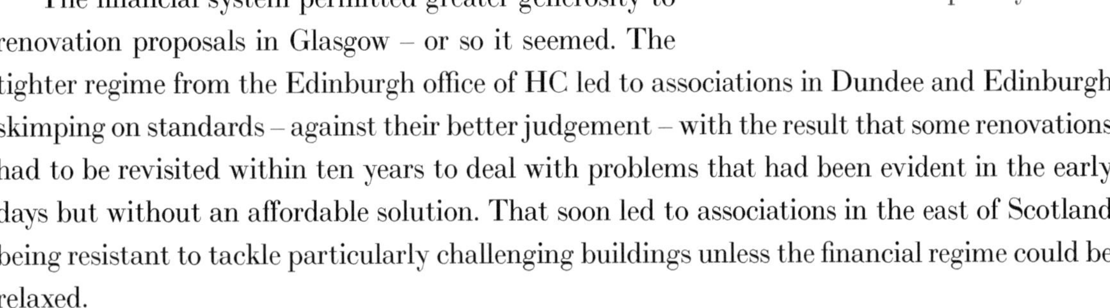

The financial system permitted greater generosity to renovation proposals in Glasgow – or so it seemed. The tighter regime from the Edinburgh office of HC led to associations in Dundee and Edinburgh skimping on standards – against their better judgement – with the result that some renovations had to be revisited within ten years to deal with problems that had been evident in the early days but without an affordable solution. That soon led to associations in the east of Scotland being resistant to tackle particularly challenging buildings unless the financial regime could be relaxed.

As the only city authority in Scotland with a concerted programme of tenement renewal, Glasgow had the lion's share of central government funding for tenement renewal. This level of funding (60 per cent for the west of Scotland) was secured intentionally by the director of the Glasgow office, Raymond Young – the former student architect activist. The west/east 60/40 split of the budget strained relations between associations in the sector across Scotland. Association leaders in Dundee, Aberdeen and Edinburgh particularly resented the dominance of Glasgow and the west in funding allocations but also in the way that funds were more generously managed.[15]

15 As director of a local association, the author represented Glasgow associations on the council of SFHA at the time and recalls regularly tense, even heated discussions about whether the east/west split – as it was known – offered a fair balance. The west's argument was based on the numbers of people living in awful conditions; the east's on the needs of older and disabled people. The Dundee argument actually aligned with the west though by virtue of being in the east of Scotland they were adversely affected by the split.

Inevitably the cost of capital works rose not least due to general inflation in the 1970s but also because of rising standards. Some associations realised early on that it was a false economy to do the minimum renovation to a building. It could end up failing to deal adequately with dry rot. It could lead to underestimating contingencies exhausted by unexpected problems on site, with cost overruns and disruption to the lives of residents. Minimum renovation built in higher maintenance costs, and could require earlier replacement of elements. By the early 1980s it had become clear that the surgical 'patch and repair' approach was less than effective and potentially more costly in time and money than a more radical approach. As mentioned earlier, some associations (notably in Dundee), declined to embark on renovation on unacceptable terms.

The failings of patch and repair were eventually recognised and the system was improved in the early 1980s, certainly in Glasgow. The revised approach was called comprehensive tenement improvement (CTI), based on standardised rehabilitation indicative costs (RIC) defined by the HC. This led to tenements being more likely to be planned to be stripped of internal elements and replacement of internal finishes, unless they were of historic importance or if required by independent owners taking part in the overall scheme. One consequence of this CTI approach was that associations ended up owning a higher share of flats in any one tenement as other owners sold up to avoid disruption and expense. But overall, costs were more predictable and the programme was easier to plan and deliver so constituted a much smoother operation, as long as the money kept flowing. A long-term benefit of this approach was to facilitate good asset management, where the life cycle of various components of the building could be estimated and planned, and replacements programmed.[16] It also enabled materials susceptible to dry rot to be removed at an earlier stage before reinstatement of finishes.

The overall programme hit the buffers in the form of a moratorium on government spending on Scottish housing in the mid 1980s, after the miners' strike and after the Falklands War. This pause lasted for around six months and had an instantly detrimental effect and repercussions on the management of the programme. It created havoc in the contracting industry and for residents. As for the agency whose financial management may have triggered the moratorium, there was no future.

Reinstatement of funding overall within a few months was based on the same model/formula but only lasted a few more years before much more radical changes were made to subsidy calculations and to the structure of regulation. Programme funding existed more or less in its original form from 1974 until 1988 when a new regime was introduced.

The new regime turned the previous formula on its head by requiring associations to set their own rents based on a government set ceiling on subsidy – with a target of less than 90 per cent and falling, compared with 95 per cent[17] plus. Although that level seems generous by today's standards, at the time it was unwelcome because it entailed higher rents, and a squeeze on management and maintenance costs. It also meant that associations had to take the business of borrowing much more seriously: the amounts borrowed were greater and for longer periods, potentially, and higher interest rates. So the banks and building societies

16 Later imposition of the right to buy on housing associations meant the loss of many of these advantages.

17 Since the late 1980s the target funding level has dropped steadily and is now less than 50 per cent and eroded by inflation.

became more important in the provision of finance to the sector, something that came to be much more significant in housing association funding in later years. And it challenged many committee members and staff to think differently about financial responsibility.

People in neighbourhoods, committees and in work

Lister is the only 'pure' cooperative to have engaged in tenement renewal and endure to this day.[18] Cooperatives were very attractive to radical policymakers in the early 1970s and explored as a possible vehicle/agency for renovation, because the cooperative model places residents in control. But that very model requires absolute correspondence between the status of the resident and the member of the ownership body on an equal footing. If the objective in Glasgow had been to acquire every single property in designated buildings for what we now call social housing for rent, a cooperative might have been achievable. But the policy appetite for effective cooperation with existing owners who wanted to remain as resident owners precluded the possibility of a 'pure' cooperative.[19] The only way to achieve tenant and owner involvement among residents was therefore to require a simple connection to the designated area/community in order to be eligible for membership of the association.

The point of a local CBHA in the Glasgow area was that people had to be local to take part. The 'model rules',[20] as they were known, allowed anyone who lived, owned property or worked in the area defined by that neighbourhood, to apply to an association for membership and a £1 share (which equates to just over £11 in 2022). This involved mostly residents – tenants or owners, but also shopkeepers, health workers, teachers and so on. Many of the early committee members were skilled workers, organisers and activists, politically engaged in different ways.

Every association, whether a CBHA or not, decides who to admit to membership, though very few applicants will have been turned down (or removed, though that was possible). Any member can stand for election to the committee (or board), which is typically required to be a minimum of seven and no more than the defined maximum (12–15 people), who then elect office bearers – the position of chairman being crucial. The committee governs the association and employs paid staff, who undertake the day-to-day operations from a local office: in the early days the office was often a local shop or a domestic flat no longer in use. Committee members would say they 'ran' the association: one could question whether this was ever really true! They took and were ultimately accountable for the key decisions – which properties to buy, how to improve, but always based on investigation, advice and recommendations from employees. It was a brave staff member who took a decision without reference to the committee, primarily out of respect for the model and the individuals, more than deference or fear. This essential structure endures to this day, though with variations and differences of emphasis around delegated powers.

18 A cooperative called Avalon was established in Glasgow in the 1970s but was later subsumed into another association.

19 In a fully mutual cooperative (FMC), all residents have the same status and are all equal members of the organisation. This is rather difficult to achieve in practice and thus extremely rare. In some instances in Glasgow, organisations were formed as non-fully mutual cooperatives (NFMCs); in effect a housing association with a higher proportion of resident involvement and control.

20 These are more commonly known today as 'Articles' or informally as the constitution.

Where associations were active on renovation work, the committee would meet every other week and make decisions based on briefings and recommendations from staff. Committee members usually stepped back from dealing with the particular details of cases: staff dealt with these working within an agreed framework, often determined collectively with other associations through a body such as the newly formed Scottish Federation of Housing Associations (SFHA) and its various committees and forums.

Each association had sub-committees set up to deal with matters such as finance, or design or management. The latter was required because of the acquisition programme which of necessity preceded redevelopment. But any delay in redevelopment could expose the association to management or maintenance problems. The sub-committee system allowed expertise to build up in different aspects of association activity and a huge amount of work to be achieved quickly. Design and improvement were always a key attraction for committee members, with fewer people motivated to take on the management of tenancies which raised thorny issues of slum conditions, repair services, rent issues including arrears, and antisocial behaviour. Although all issues related to people who were neighbours and therefore potentially known to committee members residing locally, policy guidelines and confidentiality principles were quickly established in the culture and practice of most associations. And associations had to be careful to safeguard decision-making from any perceived conflict of interest around benefits to committee members, without antagonising volunteers willing to give time and effort to an association.

The staff of new housing associations were pretty entrepreneurial at the start, almost 'making it up on the hoof' – what we might now call a 'problem-solving approach'. It was essential that they were able to work on this basis, within an agreed framework. There was no established path into working in a housing association. People working in 'housing' at that time usually worked in housing management for a council, which was not at all the same as redevelopment in a body accountable to the local community. Council staff were often seen as bureaucratic, paternalistic or judgemental and lacking in imagination or initiative. By contrast the people attracted to and successful in redevelopment work were often young, recent graduates with degrees in geography or planning, architecture or surveying but from distinctly diverse backgrounds, and not necessarily local. Commitment, pragmatism and aptitude were often the most highly prized attributes.

In 1977 an official report by the Scottish Housing Advisory committee for the Scottish Development Department (SDD), *Training for Tomorrow*, criticised the housing world for lack of professional development, qualification and competence. Anyone who wanted recognition by the professional body, the Institute of Housing, had to undergo a correspondence course and sit three diets of exams mostly about law. And the numbers of people undertaking or even completing this route were small.

Training for Tomorrow

It was only in 1980 that an innovative course in housing administration was launched at the University of Stirling, with just eight bursaries (sic!) funded each year by the SDD in the then Scottish Office. Many of the early graduates took up posts in development/renovation in housing associations and quickly progressed into management of those associations, coordination of staff teams and governance. And committees came to rely heavily on those individuals, many of whom stayed in post for many years becoming ever more senior and influential.

The Housing Corporation (and later Scottish Homes) also employed people to work on legal, technical and financial aspects of tenement renewal. They were civil servants, and mostly office-based. Some of them left HC to join housing associations in a leadership role, often frustrated with the bureaucratic constraints of HC. And associations engaged professional consultants – private sector firms which employed squads of architects, surveyors and engineers many of whom found it difficult to adjust to resident control. Younger staff who were recently qualified found it exciting to be in at the start of a movement, making a difference to people and buildings. Professional fees were considered a legitimate part of the cost of renewal. The fees were usually a standard percentage but firms competed for business, on the basis of track record, style, preferences, quality and value for money. Each association chose a professional team of consultants for each phase of improvement with a degree of churn and turnover and some in Glasgow established local neighbourhood offices.

The people working together in the movement had good relations overall – hence the sense of a 'movement' in each region. Everyone was committed to restoring the tenements to their rightful place as an important part of the built heritage, certainly in Glasgow. Some groupings within Glasgow fell out with each other, for example over employment issues. This was mild in comparison with later rifts over the emergence of stock transfer from Glasgow City Council into 'community ownership' in the 1990s. Since that episode raises many issues not directly pertaining to tenements, no more need be said about that here.

A key factor in the creation and sustenance of a 'movement' was the establishment of a network called a forum. For example, the Glasgow Forum of Housing Associations at that time would meet every month with a speaker on some pressing topical issue, and was attended by throngs of staff and committee members alike. Similarly other forums – for Edinburgh and Lothians, or Tayside Grampian and Fife – would meet to compare notes, join forces and plan campaigns.

Each forum set its own priorities and fielded representatives to the SFHA council which was itself regionally constituted at the time. SFHA had various themed committees to look at housing management, development and finance and led representation of shared association interests to the Housing Corporation and to the Scottish Office on their collective experience of aspects of the system – notably finance, and governance. Within the broad church of SFHA, representatives worked to find ways to establish and work on shared interests behind closed doors, in spite of heated public differences over the balance of spending priorities.

Party perspectives and alliances

Housing associations have long been regarded as non-aligned, a position widely and jealously guarded by the sector for decades. Since they are neither strictly public nor private entities, they appeal to the Left and to the Right: to the Left because meeting need is central to their *raison d'être* and their goals are primarily social and non-profit seeking. The Right traditionally values housing associations because they are not public or state-owned and because they harness people in communities acting rationally in meeting their own needs and thus pursuing self-interest. The association movement had to be able to work with both sets of interests to be able to function, and they rode both horses very effectively.

In Scotland at the start of the tenement renewal programme in the 1970s most cities in Scotland were controlled by Labour, though not Edinburgh. From 1979 central government was under Conservative control, with Scottish Office ministers determining policy and priorities in an era of administrative (rather than political) devolution. At a time when government was tightening control over local authority budgets, Labour politicians in local councils recognised the value of agencies such as the HC and associations being able to access additional funds from central government to spend locally, which allowed investment in housing to preserve heritage environments and enable economic growth.

The moratorium on funding set by a Conservative government in 1985 provoked a campaign between associations and councils to restore funding to complete the programme. The moratorium occurred during the second term of the Thatcher government and was considered at the time to be related to public sector cuts to which voters in Scotland were opposed. This vociferous campaign took a posse of activists to London on the basis of the cut being viewed as a political move against social housing. Arguably the explanation after the event was much more delicate, with the cut having more to do with the Scottish Office seeking to restrict over-spending by bodies other than HC and housing associations. The body responsible for the overspend ceased to exist within a couple of years, having been absorbed into the new strategic agency, Scottish Homes, which in 1988 became responsible for registration, supervision and funding of associations. From that point on, the HC's operations were limited to England and Wales.

The creation of Scottish Homes in 1988 and separation of the Scottish functions of the Housing Corporation from the GB body, led to changes in political priorities and accountability, as well as in the funding formula. The capital programme funding changes were in the same direction as in England but less extreme in terms of subsidy cuts and rent levels. Increasingly the capital programme was funded by receipts from disposing of assets (tenanted homes) previously built/owned by the Scottish Special Housing Association (SSHA).[21] These properties were increasingly bundled into area portfolios for management buy-out or 'tenant-led' transfers drawing on the experience and practice of community-based associations. Very few if any of these portfolios included the kind of tenements considered here, but it did apply to newer tenement forms.

21 SSHA was established in 1937 as the Scottish National Housing Corporation (SNHC) to build houses for economic activity. It was active post war in devising new house forms and new technologies, as well as developing housing for workers relocating in response to new factories, mines or oil facilities. In the late 1970s SSHA started to support councils with regeneration and became the second largest landlord in Scotland. It was not a housing association in the sense of this chapter – in spite of its name. It was subsumed into a new government agency Scottish Homes in 1988, which, amidst great controversy, disposed of all its rented housing to various registered social landlords. Ultimately the entire stock of the SSHA was sold or 'transferred' to associations, leaving Scottish Homes with unfunded debts. Scottish Homes was in turn replaced in 2008 by Communities Scotland, with the regulatory powers falling to a new body, the Scottish Housing Regulator (SHR).

Until 1988 Housing Corporation officials were employed by the British civil service and managed from London in financial terms but in liaison with the Scottish Office. This allowed officials to be effective in channelling pressure on the HC through ministerial influence. From 1988, Scottish Homes' staff were appointed directly to the agency – technically a non-departmental public body controlled by a board appointed by Scottish Office ministers.

For their part, Scottish Office ministers during the most intense period of investment in tenement renewal (1974–88) valued having opportunities to be seen fronting and championing housing projects at official openings on completion of projects. This was in marked contrast to the way that local authorities as social landlords, were experiencing increasing restrictions on budgets and powers which often meant tense, even hostile relationships between central and local government, with repercussions for relations between associations and councils. Ministers were regularly invited to address conferences of housing associations and were open to regular dialogue with representatives of the movement. This consistently benefitted the renewal of tenement neighbourhoods.

Legacy of resident involvement in tenement renewal

The model of community-based associations in the west of Scotland became very significant in subsequent policy and institutional developments throughout Scottish housing post 1988. The restrictions on council housing budgets created pressure between landlords and tenants for improvements and better services, and not just in Scotland. The incoming Conservative government in 1979 had introduced the right to buy for public sector tenants, which had a significant impact on council ownership, particularly of houses, and was seen as politically successful for the Conservatives. By the third Thatcher term, government thinking shifted to increase the discount available to tenants, particularly in flatted properties, and also to introduce a right to choose an alternative landlord. This was an individual right in Scotland and barely used in practice.[22] But it effectively signalled that if tenants considered changing their relationship to the public landlord, they could become 'private' while remaining as tenants. The creative response to this shift in Glasgow led to experimentation around cooperatives and community-based associations in non-traditional buildings in peripheral estates built by councils in the 1950s. Technically those buildings are tenements, though not in the sense of this book. They were walk-up flats built in concrete, with small rooms, limited insulation and heating, minimal investment in kitchen and bathrooms now past their 'sell by date' and long overdue for modernisation and upgrading. The problem was the council could not keep pace with the demand for modernisation. As pressure tightened on council finances in the 1980s the programme of modernisation all but closed and prompted some activists to protest and ultimately ambush a Scottish Office minister on a flight to London, with some success!

Taking the community-based model in inner-city tenement neighbourhoods as a point of inspiration, tenants in peripheral estates were encouraged to form 'private' organisations on

22 The main use of this right was in the Scottish Borders where managers from Roxburgh District Council created Waverley Housing which achieved individual tenant consent to acquire ownership rights over former council and SSHA housing. Another use of the individual right to choose was in Aberdeen and Aberdeenshire, with managers forming a service agency under contract to tenant-controlled associations in local communities.

cooperative principles, which could be registered and funded to acquire and upgrade their homes. Overcoming scepticism and a degree of council/landlord resistance, tenants engaged with this agenda, supported by Scottish Homes. The net result was a trial of community housing cooperatives in three estates: collective private housing with mortgage interest tax relief. This would work alongside the new variant of a traditional association model in three other estates with newly created associations. Both versions required the majority Labour councillors to accept the voluntary sale of council housing to a tenant-controlled body: the successful prior existence of community bodies in inner-city neighbourhoods helped councillors representing the peripheral estates to convince opponents of the merits of the sale.

The first six bodies were in Glasgow but the community-based model was soon picked up in other council-built/ owned estates in Clydebank, Dundee, Edinburgh, Greenock, Wishaw, Perth.

A further change in the Conservative third term was the move to terminate the five new town development corporations, which still owned tens of thousands of tenanted homes, and to dispose of all the public housing inherited by Scottish Homes from the SSHA. These were extremely controversial moves. Local councils eventually prevailed in securing rights to acquire most of the new town housing, but the Scottish Homes stock across the country was mostly parcelled into area based portfolios, with local managers engaging with tenants to create what were known as tenant-led transfers to newly registered housing associations, with some being sold to existing associations. These were rarely cooperatives but permitted tenants to become members of the associations and thus to form the committees which controlled the work of the organisation. These bodies have been responsible for the long-term management and maintenance of their assets, typically without financial support, other than to tenants on low incomes through rent support from housing benefit/ universal credit.

In this way the model of community-based associations in inner-city tenement neighbourhoods generated a plethora of local organisations accountable to members and (supervised) regulated by the state – currently in the form of the Scottish Housing Regulator (SHR). In 2020 the sector owned/controlled over 300,000 properties and tenancies, around 12 per cent of all homes in Scotland. The model of resident involvement in tenement renewal has thus been instrumental in changing the nature and governance of housing in Scotland at a time of significant transformation. At 2022, in six local authorities, the only social housing providers are housing associations or registered social landlords (RSLs).

Sustaining model and refreshing leadership

It has been difficult at times to get associations off the ground but that doesn't mean it is plain sailing once they are in place: both creation and maintenance take sustained effort, whether as a concept or as a specific organisation in a particular place.

It has long been recognised that the excitement generated by (re)development attracts people to get involved in an association. The development sub-committee is often the one that is most popular – with the greatest imperative to action. Its achievements have the most immediate impact on people: on the community and the neighbourhood, on the future of the association. Matters like management, maintenance and finance are important but don't usually hold the same appeal and that affects the vitality of an association.

Participation in voluntary organisations is not as popular in the 21st century as it was when many of these organisations were first established. Participation in governance is not top of the list of what most people want to do with their spare time. It has been too easy for associations to be governed by the same people year after year: they build up a wealth of experience and knowledge, they know each other, they are proud of their achievements. And while that is all good, necessary and useful, it can also inhibit others from joining and inhibit renewal of the organisation. The very fact that an association has solved the problems of the past should mean the need now is less pressing. But those individuals inevitably get older and it's often hard to recognise that new people and new ideas are needed, welcome, refreshing, vital.

Focusing on tenant participation can make an association lose sight of the opportunity to include others from the community – other owners, businesses, services. Associations which are mutual have to work harder at drawing in external views; associations which are essentially philanthropic and often bigger, have to work harder at engaging with the residents who are affected by decision-making.

While the vast majority of associations have flourished and become anchors for provision of good-quality affordable housing solutions, sadly not all associations have thrived. Even within the first decade there were instances of discrimination, toxic or inappropriate behaviour which prompted regulatory action to terminate the association and transfer its assets to competent owners. In the last decade, the regulator has set and pursued ever higher standards of conduct and performance and held associations to account to those standards. Not all associations have been able to demonstrate 'compliance' with the current regulatory regime, or satisfactory conduct. Some practices by individuals or associations have come to light which reflect badly on them and on other associations. Fortunately, these are a minority but there can be no room for complacency.

Governing an association today is not without challenges: it is a complex business legally and financially, responsible for the wellbeing of many individuals, families and communities. But it can be a hugely rewarding activity and requires injection of energy and imagination to keep any association alive and effective.

And in relation to those associations which are custodians of heritage tenements, many or few, it is vital that they use their experience as a platform/springboard to ensure that those buildings have a sound future, providing affordable energy-efficient homes. That is no small challenge for those who own and manage property lets to tenants. It is just as important when the association is acting as factor on behalf of different owners.

Concluding thoughts

It is difficult now to imagine how urban tenements could have been brought back to life and conserved from the 1970s without a new framework of powers, extensive public finance and committed agents acting locally. Civically minded people have played a hugely significant part in realising the potential of public private cooperation, largely through housing associations.

The 1970s witnessed a real growth in the housing association movement in Scotland, reflected in a dramatic rise in registered bodies during that decade, particularly local neighbourhood associations. The decades that followed have seen even greater growth in the number and scale of the contribution of associations across the country. Tenement rehabilitation and area renewal by housing associations contributed significantly to restoring pre-1919 tenements and to establishing associations on a firm business footing. Although tenements constitute an important foundation of the association sector, its history and legacy, today they represent a relatively small proportion of the total stock owned and managed by associations. Associations themselves have succeeded in diversifying their offer of housing, amenities and services to local communities.

Based on the Scottish House Condition Survey of 2019 we know that two-thirds of all property in Scotland is flatted. Almost 600,000 homes in Scotland are in some form of tenement, with 184,000 tenement homes built before 1919. With approximately one third of all flats in Scotland rented privately, the challenge to owners – including landlords – is to maintain flatted blocks including tenements in such a way as to safeguard the asset for future generations, use the carbon already captured in these buildings and provide much needed homes which people can afford to rent and heat.

The model of participation developed in Glasgow to deliver renovation, refurbishment and renewal resonated with people dealing with other forms of housing in other contexts elsewhere. The community-based model served to inspire and nourish the development of local capacity in urban and rural communities. It serves to remind policymakers of the extraordinary value of harnessing the ideas and commitment of people affected by decisions.

Credit for success in establishing associations is due to the competent committed people – voluntary, paid and professional – who embraced the challenge locally, but also to their partners in central and local government. They operated on the basis of courage, imagination, trust and collaboration to achieve something no one could do alone.

The strategic approach to renovation helped to combat the fragmentation of ownership and management of tenements which predominate in Scottish cities. That heritage can only be sustained in future with a similar degree of coordinated effort.

[upper] Prime Minister Edward Heath visits Glasgow to inspect the Comprehensive Development Areas

[lower] A woman in George Square who was enraged when they took down one side of George Square in 1976

Chapter 11

Repair and renewal in Glasgow from 1970 to 1990

Changes and problems in the 20th century

After the Second World War, the Bruce Plan for Glasgow proposed the demolition of most of the city centre together with a new ring road. A similar plan was made for Edinburgh. The M8 motorway in Glasgow was built, while Edinburgh's inner ring road was not. The M8 sliced its way through Glasgow, dislocating communities such as Cowcaddens and Garngad from the centre and destroying both the factories where people worked and the houses they lived in. Road widening, to make room for the car, which was seen as the future, was extensive.

Boarded up tenement collapse and decay

Slum clearance to make way for Comprehensive Redevelopment Areas resulted in further demolition as the Corporation of Glasgow invested in new system-built multi-storey developments. The Gorbals (Laurieston and Hutchesontown), with 67,000 residents, 1,100 shops and many public buildings, was demolished in stages, starting in 1956; most of the tenements were demolished in 1973, but, by then, the deck access precast concrete flats in Crown Street (Hutchesontown D) had been completed and tenants started to suffer from damp problems. A Corporation publication of the time showed a street of tenements and captioned it '*Typical obsolete tenements*'.[1] After the war, in the 1950s, was a time when everything had to be new; old furniture and old houses were discarded, so there were only very limited campaigns against demolition of tenements in most areas.

Housing Action Areas

The Cullingworth Report in 1968 sought to address the problem of slum housing by determining a housing quality standard (the tolerable standard[2]) and also offered owners enhanced improvement grants if their house failed the standard. This led to the Housing (Scotland) Act 1969 which included the Tolerable Standard and made proposals for 'Housing Treatment Areas' (HTA) of substandard housing.

In 1968, Glasgow also saw 'the great storm' in which 20 people were killed, 70,000 homes damaged and 300 destroyed.[3] Many roofs and chimneys were wrecked. The Corporation set up a 'storm damage centre'. Theo Crombie, a Glasgow Council officer, helped to tackle the substandard housing in Glasgow. The story of Glasgow's tenement improvement is described in Chapter 10, on community-based housing associations.[4]

In the mid 1970s onwards, tenements which were once tenanted and owned by landlords or factors were sold to owners, many of whom had no experience of maintaining the common parts of a tenement. The system of factoring often broke down, as it only took one owner to refuse to pay for their share of maintenance, to prevent work proceeding. Some areas became grey areas which were '*red-lined*' by mortgage providers, making selling or buying difficult.

The Housing (Scotland) Act 1974 allowed owners to apply for enhanced improvement grants from the Housing Corporation and in Glasgow led to the development of the Community Based Housing Association movement as expertise was required to administer the take up of the improvement grants. Housing Action Areas were declared for houses that

1 *Glasgow's Housing Centenary 1866–1966*: published by the Corporation of Glasgow Housing Committee.

2 The Tolerable Standard of 1987 stated that a home would meet the standard if:
  it was free of penetrating or rising damp;
  it was structurally stable;
  it had satisfactory provision of natural and artificial lighting, ventilation and heating;
  it had satisfactory thermal insulation;
  it had an adequate supply of drinking water and a sink with hot and cold water;
  it had a bath or shower within the house;
  it had a WC and an effective drainage system;
  it had satisfactory facilities for cooking;
  there was access to all external doors.

3 Repairs cost £30 million, equivalent to £150 million in 2000.

4 Raymond Young gives his personal view in his book *Annie's Loo*. Essentially a pilot project in Govan showed how a tenement flat without a bath could be improved by installing a new bathroom in the bed recess. Despite the success of the Housing Association movement in Glasgow, tenements were falling derelict and demolition was the result in many areas, given the general perception of the sandstone tenement at that time.

Grant Street and Arlington Street to go

The aftermath of the great storm in 1968

failed the tolerable standard which allowed for grants to cover both improvement and repair works, but also included some compulsion in the work.

Initially this work was carried out without decanting tenants. Windows would be simply repaired, replacing a batten rod or sill, but eventually after dry rot was found to be commonly spread throughout most tenements, housing associations arranged decant flats for tenants and owners to move into whilst the work was being done.

Backcourt improvement work 1974–86

Backcourts in the 1950s and 60s had become common areas, unfenced, poorly drained and used to house the bins. Some wash houses were still standing and some backcourts still had workshops. They were not ideal play areas for children.

The backcourts were often ruinous

Environmental improvement grants were the way of rehabilitating them. Owners had to get together with their neighbours and form a 'Backcourt Committee' and then, after initial agreement with the council, an architect would be appointed to prepare proposals and seek tenders. In Glasgow, the grants were usually 100 per cent so although there was not a problem in getting people to agree to such grants, it was often necessary to hold individual close meetings, so that everyone could have a say in the design, which would eventually be agreed at a block meeting. In other areas, no funding was provided for environmental improvements.

Annandale Square

The main problem was gaining access to the backcourts as they were often entirely enclosed. In some cases this meant that all material had to be wheeled in though a close. One contractor[5] dismantled their digger and then hauled it through the close to then reassemble it. One contractor in Govan bought a ground floor flat then demolished the walls, formed a ramp and brought machinery to pump concrete through the flat to form the bin store foundations. In Carfin Street, Govanhill, individual glass-reinforced bin stores were made in a factory which could then be wheeled through a close.

Carfin Street Square

5 John LeHarivel remembered the contractor Trunk Contracts doing this, but also many smaller contractors, such as the Todd Brothers were involved in Govanhill.

Later backcourts in the West End often had openings or pends to allow such access.

Most improvements included new metal fences to create an individual backcourt space for each close, although a few (in the West End) created communal backcourts which allowed much better space for plants and communal play areas. New metal clothes poles were included along with paving, retaining walls and steps as well as ground drainage.

Bin stores were invariably low level, usually topped with an in situ concrete roof and designed for a standard metal bin. Over time, the amount of packaging grew and with few recycling schemes, and a lack of housing management,[6] backcourts deteriorated (see Chapter 9 for more information on backcourts).

Larger wheelie bins were eventually introduced which had to be located in the street where bin store roofs had been removed, or in bin areas without a roof.

Improvement and repair work 1976–90

Outside the Housing Action Areas, improvement works were still needed. Initially, where houses had internal bathrooms, they might be upgraded and new kitchens installed, along with roof replacement works. Existing windows were simply repaired. Sometimes structural repairs were needed.

But in Housing Action Areas, it was soon realised that full decanting of residents was essential, as it was felt best to lift floors, drop ceilings and deafening, to check the condition of the floor joists for any rot. The work was very disruptive. The work to install new kitchens and bathrooms is described in the section on new bathrooms, although it would have included most of the items below.

Windows

In the early days (1976–8), existing sash and case timber windows were repaired. Parting beads and

6 More prevalent in tenements that were not managed by housing associations, as many owners and landlords did not have a system to manage waste.

All was not right in the world, as many factors did not have a community base

A Reidvale close

PLS Construction on Dumbarton Road rehab

[left] J. B. Bennets on Butterbiggins Road ...

[right] ... and in a container

batten rods along with recording sashes were common, and tenants remained in their flats, but eventually, in Glasgow, when work became more extensive as tenements were gutted, tenants were all decanted. All the windows were also replaced, usually with timber double-glazed, pivot hung sashes with a central fixed transom. These have not weathered well and most have since been replaced (see Chapter 7).

In Edinburgh, in listed buildings and conservation areas, the sash and case window remained and was repaired.

Many landlords and some housing associations have started to use UPVC windows which unfortunately have been recommended by the Energy Savings Trust. From an owner's point of view, UPVC windows appear cheaper to install and require no maintenance. They are often installed into the old timber case of the original window, which makes the work less disruptive from an owner's point of view. However, it results in reducing the glazed daylight area by 10 per cent because the plastic frame protrudes. Also, the old timber case still remains and may eventually become rotten. The timber sill is usually trimmed back and a plastic plate applied at the sill which may create dampness issues. UPVC is not recyclable and leads to landfill waste and the plastic material can be made with toxic materials which eventually leach out. Far from being 'maintenance free', plastic windows are nearly impossible to maintain. Ironmongery replacements can rarely be matched and, if the plastic is cracked, it is difficult to repair. In a fire, toxic gasses can be released.

Doors

Surveying a roof in 1978 (note the lack of harness and hard hat!)

Detail of roof

Original internal doors were usually panelled doors, with knobs and latches, faced with architraves. If retained, owners might have sheeted over them with hardboard (to avoid dust or just to 'modernise'). Invariably the internal doors would be renewed with ply of hardboard faced flush doors and handles and latches replaced the older knobs. Front flat doors would be replaced with solid core fire doors[7] with a door closer. This often led to the loss of the original etched glass or stained glass in the door, although some were saved in listed buildings. The storm doors were usually retained, but any fanlight above the flat door had to be replaced with Georgian wired glass to make it fire resistant.

Roofs

Roofs were stripped of slates and old flashings; after the rafters and sarking were repaired, the roof was felted with a bitumen felt. It was then simply battened to receive concrete tiles which initially might only be nailed every third course. Concrete tile ridges were bedded on cement mortar. Counter battens came in around 1979 to allow ventilation under the tiles.

Today we would use a vapour permeable roofing felt and then batten and counter batten the roof before fixing any roof tiles, with each tile clipped securely with stainless-steel clips.

In many cases the rafter and joist ends would have been affected by rot due to overflowing gutters. New rafters and joists were spliced into place and set onto new treated wall plates set on DPCs. When front moulded gutters were replaced, they were usually set onto a bitumen undercloak. Now, in the Glasgow area, it is best to use cast-iron gutters and install a lead undercloak. In the past, glass fibre moulded gutters were used but didn't perform well because of snow loads. Extruded sheet aluminium gutters with offsets at bays were joined with putty[8] or mastic and sections of flat aluminium plate were used inside the gutter. On curves the sheet aluminium could be welded easily so it was cheaper than using cast aluminium. The result is functional, if not as elegant as a curved cast-iron gutter, which can still be sourced for most curved bays.

Rear gutters were often replaced with UPVC deep flow gutters connected to rafter brackets.

Flat roofs in Edinburgh were recovered with bitumen felt, often prone to being blown off the roof. Lead flat roofs were generally left in place. Builders would often repair a crack in lead with a proprietary product called 'flash band'.

7 Early fire doors relied on a deep check, later ones had to have intumescent material set into the edge of the door and also smoke seals.

8 As well as putty, polysulphide mastic or 'Plumber's mait' was used.

In the 1970s and 1980s, lead flashings, if leaking or missing, might be replaced with Nuralite or Aqualite, proprietary bitumen flashings. The only advantage these had was that they would not be stolen, as lead theft was then a problem.

Cement render on a rear chimney

Stonework

Where stonework was defective, it was repaired with cement mortar, the decayed stone would be cut away and then a plastic repair made using two coats of cement mortar, often coated with Linostone or finished using a coloured mortar. If the repair was deep, copper screws would be plugged into the stone and then encircled with copper wire to allow the cement mortar to bond to the stone. Invariably these cement repairs have failed and caused more damage to the stone.

The effect of cement coatings on sandstone

In Edinburgh defective stone was often *indented* with real stone to match the existing stone, but only in listed buildings as it was possible to get an extra-over grant from the Scottish Development Department. Otherwise, for unlisted tenements, cement mortar repairs were used, as in Glasgow.

Rubble walls at the rear were usually repointed in cement mortar – again, not the best option as lime mortar is needed to ensure the stone can dry out through the mortar beds.

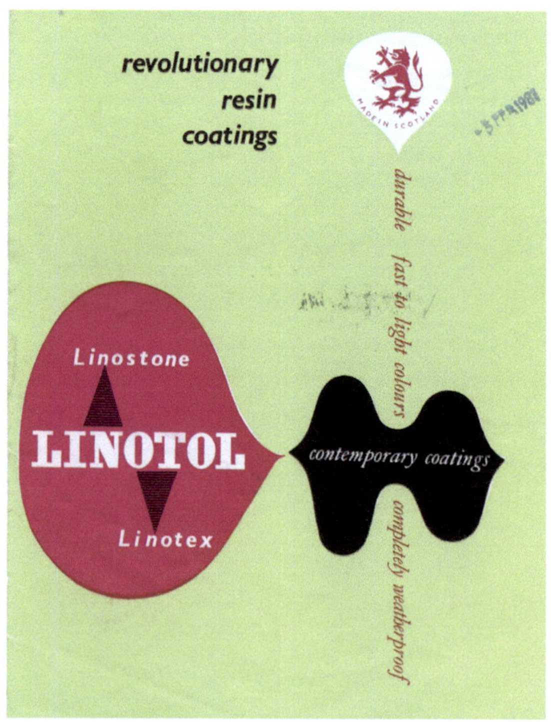

Linotol was a trade name

Stone cracks caused by metal corrosion

One problem that has become increasingly problematic in Glasgow is that embedded cast-iron plates and iron cramps (called dogs) and not always wrought iron, are corroding due to wetter weather, causing the adjacent stonework to crack. This can sometimes occur at the corbelling of oriels and at bay and oriel window lintel joints, resulting in small diagonal cracks caused by metal corrosion of corroded dogs (see Chapter 3, Bays and oriels).

Exposed cast-iron lintels under corbel stone in Glasgow

Around 1900, oriels were tied back into the tenements with thick cast-iron plates which often start to corrode and damage the adjacent stone.

In Edinburgh, around the 1890s, exposed cast iron lintels[9] were often used on the frontage at ground floor windows and close openings. These have not caused a problem, probably because the east of Scotland is a lot drier and the cast iron has only corroded on the surface and does not affect other stones. It is thought they were installed when the tenement was built on soft ground or near a stream, in the belief that any structural movement would cause most damage at the ground floor openings.

In the 1980s and 90s, little was done to bays and oriels; cracks may have been superficially pointed up, but if there was clear evidence of movement, steel ties might be strapped around the bay and tied into the floor. As the stonemasonry skills were not available, some oriels had GRC formwork wrapped around their base to disguise any steelwork or rough stone.

Stone fall

In 2019 there were nearly 180 reports of falling masonry from tenements in Edinburgh and Glasgow has a similar problem, often caused by embedded metal corroding and cracking stones. Owners need to ensure they have common insurance which covers this but more importantly they should regularly check their building for movement and cracks.

Converting old shopfronts into houses

When old shop units ceased to have any trade, housing associations would often buy the unit and convert the shop into a ground floor house. Sometimes this required structural work to support the outer wall. In the case below, a bay had to be converted into an oriel to allow the change to occur.

A glass reinforced concrete (GRC) oriel corbel formed where a ground floor bay has been removed

Upgrading shopfronts

Very little was done to shop units unless the structure of the tenement required major work which required the shop unit to be upgraded. Most of the renewal to old shopfronts has been carried out more recently under Townscape Heritage Initiatives or with a CARS (Conservation Area Regeneration Scheme) (see Appendix 1).

9 It is most likely that these cast-iron front lintels were used because of the risk of ground movement.

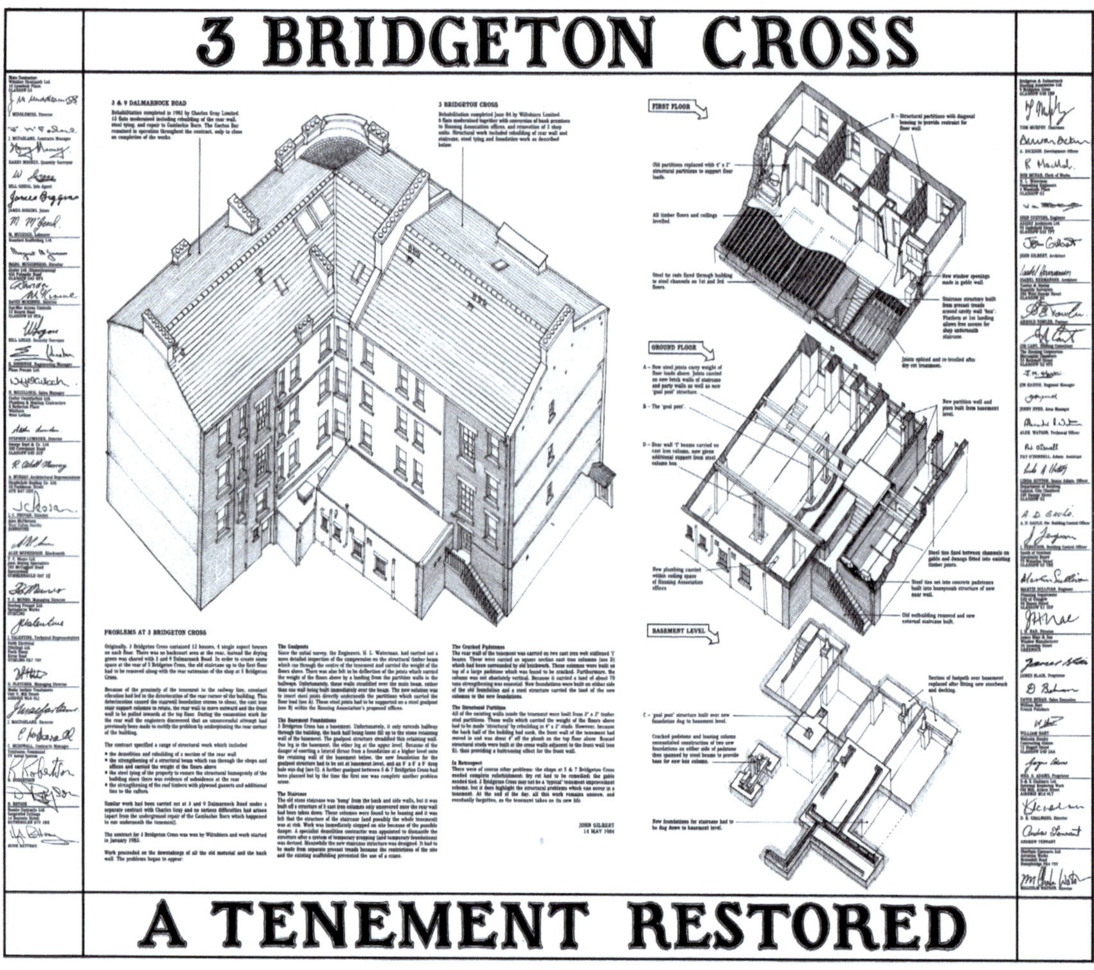

Drawing depicting structural work at 3 Bridgeton Cross, Glasgow, and companies involved

Structural problems in tenements

Settlement continues to be a problem, particularly if a tenement has been built on made up ground. Differential settlement between the main outer walls and the internal walls is common as the loads, and foundations, can be different. Cracking can develop in close floors at the outer wall or in the outer corners of the rear close stair wall. Here the slender internal brick wall is usually not fully keyed into the outer stone wall, so cracking develops.

In some cases the outer wall structure was so bad, the stonework had to be taken down and rebuilt. One tenement at 3 Bridgeton Cross had the back wall, stairs and all internal load-bearing partitions rebuilt with a new steel goalpost structure to support the upper floors. The drawing was signed by the main contractor, subcontractors and suppliers to show the number of people involved in tenement refurbishment.

Movement of foundations

Despite the slender stone footings which were used in Edinburgh New Town, there has been no major sign of foundation movement since the tenements were built, although some localised subsidence has occurred. This was largely dealt with in the 1980s by inserting a cementitious grout into suspect areas.

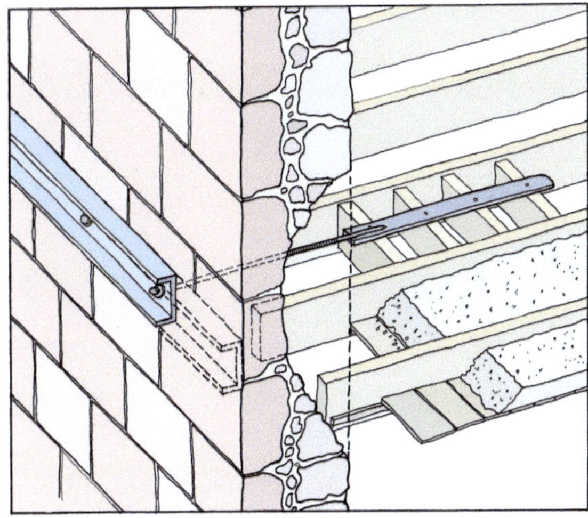

[above] Cementitious grout being injected under a foundation

[upper right] Steel channels to provide stiffening to a tenement block

[right] Steel channel bolt fixing into timber dwangs fitted between joists

Sandstone was usually quarried locally and when local quarries were exhausted, they were then backfilled with the town's rubbish or uncompacted material. Eventually the site might be built on. This is what causes most problems for foundations, so knowing the whereabouts of old quarries or backfilled areas can help to diagnose a structural problem.

As ground conditions varied, so did the foundations, so sometimes large stones or concrete foundations were not even used. In Glasgow in Partick, it was discovered in the late 1970s that a tenement in Rayburn Street had been built on compacted sand, with no signs of settlement.[10]

Glasgow was also beset with shallow mine workings, using the stoop and room technique. The coal would be mined by creating rooms with the ground held up by timber props and columns of coal. Naturally enough, these eventually gave way and subsidence resulted.

It was a big problem in Glasgow in the Govanhill area, when tenements were being rehabilitated in the 1970s and early 1980s. Then the engineering approach was to encircle the tenement structure with steel channels at each floor level, tying the tenements together so they would sink evenly. This was rarely an effective move as much of the steel channels was

10 A leaking water pipe would change that in a year, washing away the sand and causing settlement. There was also a problem of settlement under bays on the south side of Glasgow as underground burns or rivers washed away the subsoil as recounted by architect Matt Bruce.

Cementitious grouting to stoop and room mine workings

not galvanised and once corroded can do more harm to the stone.[11] In some cases these steel channels are covered in a steel mesh and covered with a cement coating to mimic a large string course. Also it did not solve the problem of the risk of subsidence from the mine workings. The steel would be secured to the wall by inserting steel rods through the external wall, then welding to a flat steel plate which was secured to new timber dwangs fixed between the existing floor joists.

In the 1980s a programme of consolidation[12] was approved where a cementitious grout was pumped into the voids of the old 'stoop and room' coal mines in order to reduce the likelihood of further foundation movement.

A more localised problem can occur if a tenement has been built over an old mine shaft. Such shafts were rarely sealed correctly. Instead some large timbers would be laid across the shaft and the ground infilled, only to then be covered by a tenement. Over time the timbers would rot and the ground immediately above would fall into the shaft creating local subsidence in the tenement foundations.

Fire

Tenements are vulnerable to fire and, although serious fires are rare, when they do occur, the whole building can collapse. The only way of rebuilding is to ensure that all owners of a tenement contribute to a common insurance policy which covers fire risk for the whole tenement. Albert Cross in Pollokshields (fire in 2019), Glasgow, was recently badly hit when two fires occurred at different times, leading to massive damage to this unique conservation area. Unfortunately, insurance cover (as it usually is) appears to be insufficient, so any rebuilding will likely be compromised.

Although there is not yet a requirement for fire suppression systems in older tenement flats (though there is a requirement for fire alarm systems), it is still essential to ensure that there is fire protection between each flat and to check that ceilings of shops underneath are sufficiently fireproofed. Also, that front doors are fire doors and that integrated smoke alarms are installed in each flat. The stair must be kept clear of any flammable material and anything that would hinder an escape.

In the loft, one should check the party wall to ensure that there are no holes in it to the adjoining tenement. Party walls were required to be a foot higher than timbers in the roof to provide a fire break between tenements. These walls are the visible 'skews' that form the fire break between each tenement.

11 Sometimes these steel channels were boxed in and rendered with a cement render coating.

12 Mineral consolidation was carried out in the 1990s to pump a cementitious grout into the old mine working voids left from 'stoop and room' coal mines.

[above] Albert Cross in Pollokshields, Glasgow before fire...

[right] ...and after

Chimneys

Chimneys would often require to be rebuilt in brick with new copes. They were invariably in poor condition, and many had to be rebuilt. The brickwork was usually rendered with a smooth cement mortar and lined to mimic stone coursing then coated with 'Linostone'[13] or a coloured mortar would be used. In later periods a metal mesh was used (twil-lath) and a coloured mortar applied over the face. Facing bricks were also used which did not require coating.

Stone chimney copes would be replaced by in situ concrete copings, often without a drip. The copes would be formed by creating a timber shutter and forming the concrete cope in situ. In situ concrete copes rarely have a proper throating, so water may not be thrown off the face of the stone; if it is a very exposed area the stonework can suffer.

Mutual chimney with stainless flue liners, in poor condition by 2018

Because flue bridges were often defective, stainless steel flue liners were sometimes installed from each fireplace. Some housing associations removed stone chimney stalks altogether and allowed stainless-steel flues to rise out through the roof. If these flue pipes were not insulated adjacent to timbers, the pipes could become very hot.

13 Invariably cement mortar and Linostone treatment have failed.

Wallhead chimneys were also often lowered as flues were no longer used, although the flues would still be vented to the room. Chimney pots would be capped with terracotta vents[14] or stainless-steel vent caps.

Use of flues today

After the Clean Air Act in 1956, coal fell from popularity, although 'smokeless fuel' could still be used. It became common for fireplaces in bedrooms and the old stoves in kitchens to be sealed up. Ideally, they should have been vented to help dry out the chimney head and ventilate the house. However, these are often sealed given the need to draughtproof houses, although care should be taken in doing this as ventilating the disused flue can help the chimney dry out as well as providing necessary ventilation to a room.

Flues and chimneys continue to be used with gas fires and sometime gas boilers, although, when this is done, the flues need to be smoke tested to ensure the flue bridges are intact. If they are not, then stainless-steel flue liners are often installed.

Modern balanced flue gas boilers are usually vented via a balanced flue out the back wall. This requires cutting a hole through the stonework with a core cutting drill; this is rarely well done and when installed on front walls can detract from the appearance.

With the need to combat global warming, gas heating will eventually be banned and tenement flats will need to be better insulated and heated by zero carbon appliances.

Rotwork

As the west of Scotland is more affected by increasing rainfall due to climate change, incidences of rot in embedded timbers is becoming more common in the west as well as stone walls not having time to dry out. Some of these repair issues are discussed below, along with the work done in the tenement areas in Glasgow over the last 45 years. Repair work in Edinburgh was different.

Inevitably old tenements suffered a lot of rot problems, not always seen on an initial inspection. As timber was bedded into the outer walls, floor and ceiling joists, wall plates and timber safe lintels are all susceptible to becoming damp; any moisture in the stone can soon lead to decay of such timbers and often travels inside the unvented deafening material to cause dry rot. At the start of most projects, after the initial stripping out of defective ceilings and walls, a survey would be carried out to ascertain the level of rot in the tenement.

One English based surveying firm[15] uses trained rot hounds, Labradors that can sniff out dry rot, but not wet rot, and are successfully used along with a surveyor with a borescope, to investigate hidden dry rot. Embedded timbers such as bressumers, balks and safe lintels would be drilled to check if they were affected. This often led to replacement of floor joists, bressumers, deafening and flooring as well as old lath and plaster linings. The treatment for dry rot would have been to drill the affected walls and irrigate them with a dry rot fluid. Replacement of safe lintels was normally with precast concrete lintels and for large timber beams, rolled steel joists were also used. Treatment specification was laid down by the rot specialist as they were required to issue a guarantee for the treatment which was required by housing associations and mortgage providers.

14 Sometimes referred to as 'elephants' feet'.

15 Hutton and Rostron Environmental Investigations Ltd.

[above] Dry rot on underside of flat roof

[upper right] Repair after rot treatment carried out

[right] Dry rot tendrils on a stone wall found behind plasterboard

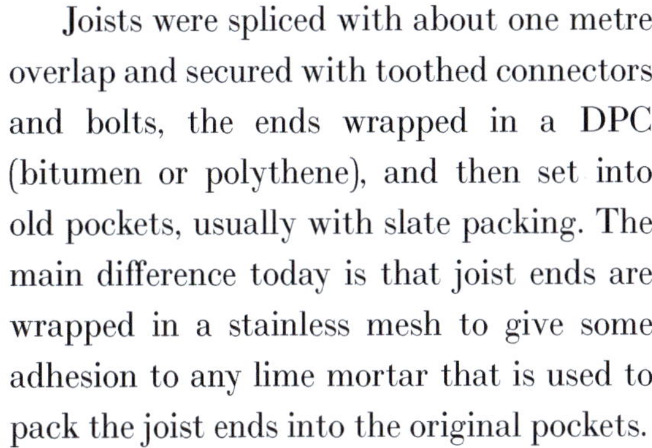

Joists were spliced with about one metre overlap and secured with toothed connectors and bolts, the ends wrapped in a DPC (bitumen or polythene), and then set into old pockets, usually with slate packing. The main difference today is that joist ends are wrapped in a stainless mesh to give some adhesion to any lime mortar that is used to pack the joist ends into the original pockets.

In fact, wet rot and dry rot can be treated by applying heat for a long enough period and, provided the source of moisture has been repaired and the wall dried out, the rot will die off. In tenement improvement work in Copenhagen in the 1980s they enveloped the whole block with a membrane over the entire scaffolding of the block, then heated

Collapse of front due to rot: Albert Road

Front bowing out: Albert Road

[above] The joists at the ends have succumbed to rot

[right] Now rebuilt but without the bays

the tenements to 55°C and kept it warm for at least three weeks. This eradicated all the hidden rot in the walls of the building and avoided the use of toxins.

Because tenements were built with floor joists embedded in outer walls, a wetter climate means they become more vulnerable to rot. Given they provide a timing action to restrain the front wall of a tenement, if not dealt with in time, rot can lead to catastrophic failure, as occurred at a tenement in Albert Road, Glasgow where the front wall had started to bow out. Residents were decanted and eventually the whole of the front facade including both bays had to come down. From inspection one can see that the joist ends that tied the outer wall together were badly rotted, allowing the wall to move out. Thankfully the front facade has been rebuilt in stone, although the bays have not been reformed.

Damp-proof courses (DPCs)

Invariably many external and close walls were treated with a DPC which was usually an injected silicone. Despite many tenements having been built with Caithness slabs as damp-proof courses, the pavement level had risen, or the building had sunk, so they were no longer functioning. Also ventilation of the solum was rarely working. After a 'specialist rot surveyor' would report on a tenement's faults, invariably they would declare rising damp, even if they had not inspected to find the old DPC. Injected DPCs were often used. Inside a house this might mean replastering the walls with a cement render, although most close walls were strapped and then insulated and lined with plasterboard. Other methods were also used with varying effectiveness, such as 'vent plugs' which provided additional ventilation to a wall, and electrical systems which relied on a current being passed through a wall. When solums were low, the ground floor walls might be treated with a corrugated bitumen backing which tried to

Slate DPCs were used in tenements after 1900, often covered by earth or coated with cement

seal the wall, before it was replastered. Damp in walls would sometimes remain well over three feet above floor level, due to the wall not having enough time to dry out (or if the plaster was hygroscopic).

Floors and ceilings

Floorboards were removed, then the deafening and deafening boards and lath and plaster ceilings dropped. The floor and ceilings were replaced by fitting new deafening boards between the joists, then voids filled with dry sand or a mixture of dry sand and plaster. In later periods this changed to using limestone chippings.

Such work was rarely a solution to fully reducing sound transmission and it did not provide sufficient fire protection, so it was then changed with the joist void filled with a glass wool acoustic quilt and two layers of plasterboard on the underside of joists for fire protection. Eventually a lowered independent ceiling spanning between walls was created to support a new ceiling and the void space created, then filled with an acoustic quilt rather than filling a quilt between the original joists.

Today the aim is to reinstate the original sound deafening between the joists as this provided a mass that helps to reduce the sound impact. Ceilings still need to be protected with 30mm of plasterboard to provide fire resistance between flats; the boards are usually plastered with a bonding coat and skim coat of plaster as filling small voids helps to reduce sound transfer.

Where the deafening and existing ceiling is retained but the ceiling has cracks, it was common to fix timber branders[16] to the underside of the ceiling and then plate the ceiling with plasterboard and skim coat it. Also, where the existing floor was retained, an additional floating floor was sometimes added over it, set onto resilient battens which helped to reduce impact noises and also allowed central heating pipes to be installed easily, although it did require adjustment to the skirtings and door heights.

Noise between flats

Noise between flats[17] is a recurring problem, particularly with the fashion to have bare floorboards or laminate flooring, when sound is transmitted by footfall.

Noise is transmitted through walls and floors and can be either airborne or impact.

Any new or refurbished flat will have to undergo a sound test under part E of the Scottish Building Regulations. Typically, such a test would provide two airborne wall tests; two airborne floor tests and two impact floor tests. If the test is passed, the flat is approved for occupation. However, flats refurbished before 1990 were not required to have sound tests, and for existing tenements and flats there is no requirement for sound testing.[18]

When flats failed a test, it could be for any number of reasons, although the testing process would show if the problem was airborne or impact noise. For example, a steel support beam within the depth of a floor could lead to failure of the test. To avoid such problems, sound tests might be carried out early, both before work started and before the end of the contract. A

16 Usually about 45 × 25mm battens screw fixed through the old plaster ceiling into the joists and sometimes packed if the original ceiling needed to be levelled. Often referred to as a *'brandered'* ceiling.

17 Refer to www.underoneroof.scot for advice on this topic.

18 Sound testing in Glasgow started around 1987.

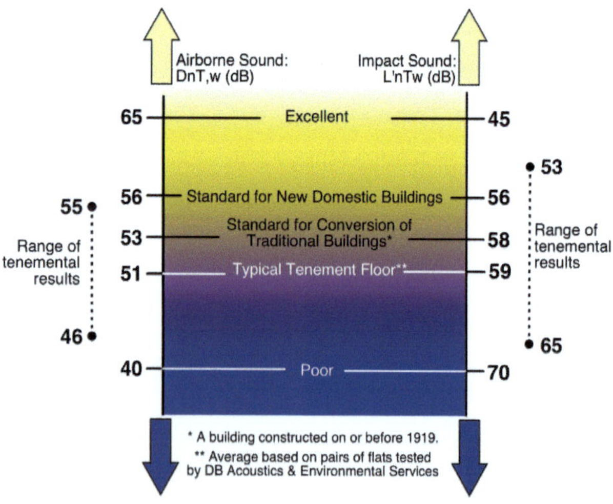

Diagram showing airborne and impact sound limits found in tenements

The problem!

From the CIRIA report

typical remedial solution used a product which comprised hardboard over carpet felt with lead between, which was nailed to the floor.[19]

In Victorian tenements mostly in the west of Scotland the space between the joists was filled with deafening boards (see Chapter 3 photo 3.11) and then a mix of fuel ash and lime was laid over the boards to a depth of about 70mm. When intact this works quite well but sometimes the space between wall and joist is not backfilled and if old cornices have been removed, the sound can travel through this void. The whole construction of the original floor and wall junctions contributes to the effective sound reduction, with large heavy cornices and lath and plaster. Such materials were often lost during rehabilitation.

In the tenements of Georgian Edinburgh it is more likely that the floor construction is slightly different. There is still deafening material between the joists, but the ceiling below is hung from ties and framing, so it is not fixed directly to the floor joists. This provides slightly better soundproofing.

However, if the original sound deafening has been dislodged, or even removed, then this can have a big impact on the noise transfer through the floor. The deafening is sometimes removed by plumbers to fit pipes under the floor, by electricians to feed cables through, or after a flood. Often it is not replaced. Sometimes it is replaced with glass wool quilt in the mistaken belief that this will attenuate sound. It does not.

Mass is important as the mass of a floor helps to reduce impact noise.

Before 1980, radios and hi-fi systems were not as powerful as now, and fewer people worked different hours, so sound insulation wasn't so much of a problem. People also used carpets a lot. As old deafening was compromised, music became louder, with residents frequently working at different times; the trend away from carpets exacerbated the problem.

In the 1980s and 90s, tenement refurbishment carried out by housing associations, often involved dropping all the ceilings and removing the old deafening. Independent ceilings

19 It surprisingly worked!', John P LeHarivel.

which were timber studs fixed to the side walls (and not suspended from the joists above) were often used to combat noise transfer, often with a glass wool quilt above. A report by CIRIA[20] in 1987 was influential in improving the sound insulation in tenements.

Flanking noise can occur both with airborne noise and impact noise. Sound is transmitted in a wall and carried up or down that wall through vibrations. In a tenement, the weak parts of a party wall are at the presses and bed recesses as the wall behind a press is usually quite thin. Walls adjacent to closes are often only 4½ inches (114mm) thick, so sound from the close can transmit to rooms next to the close.

Asbestos

Where original walls and ceilings (and stair soffits) were retained, they might be decorated in woodchip paper to disguise unevenness. However, more often they received a coat of Artex, which was a white textured coating that covered up old cracks. Artex usually contains 1–4 per cent of chrysotile asbestos which is also known as white asbestos. It was used extensively in the 1960s through to the 1990s. Not all Artex contains asbestos as 'asbestos-free' Artex was available in the 1970s but all will need to be tested.

Asbestos ceiling tiles were also available and often applied by residents to ceilings with a paste.

Asbestos cement boards were also used in some areas, usually at fireplaces of boiler installations. Although not used in tenement refurbishment, if work was done in a house before 1970, then there is a chance that some boards may be present.

Asbestos cement gutters and vents were also made, but little used in tenement construction.

At present, if any owner is planning to refurbish their flat or tenement, they first must commission an asbestos survey of the affected areas to include in the building's safety plan (unless the work is for minor decorative tasks).

Internal and close wall insulation

Where wall linings had to be removed due to rot, a new stud framework was built, usually using 2 in. × 2in. studs and glass wool insulation inserted in the space between the studs. A vapour barrier was then applied and then a sheet of plasterboard which might be Ames taped. The timber studs were held off the outer wall by vine ties. Thermal boards were also used on external walls.

Walls adjacent to close walls, at the close stairwell, were usually single brick. They might be insulated in a similar way but using a foil backed plasterboard which provided a vapour check. If space was tight a thermal insulation board might be fixed to timber battens secured to the close wall. In slightly later periods, these thermal boards might be bedded into place with dabs of plaster. The thermal boards usually were made from polystyrene or polyurethane and bonded to a plasterboard.

The facings and linings around the windows were often replaced when the new windows were installed, but rarely insulated. Little or no account was directed at draughtproofing.

Loft insulation

Insulation was applied to most loft spaces in the 1970s and 80s, using glass wool (or

20 CIRIA Practice Note 46 1987, 'Sound Insulation of timber floors in rehabilitated Scottish Tenements'.

sometimes blown cellulose fibre or even Rockwool insulation) although only to about 100 mm thickness. In most cases additional insulation added under home energy grants has since been applied. There was a time when concern was expressed about lung disease caused by the small fibres in glass fibre and in the 1990s some manufactures made a change to fibre sizes.

Sundry problems

Other interventions that were common around the 1980–90 period which should be looked out for include the following:

Skirtings and door facings were all renewed, and door frames also altered because new flush doors were often a different size to any original door opening. H-frames, previously discussed, had to be removed and new precast concrete lintels formed into repaired brick walls. As new bricks were often used, the coursing did not always bond in with the existing coursing.

Windows were also removed and replaced with new timber pivot windows.[21] These replacement windows were poorly made and rarely lasted. Most have since been replaced with new timber fully reversible windows and double-glazed. In conservation areas, sash and case windows were used although sometimes fully reversible were accepted if they mimicked the sash and case profile.

Skew flashings were usually replaced with a bitumen material such as Nuralite although if lead was used it was sometimes painted black to deter thieves. The skew flashing is a common cause of damp penetration, because water is focused here. Lead flashings are often poorly laid with not enough allowance for expansion and poor clipping of cover flashings. Mortar flashings avoid the need to have a secret gutter but tend to do best in the east where there is less rainfall.

Close and close stairs. As they were usually worn they would be resurfaced with a granolithic mix or an epoxy concrete mix; a proprietary product was called Stonehard. In later periods they might be recovered with Veitchi floorings Linotol coating, which came in a variety of marbled colours and was appealing to residents as it appeared easy to clean. After a corner hanging stair collapsed in Woodlands Road, the council inspected other stairs and decided to add steel T bars to the underside of all hanging stairs in Glasgow.

Close stair balusters and handrails were restored as far as possible. It is still possible to obtain replica cast-iron balusters to replace broken ones. External stairs are more problematic. Long stone treads need be set onto cast-iron runners which span across dunny areas (in older tenements, stone arched supports were formed). The worn stone can be carefully indented or replaced. The railings, which are often cast iron, are vulnerable to cracking and, over time, poor repairs may have been carried out. Such handrails are important as they protect the public from falling into lower basement areas. Some owners in Glasgow have been sued by a student who was badly injured when he leant against a cast-iron railing and fell into the dunny below.

Close walls sometimes had to be replastered. This was usually done with a cement render and other walls with a renovating plaster.

21 The early pivot windows were of poor quality and most have now been renewed with new double-glazed fully reversible timber windows.

Stair collapse in Woodlands Road

This led to steelwork being installed under such hanging stairs

[left] Hot water cylinder with immerser

[right] It would also be vented to the roof (a failsafe device). Breached 'coffin' tanks in loft (each one connected to another)

Electrics. All the electrics would have been replaced using PVC covered cables and installing new consumer units, although RCDs[22] were not fitted originally. The provision for sockets would now also be much increased. TV cabling would have been to a terrestrial communal system. If you are the householder of a flat, the electrics should be tested every 10 years. If you are a landlord additional safety checks are required every three years.

Water supply. This was invariably a lead supply coming from a water toby[23] in the pavement. It fed large common storage tanks in the loft which were made from 30mm thick wood carcass with tongued and rebated edges and then lead lined. There would be one tank for each of the four flats directly below. The feed was terminated with a floating ballcock and there would be a lead overflow pipe flowing to the outside below the gutter. Originally the tank would have had lids and been vented to the roof. The water supply pipes were renewed with polyethylene, although copper was originally used. The water tanks were removed and the lead resold by the builders as 'bunce'.[24]

The tanks were replaced with glass fibre coffin tanks breached together, usually four tanks per stack, and then insulated and covered with lids. Supply was still controlled by a single ball cock but there would be an additional stop valve to shut off the water. The overflow would also be renewed in plastic. Plastic water tanks eventually came into use and sometimes the old lead lined tank was retained and just relined with glass fibre material. Individual cold feeds went from the breached tanks to each flat where they supplied water to the hot water cylinder, toilet cistern and other fittings, although the kitchen sink was fed directly from the rising main.

Gas and heating. Any gas fire that remained would require to be tested on removal and the original flue chamber revised to the Gas Board's design for a square flue box. Also sufficient trickle ventilation was required. Each flue would be smoke tested and if the flues failed then a stainless-steel flexible flue liner would be installed. Gas fired back boilers made by Baxi were

22 RCD stands for residual current device, while RCB stands for residual current breaker.

23 Tobies are under pavement stop valves located in the pavement at the front of a tenement. The whole water supply can be shut off by inserting a water key into the toby and giving it a quarter turn.

24 Booty – material that can be removed from the building site and sold.

sometimes installed in front living areas. These would have a circulating gravity fed pipe to and from the hot water cylinder which might also supply some heat to a radiator. Originally the kitchen would have had a cast-iron stove for cooking and this was often linked to heat the hot water cylinder. However, nearly all these stoves had been removed by 1970. Gas lights were used to light the closes in Glasgow up to the 1960s – the last gas light turned off in 1971.

Hot water cylinders were replaced with new copper cylinders and electric immersion heaters. Hot water cylinders were originally found in the old kitchen area in the outer corner above the press. They were often supported on a steel angle frame and by a steel rod hanger fixed to the underside of the joist. The new cylinders were often relocated above the lowered ceiling formed above any new bathroom. In some locations space was so tight that breached horizontal tanks had to be used above ceilings. When installed in narrow wall presses, thin vertically paired Doublo tanks were used (in lofts). It was only after 1990 that the larger combined open system tanks were used.

Drainage. A new drainage system was usually required. Inside the tenement this would usually have been a PVC soil pipe passing through each intermediate floor and sometimes fire stopped, although not always. These internal soil pipes might drain to join into the centre of the close where a new manhole would be built in brick with a cast-iron lid, although sometimes the soil pipes travelled under the ground floor to the street.

A rear disconnecting manhole would be formed at the backcourt to link any rear downpipes into the drainage system.

A drainage problem can best be obtained by a CCTV survey of the drains to ascertain their condition and where any blockages occur. Plants are a common cause of drain blockage as plant roots seek out a source of water and can drive through narrow gaps at pipe joints, eventually growing and cracking the pipes. All rain water pipes should have an access eye at the base of the pipe,[25] which allows a CCTV camera, or rodding set, to be inserted.

More commonly, cast-iron downpipes are neglected and don't get painted, particularly at the rear of the pipe. If water spills out of the downpipe it can saturate the wall at a point where timber may be embedded, so ensure rainwater gets away from a tenement as swiftly as possible.

Kitchens and bathrooms

New kitchens

These were usually formed in the space originally provided for a WC which was very narrow. The kitchen was formed into a galley kitchen opening out into the rear room which became a living area and the front room became the bedroom. If there was an old narrow bathroom, this was often converted into a kitchen and the new bathroom installed in one of the bed recesses. Internal kitchens without daylight were not approved.

When there wasn't an adjoining room, the kitchen might have been installed in the front bed recess with a low-level counter to 'hide it'. Ventilation ducts would still be installed and the waste connected to the soil pipe in the bathroom. A storage cupboard above would sometimes house a hot water cylinder, if there was room.

25 If there are no access eyes, and you have cast-iron downpipes, your plumber can core drill a hole into the pipe and fit a removable rubber stopper to form an accessible access eye.

[left] A galley kitchen opened into living space and installed where there would have been an old bathroom

[right] In the front parlour bed recess

New bathrooms

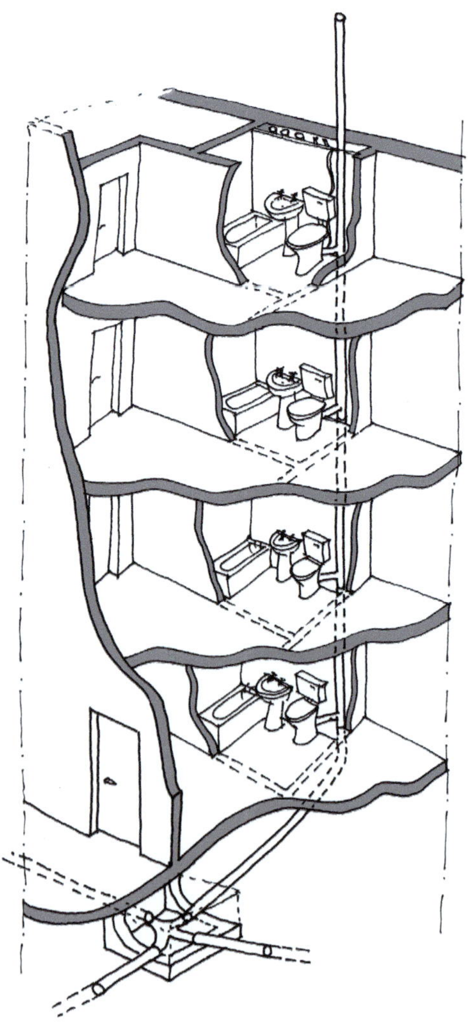

Bathroom stack in bed recess

If you want to know the full story of how tenements in Glasgow were saved by installing bathrooms and kitchens, it is well worth reading Raymond Young's book *Annie's Loo* (2013), as mentioned previously. It started around the time Raymond Young decided to do a thesis on design participation at Strathclyde University's architecture school when Jim Johnson was his tutor. Raymond had been living in Govan and doing voluntary work with the New Govan Society Task Force. In 1971 a tenement improvement project survey in Govan showed that people wanted to stay in their houses rather than be decanted as happened when the council improved the Old Swan area. In 1969, Peter Robinson, also a Strathclyde student, showed how tenements could be improved by installing new bathrooms in the bed recess space. A mock-up demonstration flat was prepared at Gourlay Street which was a Christian Action Housing Association flat. A year later a residents' association was formed in Govan and a department of the university created called ASSIST (Architectural Students Improve Scottish Tenements) with an office in the main street of Govan. Raymond was appointed with Jim Johnson as director along with a part-time secretary.

Eventually, after many meetings, plans were prepared with the residents to improve their flats and use improvement grants. The process was driven by the community, a different approach to the 'take it or leave it' approach previously offered by the council. The first project allowed residents to see what was possible. Eventually Govan Housing Association was set up in 1971 as the first community-based housing association in Scotland, and, with funding channelled through the Housing Corporation, more community-based housing associations grew to help improve their areas using improvement and repair grants.

Although the plan proposals look quite simple, what was different was the process that was involved in getting plans agreed with each individual resident, as this was essential in tenements where a layout of one flat might affect what happened in the flat below. As housing associations have the power to buy out owners, it was sometimes possible to do away with single ends and amalgamate these small single aspect flats into the two adjacent flats. After Govan Housing Association was set up, Govanhill and Denniston formed their own housing associations.

Bathrooms were not always installed in bed recesses, but this was the only location possible in many flats as old bathroom spaces, which were very narrow, were converted into *galley kitchens* off living rooms. In some flats only a shower installation was possible. In out of the way places, an air admittance valve was used as the termination would be far distant from the supply (a 'loop vent' or 'air cushion' as it was commonly called).

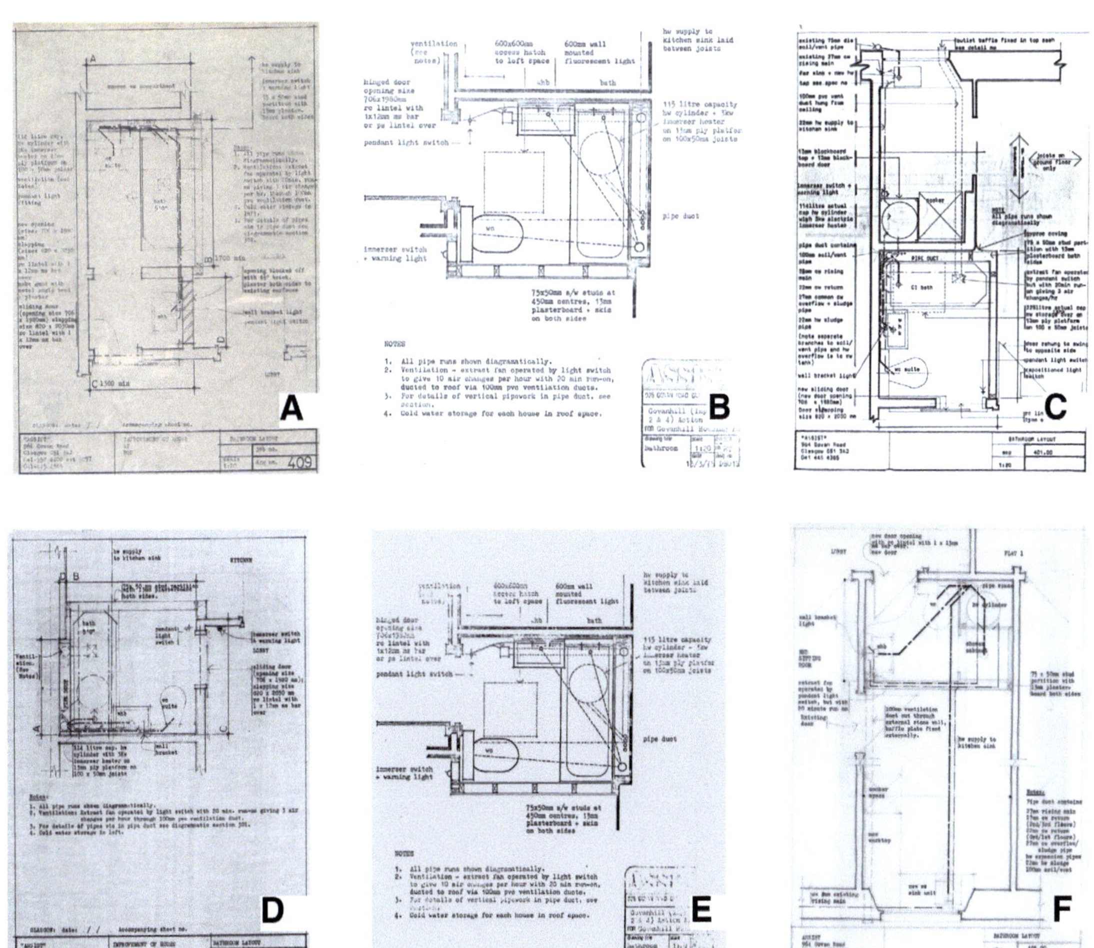

Bathroom plans provided by ASSIST Architects

When installed in bed recesses, one above the other, it was easy to plumb. A single UPVC 100 mm soil pipe which would be vented at the head, would drop through each floor (fire stopped at each level) and connect to each individual WC as well as taking the waste from washbasin and bath or shower. In the solum, the drainage run might connect to the main drain under the close after a manhole was built to take the new soil stacks. Sometimes it might go under the ground floor flat and through the front wall, then connect to a new manhole in the garden or pavement.

The space in the bed recess was formed by building a timber partition across the bed recess area (see Chapter 7 page 191) and forming a new door from the hallway. A lower ceiling was also formed creating a high-level cupboard accessed from the old kitchen. This was floored and often contained a new hot water cylinder.

A standard sized bath of 1700mm could often be installed although sometimes a smaller 1550 mm bath had to be used. Only rarely were showers provided and never over a bath. A single row of tiles was installed around the bath edge and at the washbasin. The bath was boxed in with ply panels or tongue and groove planks.

Many flats were small, so the door into the bathroom often had to be formed as a sliding door to save on space.

The only lighting provided was a ceiling mounted fluorescent light controlled by a pull cord. This also controlled the extractor fan which would run on for 20 minutes (until later ones were developed with humidistat fans).

Ventilation systems

As internal bathrooms were not daylit, they had to be ventilated. This was done by installing a vent pipe in the same duct formed to house the soil pipe. Housing associations often took different approaches:

- A common duct system called a Twinfan, which had a large common fan in the loft which pulled the air up and smaller individual fans which would 'push' the air up the duct and be controlled by the tenant.
- Individual paired ducts with a fan that pushed the air up the duct. Two flats (no more) could share the same duct and duct connections made using a shunt duct system which made connection to the extract duct a floor above to limit noise transfer.

[left] Nu-Twinfan

[right] Air extractor from everyone's bathroom

In both cases condensation within the ducts was an issue so condensate drains had to be introduced usually draining into the bath overflow.

Top floor flats could be vented by a single duct using a passive stack system, often with a humidistat-controlled vent grill.

Environmental work

In the 1970s, there was a programme of environmental improvements which gave out grants to allow backcourts to be improved, creating new railings and fences, landscaping and building new bin stores along with paths and better drainage. Old factories and workshops in the centre of backcourts were often demolished during this period.

The grants were provided by the private sector of Glasgow District Council and were for 100 per cent. Residents were expected to form residents' associations to then appoint an architect to prepare plans and manage the contract. Most residents decided on forming individual backcourts to create defensible spaces around their own close, although in some backs a communal area was formed and landscaped. Sometimes a central area was made into a communal area or playground, although most residents tried to avoid playgrounds as they were concerned about antisocial behaviour. Trees, which were proposed, were often declined because residents wanted to avoid leaves!

Drainage was improved in each of the closes, and retaining walls, steps and paths were built, along with drying areas and low-level bin stores to contain the old metal bins. Metal railings were added between closes and new gates were formed into backcourt areas (where such openings existed). Over time, the bin stores have not kept up with the need for recycling and the use of taller plastic bins, and some backs which are totally enclosed have had to resort to large street bins.

Stone cleaning

Stone cleaning was funded by environmental improvements. It is regrettable how much of Glasgow has been cleaned, as the damage caused is extensive. We started to grit blast tenements, and then people complained that they felt unwell, so wet grit blasting came in. But operatives were rarely well trained and the stone surface was often over-blasted, causing much damage to the ashlar faces and arises.[26] It was still the same blasting techniques used, but it was much safer. In some tenements, they did chemical cleaning. Some of it was to remove the old paintwork. This too was annoying.

Today we have steam heated water droplets, largely to remove paint or Linostone, using a much less invasive system which employs steam under pressure,[27] commonly referred to as the DOFF system.

Despite all this, there is, nevertheless, something to be said for cleaning buildings. It can give them a new lease of life and, finally, tenement refurbishment has caught up.

Ashlar sandstone with marks left by excessive use of grit blasting

26 An ariss (aris) is the sharp exposed edge of ashlar stone.

27 The 'Doff' System is used to clean paint and stains off stone. It is a steam-based system, heating water to 150°C which helps to remove the paint. As it does not involve grit, it does less harm to the stone. For conservation cleaning of stone details, laser cleaning is possible but very slow.

Stone repointing

The lifetime of stonework is dependent on the pointing. Most of the pointing would generally have a life expectancy of 70 years.[28]

If your tenement has not been repointed before, all rotten or deteriorated mortar should be carefully cut out and raked from the building. Some of this will involve the loosening of a number of smaller wedging stones, or pinning stones. These must be set aside and reused when the new mortar is prepared. The walls are washed with water wash, and the core face of the stone well watered. A lime mortar, preferably matching the original, should be mixed and ready for pointing. The loose stone should be raked into the open joint to ensure the remaining stonework has enough strength for the original building.

Where stone ashlar is evident, pointing may be required. If the lime mortar is present, it may be used or cut out to allow a lime mortar to be injected in the void. New putty lime should be made and this should match the original. This was rarely done in the 1970s to 1990s.

The conservation of tenements

We have long passed the time when old tenements were considered obsolete. Despite the considerable demolition that followed the Comprehensive Development Areas and planning from the 1950s and 60s, tenements have remained as a form of housing suited to modern urban living. The carbon savings in keeping older structures is considerable in comparison to demolition and new build. Older stone tenements have largely been upgraded, but much still needs to be done to upgrade insulation and heating systems, as well as fabric maintenance.

Most of the recommendations and approaches that exist in conservation areas also apply to unlisted Georgian and Victorian tenements that are not within conservation areas. In particular, attention should be paid to how we maintain the stone fabric, the need to ensure gutters and downpipes are fit for purpose, keeping the fabric dry and able to dry out to prevent rot and how we upgrade the external walls, floors and ceilings with insulation. Existing older stone tenements tend to be in inner urban areas, so they are close to services and public transport. Because of this, they are often favoured by the young and old who need public transport.

Scotland has 600 conservation areas, many designated in the early 1970s and then later in 1984 (see Appendix 1 for conservation areas today). They are not all tenement areas, but also cover historic land, battlefields, public parks, landscapes and infrastructure such as railways and canals. Glasgow has 25; Edinburgh has 49; Dundee 17; Inverness and Aberdeen 11 each; Inverclyde and Renfrewshire each with 8.

Those tenements that escaped demolition and were not part of a Housing Action Area managed to survive and some early conservation areas and listed buildings were lucky enough to get reslated under any improvement works, but in Glasgow that was quite rare at the time.

Although conservation areas are designated by councils after an assessment, they very much rely on the responsibility and goodwill of the inhabitants. In such areas, before 2012,

28 Ingval Maxwell 'Stone: the changing perception of traditional build', in *Materials and Traditions in Scottish Building*.

New slating being prepared for Greenhead Street tenements in Glasgow

the replacement of traditional features could be made without formal approval unless they had specific protection with an Article 4 direction.[29] Since that date, all changes require planning consent.

East Pollokshields was designated a conservation area in 1973 (extended in 1984), but planning controls have proved limited, given the amount of UPVC windows[30] and concrete tiles on roofs; planning constraints were not enforced.

If someone installs a UPVC window without permission, the council can ask the owner to apply for retrospective approval but when it is submitted and the council refuse permission, the owner can appeal to the local review body, a sub-group of the planning applications committee. Local councillors then often approve the windows and the more people that have them, the harder it becomes to enforce the conservation requirements.

29 An Article 4 direction can cover a variety of minor works which might include replacement of windows or doors, erection of fences or installation of satellite dishes.

30 One suspects that councillors living in houses elsewhere with UPVC windows wondered why East Pollokshields should be denied them. Enforcement was poor in East Pollokshields, unlike Edinburgh.

Pollokshields conservation area with UPVC windows

Despite these changes over the past 50 years, the sandstone tenement has survived and is now valued, although much still needs to be done to ensure there is a good maintenance system for tenement owners and a legal system that ensures owners carry out regular inspections, form stair associations and contribute to a sinking fund and of course, pay for their common insurance.

Carry out your own survey

Chapter 12
Tenements fit for the future

Inspect it yourself

The climate in Scotland is changing, showing a strong trend towards an increase in rainfall and extreme weather events, with milder but wetter winters and hotter summers.

About 40 per cent of Scotland's total carbon emissions come from domestic energy consumption and almost 20 per cent of all buildings are of traditional construction. Improving the energy efficiency of these buildings will be necessary to meet carbon reduction commitments required to slow this trend, and ideally prevent it. While we wait for our government to cease mining gas and oil from the North Sea and stop subsidising the fossil fuel industry, we can, as responsible owners or landlords, take action by improving the insulation of our homes.

We discuss here the different approaches to tackling energy efficiency, although it is essential that, before trying to improve insulation and draughtproofing, tenement owners should ensure that their tenement is in good repair. To do this, we think that the best approach is to form an owners' association[1] for all the owners, with a common bank account, and commission a full survey of the common parts of their tenement at least every five years. From this, a plan of action can be developed: to ensure the roof is intact and that water can be effectively drained off it particularly when there is a deluge; to check the fabric – stone walls that are wet lose much of their insulation value, so it's important to ensure the stonework is well maintained, protected and correctly pointed; and to check that the gutters and downpipes are adequately sized to cope with the increase in rainfall. Once you can be confident that the fabric is in good condition, then owners can be assured that their energy-efficient improvements will help to maximise energy saving and reduce carbon. You can start by inspecting your building with a neighbour. Help in identifying defects can be found on www.underoneroof.scot.

These following measures are considered to be suitable for all pre-1919 tenements, although, as each tenement is different, you would be best advised to appoint a professional advisor such as a building surveyor or an architect. There is now an SQA[2] award available to

1 Also referred to as a 'stair association'.

2 The Scottish Qualifications Authority (SQA; Gaelic: Ùghdarras Theisteanas na h-Alba) is the executive non-departmental public body of the Scottish Government responsible for accrediting educational awards.

professionals on 'Energy efficiency measures for older and traditional buildings'. Those who have completed this course should be better placed to advise on potential measures.

In clause 6.2.8 of the Scottish Building Standards,[3] the conversion of historic, listed or traditional buildings is discussed in detail. It should be emphasised that this clause relates to all buildings of traditional construction and not just those which are listed or of perceived heritage value. This section of the building standards notes that '*whilst achieving the U-values recommended previously should be the aim of any work, a flexible approach to improvement should be taken based on investigation of traditional construction, form and character of the building in question and the applicability of improvement methods to that construction.*'

Historic Environment Scotland has published *Fabric Improvements for Energy Efficiency of Traditional Buildings*, which is well worth reading.[4] It explores a range of potential energy efficiency measures, what their efficacy might be and how challenging they are to install. They have also produced a series of refurbishment case studies[5] which review the effectiveness of the work undertaken, carried out by an independent firm of surveyors.

Reducing heat loss in a tenement

Retrofitting tenements with insulation lowers their energy demand and reduces both CO_2 emissions and energy consumption, thus leading to lowering the environmental impact of our homes. It may be more effective to improve windows, glazing or draughtproofing before venturing to increase internal insulation of the wall. However, recent research on the U-values of stone walls suggest they may not be as bad as EPC[6] generic estimates of such solid walls. There is also evidence that, beyond a certain thickness, the benefit gained from insulation begins to drop off, whilst the risks associated with installing internal insulation to solid walls increases. The most effective way to improve the fabric's capacity to retain heat is to add a continuous layer of insulation. Nevertheless, thicker layers of internal insulation can also increase the risk of interstitial condensation and timber rot if not installed correctly (external wall insulation does not increase such risks). Also, any tenement flat will gain more insulation if all the surrounding flats are upgraded, along with floors, lofts and close walls, rather than insulation confined to one flat. Energy Saving Trust have more details on their website (https://energysavingtrust.org.uk).

A traditional stone tenement can be insulated externally or internally. Externally insulating our stone tenements is rarely appropriate for sandstone fronted buildings, although it may be considered for rear and some gable elevations. The advantage of externally insulating a wall is that it is much less disruptive to residents and by applying insulation outside, the thermal mass of the stone walls can usefully retain heat, as long as it is warmer than the outside; other benefits are that it protects the existing structure and walls and it is easier to mitigate the effects of thermal bridges. Two drawbacks are:

- the roof eaves will usually need to be extended
- downpipes and soil pipes must be adjusted so that the insulation avoids thermal bridges, also known as cold bridges.

---

3 Building Standards Technical Handbook 2022: domestic.
4 *Fabric Improvement for Energy Efficiency in Traditional Buildings; Short Guide 1* (2012).
5 *Climate Change Adaptation for Traditional Buildings: Short Guide 11* (2016).
6 EPC Energy Performance Certificate.

Internal wall insulation preserves the exterior appearance of the building, which will almost always be a requirement of the local authority's planning department for the primary elevations of the building. The downsides of insulating from inside are that achieving a continuous insulation line can be challenging, which means that thermal bridges are generally not fully mitigated. It can also reduce the internal floor area, and is more likely to necessitate the removal of existing fixtures and fittings.

There are also building fabric risks when installing internal insulation, primarily the accumulation of moisture within the wall and the damage this can cause, such as decay of any embedded timbers and frost damage to the stonework as well as biological growth. This can happen for two reasons: firstly, the reduced permeability of some insulants can impede the drying out of the wall by preventing it from drying towards the inside; secondly, when the wall temperature is lowered, because of the insulation, it will get colder and move the dew point into the wall, thus increasing the risk of frost damage which can cause damage to the stone wall (or any timbers in the wall). To avoid these risks, it is important that vapour open insulations are utilised, and that any proposed measures are first checked by a professional. Wet walls are much poorer insulants than dry walls.

Draughtproofing

A lot of energy can be lost through ill-fitting doors and windows, but also at the junction between the window and the opening. This gap is normally filled with linseed oil sand mastic or a polysulphide mastic, but this can break down, so, as well as pointing the gap with mastic on the outside, it's better to try and seal the window case or frame to the stonework with a very sticky tape on the inside. Some gaps can be backfilled with sheep's wool or hemp. Clearly this is only possible when windows are first installed, otherwise any tape sealing may have to be done between the window and over the existing wall finishes.

Gaps at skirtings can be improved by sealants, although they rarely last. It may be better to apply a small moulding or additional skirting with a rubber draught strip between the existing skirting and the floor.

Ground floors can be draughtproofed at the same time they are insulated, although it is best to remove the existing floorboards before laying any vapour permeable membrane over any insulation and ensuring it is taped at all edges (see insulating ground floors below).

Open fireplaces that are no longer in use can be draughtproofed with sheep wool insulation, which still allows vapour transfer, provided the fireplace is no longer used.

Most flat entrance doors will already have smoke seals, so they should be draughtproof, but these should be checked for effectiveness.

Similarly, windows that are leaking should have their draught strips replaced and sealed up with sheep's wool or an insulant.

Indoor air quality

Adequate ventilation should be maintained to ensure the health of the building and its occupants. This might mean using existing chimney flues and other traditional features to reintroduce natural ventilation routes. Typically, a pre-1919 tenement achieves around thirteen to fifteen air changes per hour (when pressurised to 50 pascals); improving the air-tightness of properties helps to improve their energy efficiency and comfort levels.

A recent Historic Environment Scotland (HES) case study[7] achieved an air leakage reduction from eighteen to eight air changes per hour, using the measures described in their guide. This degree of airtightness challenges the assumption that old buildings must be draughty. A balance needs to be struck between improving the airtightness of a building and reducing airflow to the point at which poor internal conditions result.

Where high levels of insulation or draughtproofing are included in any refurbishment, additional ventilation should be introduced, ideally with heat recovery to minimise the energy lost in refreshing the indoor air. The most efficient of these systems is MVHR (mechanical ventilation with heat recovery); however, to be really effective, the airtightness will ideally be improved to less than five air changes per hour (ACH).[8]

Internally insulated external stone walls

Traditional buildings are often referred to as being constructed of breathable construction. This acknowledges the fact that the materials used for their construction have the ability to absorb and release moisture.

There is little agreement on the best method of internally insulating sandstone tenements, but two main approaches are vapour tight and vapour permeable systems. Both will perform if properly applied although since the original structure is vapour permeable we think this is the best approach to take. The illustrations below show the main approaches that can be

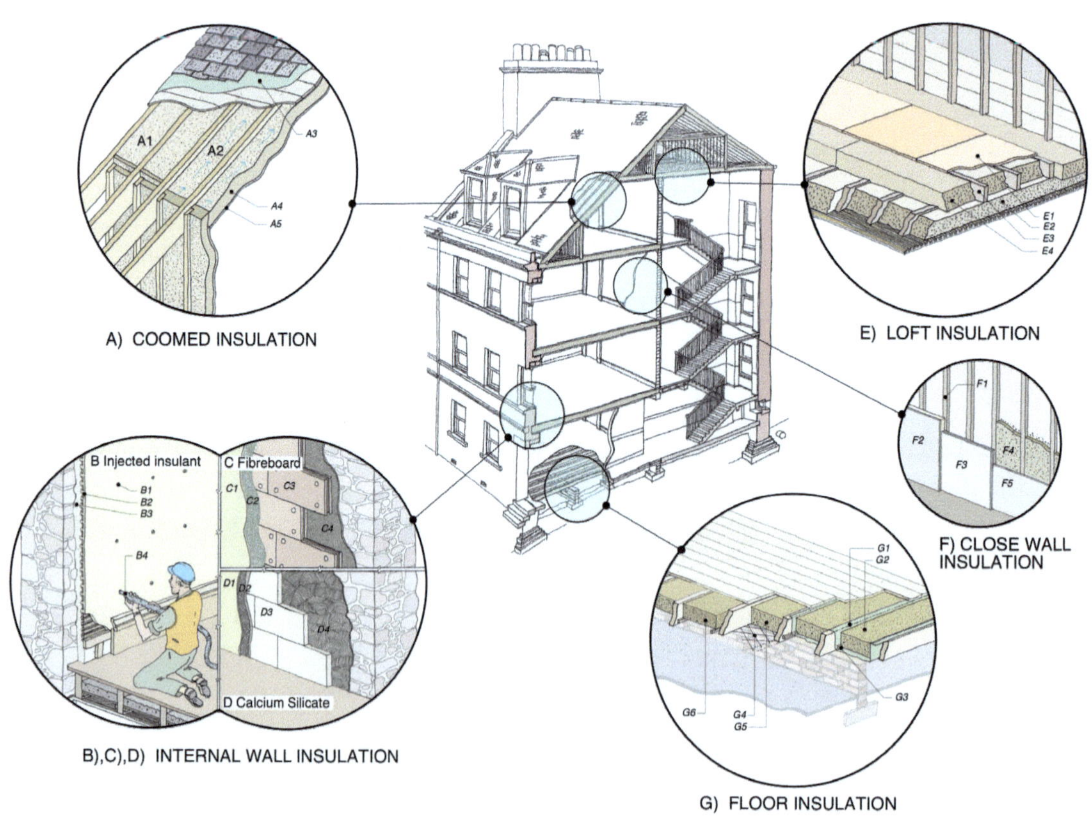

7 Historic Scotland *Technical Paper 6: Indoor air quality and energy efficiency in traditional buildings*.

8 Any heat recovery system should be carefully designed and installed by competent professionals/ installers.

taken when insulating lofts, coombed ceilings, internal walls, close walls and ground floors. Some of the main systems are described below.

A) Coombed insulation

A1 Provided the roof membrane is vapour permeable, the void between rafters can be filled with a vapour permeable insulation such as wood fibre, hemp, cellulose.
A2 As many roofs will still have bitumen felts which are not vapour permeable, it is advised to leave a 50 mm air gap to ventilate the void between insulant and sarking. Wood fibre or other vapour permeable insulants may be used.
A3 A vapour permeable membrane allows the water vapour to disperse.
A4 Wood fibre insulation cut to form a tight fit between rafters.
A5 Insulant held in place by a wood fibre board which interlocks and is secured to each rafter.

B) Insulating an existing lath and plaster wall

See main text for issues regarding this process.

B1 Holes are drilled about one metre apart and between each vertical stud, then either a cellulose fibre or a water-based vapour permeable foam is pumped into each hole (where the foam expands slowly).
B2 Existing lath and plaster wall.
B3 Insulant fills the void behind the lath and plaster.
B4 Operative pumps insulant into holes formed in existing wall, which then appears at other holes and can be trimmed off before redecorating.

C) Fibreboard insulation applied internally

C1 Lime plaster.
C2 Fibreglass mesh.
C3 Wood fibre insulation batts interlocked, secured with fixings into stonework and set onto ...
C4 3–6mm vapour permeable adhesive bonding coat formed over existing stonework.

D) Calcium silicate insulation applied internally

D1 Lime based finishing plaster applied over ...
D2 Reinforcing mesh set onto plaster base over ...
D3 Calcium silicate batts, stagger bonded and set on adhesive layer ensuring no air pockets between stone and batts.

E) Loft insulation

E1 Ply catwalk set onto new timber joists set across existing joists E3.
E2 Vapour permeable insulation laid tightly between joists avoiding any gaps.
E3 Timber to support the catwalk.
E4 Additional rolls of vapour permeable insulation laid across existing joists and under catwalk. Total thickness recommended is 350mm.

F) Close wall insulation

Although many insulations can be applied directly to an existing plastered wall, in this detail we show insulation applied to vertical timber straps (thickness may vary).

F1 Vertical timber straps fixed to existing wall.
F2 A thermal laminate board which may or may not be vapour permeable.
F3 Plasterboard or magnesium silicate bonded to aerogel: these thin boards can save space and still be effective insulators.
F4 Cellulose beads blown between the studs then smoothed off before applying a clay board and lime plaster (or other vapour permeable board).

G) Ground floor insulation of a suspended timber floor

G1 If floorboards can be removed, the best method is to secure a vapour permeable membrane over the joists forming a carrier for the insulant ensuring it is tight by fixing timber battens to hold the membrane to the joists (**G3**).

G2 A vapour permeable insulation such as hemp, wood fibre, held in place by the membrane which can also help with draughtproofing.

G4 Alternatively the insulant can be secured, provided there is a crawl space in the solum. Here, wood fibre batts (**G5**) are held in place by galvanised wire secured to the underside of each joist with stainless steel screws.

G6 Vapour permeable insulation batts may be secured by fitting timber straps on the side of each joist, but this is more difficult!

Vapour tight systems applied internally

These prevent moisture from entering the wall from inside (mainly water vapour). They are usually insulants like polystyrene, polyurethane, mineral wool, which all use a vapour control layer to restrict the movement of water vapour. In traditional stone tenements, where rain is wind forced through the stone from outside, vapour tight systems will result in moisture becoming trapped in the interface. We do not recommend these fixings. Popular systems are as follows:

Thermal board: an insulation of varying thicknesses, bonded to a plasterboard. Thermal board is often secured to the wall with plaster dabs. It can also be secured to timber battens fixed to the wall. The insulation may be polystyrene or polyisocyanurate board.

Stud and insulant: a timber stud is secured to the wall and infilled with mineral wool (varying thickness). A vapour barrier is added over the insulation then plasterboard fixed to the battens and skim coated.

Such systems prevent moisture in the wall from drying out, so it's best done in the summer, although some insulation companies state that their particular insulation and render system is vapour permeable. It is also possible to form a stud frame, secured to a stone wall with vine ties (allowing for a small vented gap), then filling the studs with mineral or glass wool, then applying a vapour barrier then a plasterboard and skim coating the board.

Internally applied insulation to stone or brick walls

These are often proprietary systems using polystyrene or Rockwool batts which are screwed back to the face of the masonry. The batts are stagger bonded and a mesh is applied over the insulation then two coats of render are applied. The renders are sometimes described as vapour permeable but as they are often applied to an existing cement rendered fabric, it is debatable how vapour permeable they are.

In both cases, if an external insulation system is applied to an existing stone wall, it is likely that you would need planning permission first.

The interface between insulation and existing wall should allow a flush tight sandwich. Therefore, the insulation should either have a flexible back face to squash against the wall and fill any gaps, or else a parge coat should be applied first to fill gaps to a smooth plumb finish.

A more robust finish for ensuring long-term weather protection could include battens and a render board or other board finish. This will provide a ventilation gap or cavity between the new external finish and the external face of the insulation, thus protecting the insulation from regular wetting and consequently maintaining optimum performance.

Vapour permeable systems

These facilitate the drying of the external wall by allowing it to dry out both to the outside and to the inside. The sort of insulation used would be a vapour permeable insulant with a vapour permeable finish. The following systems may be used:

Insulated lime render system: can be applied internally up to a thickness of 50mm and should improve the wall U-Value by about 50 per cent. Also a cork based render can be applied in two coats to internal walls and then over skimmed with 2mm of skim coat. This has the advantage of saving space and is suitable for internal walls which are quite flat.

Lime with cork and lime with hemp: both provide vapour permeable insulation; the insulant is secured to the walls internally fixed with a 3mm adhesive coat then a mesh applied with lime plaster. Cork slabs vary in thickness. A lime plaster system is used to bond a glass fibre mesh to the cork and then the finished lime plaster is applied. The overall thicknesses available are 50, 80 or 100mm.

Studs and blown cellulose or wood fibre: studs secured to wall then cellulose blown between the studs. Covered with a vapour permeable board such as calcium silicate board or clay board and finished with a lime plaster.

Wood fibre board: rigid boards of wood fibre are fixed to the wall, typically over a lime plaster parge coat; the system is finished internally with either a lime plaster finish applied directly or a battened service void with clay board over. The parge coat should fully fill any pockets or gaps in the masonry face to provide a flush plumb surface against which the wood fibre is installed.

Aerogel (trade name Spacetherm) is vapour permeable and when applied to walls is usually covered with a calcium silicate board which is also vapour permeable.

Calcium silicate boards (25 and 50mm) have an insulation value themselves. They can be glued to an existing wall and coated with a lime plaster. 50mm thick board will reduce the U-value by about 50 per cent. Parge coats and bonding coats are recommended; sometimes the board may be fixed with a glue-based adhesive.

Injected foam behind existing lath and plaster wall

This allows the retention of existing wall linings and minimises disruption to historic material, as well as reducing cost and disruption to the occupants. It is therefore necessary that a vapour permeable material is used to allow movement of water vapour through the fabric. However, a word of warning: some insulants can decay PVC cabling, and all electrical cabling that is buried under insulation needs to be de-rated, usually needing a thicker cable or requiring cables to be conduited. So, it is best to avoid injecting an insulant behind lath and plaster where there could be electrical wiring.

The type of insulation used by Historic Scotland was **cellulose beads**, which can be injected under pressure.[9] This does not degrade PVC wiring, but installers will often use a spray foam insulation such as **Icynene** (other open cell spray foams are available). The foam should be vapour open and expands slowly. However, once applied, it cannot be removed and some

9 See HES Case Study No 2 for details.

Injecting foam, wood fibre and calcium silicate

foams may be toxic. It is also rarely fully vapour permeable so it is best avoided in existing lath and plaster walls or between rafters. There have been only a limited number of pilot projects using it.

When inserting or blowing material behind lath and plaster wall linings, it is important to ensure that the masonry can stay dry by retaining or improving water vapour permeability. The role of external maintenance is necessary in achieving this, which includes the proper functioning of rainwater goods, such as gutters and downpipes.

Modern external finishes such as cement render may compromise water vapour permeability.

HES refurbishment case study number 20 has shown where open cell foam has been used, a U-value improvement from 1.1 to 0.42 was achieved, although we are unclear about the long-term effects.

Polystyrene beads injected behind lath and plaster have been trialled by Historic Scotland, although there was some concern the bead did not fully fill the cavity and in time might slump. The beads are first coated in a water-based adhesive to ensure the beads set into place once injected. The beads can also degrade any PVC electrical cabling and cabling would again need to be de-rated. HES case studies have shown the use of bonded polystyrene beads in this way resulted in a U-value improvement from 1.1 to 0.32.

Perlite beads were also poured into the wall cavity from above, but as it is a loose fill, it is rarely appropriate as the perlite can disperse into other voids uncontrollably.[10]

Prior to installing any of these methods care should be taken to find and stop any gaps in the surrounding fabric which might allow the insulation to fall out and/or lead to slumping.

Externally applied vapour permeable insulation to stone or brick walls

HES have trialled the use of a calcium silicate board in this situation, although wood fibre-based products and other vapour permeable insulation materials can also be directly fastened to a mass masonry wall (refurbishment case studies 3 and 6[11]).

Calcium silicate board is available in a range of thicknesses to suit site conditions and the degree of thermal improvement sought.

10 See HES Case Study No 7.

11 HES Refurbishment Case Study 3 and Case Study 6.

How not to apply external wall insulation. Here it is not vapour permeable, does not go below ground level, and the downpipes and gas meters are left in place to form cold bridges if it is applied to a rear elevation

Where an external wall is uneven, a parge coat of lime plaster is formed and the calcium silicate boards are secured to the parge coat with a vapour permeable adhesive. This method can improve a wall's U-value from 2.1 to 1. As noted above, a finish on battens providing a ventilated cavity to the external face of insulation will result in better weather protection of the insulation for the longer term.

Some externally applied insulation is not vapour permeable. Insulating externally will allow the original wall to retain energy as a thermal mass, as well as protecting the existing wall from the weather. Vapour tight systems are usually insulants like extruded polystyrene (XPS foam), expanded polystyrene (EPS), polyurethane (PUR foam), and polyisocyanurate (PIR foam) followed by a render mesh and render system, which may or may not be vapour permeable. We try to avoid these systems. A lot of municipal stock is using these, as can be seen from photo 12.4, where there is plenty of scope to reduce the below the ground insulation and the soil pipes would form cold bridges.

Floor insulation types

Suspended timber ground floors

Insulating ground floors can be difficult: if it is a timber floor, the solum space under the floor may be quite low and difficult to crawl into, let alone secure insulation between the floor joists.

If the solum is deep enough, then insulation can be secured between each joist. One can use a range of insulations, such as glass wool, recycled insulants, wool, hemp, wood fibre. The problem is how to secure them in place, and avoid gaps in the insulation layer. Contractors may use a plastic mesh secured to the underside, but the insulation is prone to slump over time towards the centre as the mesh stretches. Instead of a mesh, a vapour permeable membrane can be used. This has the advantage of reducing drafts from the solum – it too can stretch, so a taut install is important. Alternatively, a galvanised wire can be secured between the joists, but it needs to stretch back and forth in order to support the insulation.

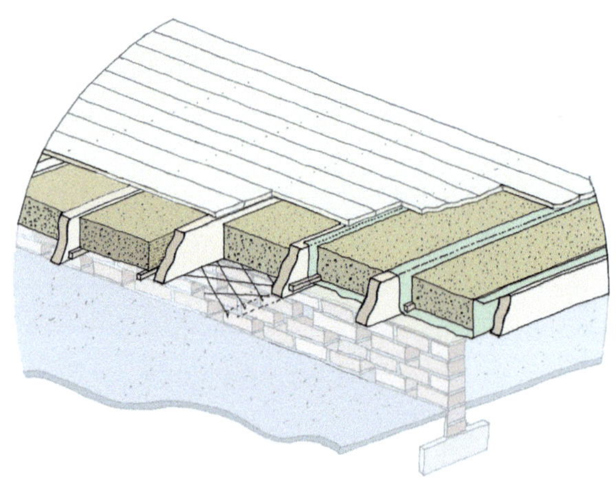
Suspended ground floor insulation type

If rigid wood fibre-boards are cut to fit between the joists then they can be supported on timber battens secured to each side of the joists although it is best if the floorboards are lifted first.

If the solum is deep enough, then access from below will allow insulation to be secured between each joist and membrane fitted to the underside. However, if there is insufficient crawl space and if the existing floorboards can be lifted, then the work can be done from above by placing a vapour permeable membrane over the joists, securing the sides to the joists with small battens to ensure the membrane is fixed in place and tight, then the wood fibre insulation is placed between the joists. The membrane will provide a certain amount of draughtproofing although one can also add a further membrane airtight vapour open over it all before replacing the floorboards (an airtight membrane is something that will allow air to circulate).

If the solum space is very narrow and the floorboards cannot be lifted, there is one company that uses a robot device that sprays a water-based foam or polyurethane foam insulation into the space between each floor joist; this can work best when the solum space is so narrow that human access is impossible, but it needs to be at least 12 inches and ventilated. The installer should also check the conditions of existing timber structure (moisture levels) and confirm that the insulation material will not degrade any PVC electrical cabling.

Should insulating between joists prove impossible or inadvisable there is the option to lay new rigid insulation over existing floors, followed by an airtightness membrane (taped at all joints and to adjacent materials at the perimeter), with a new finished floor laid on top. This option will require altering of door bases and the base of surrounding joinery.

Solid concrete or stone floors

An old solid floor is often a source of cold and discomfort and many solid floors are made of uninsulated concrete which can cause dampness in traditional walls. If you have a stone floor such as Caithness stone slabs, then it is probably best to leave it alone and rely on having mats where you want to stand.

For a concrete floor, assuming the floor level cannot be raised too much, an aerogel material can be used with a bonded calcium silicate board. These are simply laid over the existing slab and stagger bonded, then the joints are taped. A carpet or linoleum finish is then applied.

In HES refurbishment case study number 6,[12] the use of 30mm aerogel board gave an improvement in U-value from 3.9 to 0.8. Aerogel board can be supplied in various thicknesses and then cut to size and fixed with adhesive. The base of doors will usually require to be trimmed, resulting in some loss of fabric. Skirting boards should be removed and reinstated a little higher.

12 Historic Environment Scotland Refurbishment Case Study 6.

Lofts, flat roofs and coombed ceilings

Loft insulation (cold roof)

When insulating a roof, it is important to assess what the fabric build-up is above the sarking. If there is a vapour barrier present (such as a bitumen under slate felt), consideration should be given to providing ventilation to the attic roof space. Without ventilation in the attic, mould can grow on the underside of the string and rafters.

However, if your roof was retiled or reslated and a vapour permeable membrane was used, the ventilation should be sufficient as the 'penny gap' between sarking boards should provide enough ventilation.

For loft insulation and for most open fibre materials, such as sheep's wool, 280mm is recommended.[13] How this is laid out depends on whether the attic space is to be used. If the space is not used, then the material can be laid between the joists and a second layer laid over the joists at right angles to this. This approach creates what is termed a cold roof. If a deck or flooring is required, then the additional height is achieved by adding an extra joist to carry the walkway above; this is generally the approach taken by most insulation installers.

Any insulation without ventilation can damage your roof (except those with breathable insulation).

In one project, sheep's wool insulation improved the U-value from 1.4 to 0.2, and in a separate case study, wood fibre insulation between ceiling joists resulted in an improvement from 1.5 to 0.2.

Cellulose insulation can also be blown into roof spaces, but it can be difficult to ensure the depth of insulation is maintained because any air movement (most often at eaves) can blow the insulant so that it is thinner at some locations.

Insulating flat roofs

Insulating a flat roof presents a number of technical challenges. The methods and materials used should be carefully considered prior to any work taking place. The form of insulation will depend on the type of roof finish. If it is a flat lead (or zinc or felt) roof, then there should be a ventilated space under the sarking that supports the roof finish, otherwise condensation can develop.

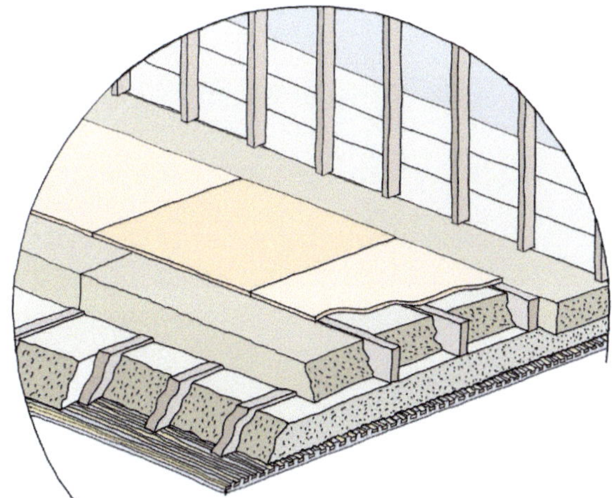

Loft insulation showing catwalk and double thickness of insulation batts

Some warm roofs are formed by retaining any existing felt or asphalt, or laying a vapour barrier over the roof initially, then laying a closed cell insulation such as a PIR board and applying a

13 Although it is possible to lay a vapour permeable membrane over the insulation, to increase airtightness, it is rarely done because of the need to trim the membrane where all the timber ties form the roof structure. Tight fitting insulation is preferred.

liquid resin finish (called liquid plastic). There are several other systems but it is best to take your professional's advice if improving insulation in a flat roof.

Coombed ceilings

Coombed ceilings can be insulated by inserting a vapour permeable insulant such as wood fibre between the rafters and leaving a 50 mm air gap between the insulation and the existing sarking.

Provided a vapour permeable roofing felt has been used over the outside of the sarking and the rafters are plated with a vapour permeable wood fibre board (of calcium silicate board), it is possible to blow cellulose fibre into the void between the rafters.

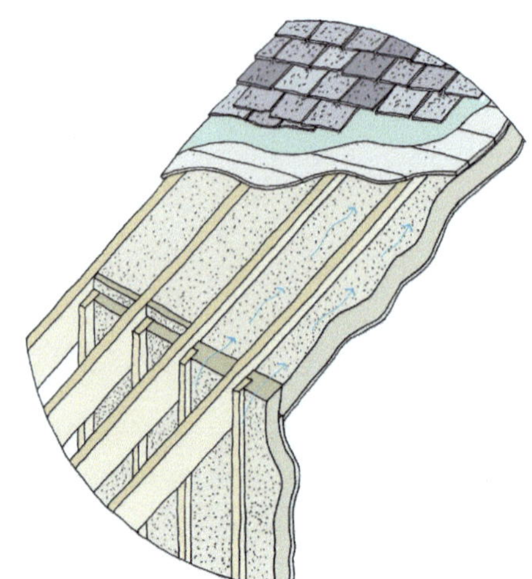

Coombed insulation. There are different options depending on roofing membrane

Shutters, curtains and blinds

Traditional options for reducing heat loss through windows, such as blinds, curtains and shutters, can result in a significant reduction in heat loss with no impact on the existing window fabric. Shutters alone can reduce heat loss by 51 per cent, with a HES trial achieving a U-value of 2.2. A combination of these options can reduce heat loss by as much as 62 per cent, only 1 per cent less than through the installation of secondary glazing. Commonly found in older buildings, timber shutters can reduce heat loss through windows by up to 51 per cent.

The thermal performance of shutters can be further improved by applying a thin layer of insulation, such as aerogel blanket, which in a HES case study resulted in a 60 per cent reduction in heat loss.

Full-length, lined and well fitted curtains can control draughts and reduce heat loss by up to 14 per cent.

Glazing and windows

Secondary glazing

Secondary glazing is one of the most effective methods of improving thermal performance, reducing heat loss by 63 per cent to a U-value of 1.7 (single pane secondary glazing). Used in conjunction with other methods, such as blinds and shutters, a reduction in heat loss of over 75 per cent can be achieved to a U-value of around 1.0.

Secondary glazed systems can be formed in aluminium frames to mimic the pattern of the window opening. This allows the secondary glazing to be easily opened to allow the external window to be cleaned or for ventilation.

Another cheaper solution is a plastic sheet held by magnets to the frame. This is much cheaper but it means you cannot easily open the window except by removing the secondary pane.

Another simpler and cheaper solution is to install a thin glass pane formed to the exact size of the existing glass and fix it to the existing glass with a magnetic strip. This forms a small cushion of air between the two panes and still allows the sash to open. There is some heat loss around the edge of the glazing, but it is still an improvement over single glazing. Such a system should be able to achieve a U-value of around 2.4.

If your tenement is listed or in a conservation area, any window replacement will require planning permission and possibly listed building consent.

Improved glass

Where appropriate, glass in sash and case windows can be replaced with types of window glass that has better thermal efficiency than a single pane of glass. New panes of this type come in many forms and they can be retrofitted into the existing window frames. This minimises the impact on the character of the window, although it does involve the loss of original glazing. There are two main types: Slimlight glazing, or Histoglass use a much smaller void filled with Argon gas, so that the overall thickness is about 10mm and can be installed in traditional sashes. It is recommended that the space between the inner and outer panes should be at least 7mm to provide better results. Even better is much thinner vacuum glazing which relies on small spacers to hold the panes apart and allows a vacuum to be formed in between the two panes. This has an even better insulation value but is likely to be more expensive.

Upgrading windows

Existing timber sashes can be draughtproofed by removing the sashes and forming a small check in their sides and rails to receive a draught strip. The strip should be continuous round all sides of the sash. This effectively reduces draughts at the sashes, although there may be more air leakage at the junction between the actual case and the stone wall. In such cases, the wall linings or shutters need to be removed to allow an adhesive tape to seal the joint or any gaps can be plugged with a wool or Oakum fibre. The gap between window and stone should be backfilled with linseed oil sash mastic. Timber sash windows can be replaced with double-glazed sash and case windows which are draughtproofed. Some owners may opt for cheaper solutions such as UPVC windows (see Chapter 7).

Heating options

It is easy in the energy efficiency discussion to focus entirely on the measures required to increase insulation and reduce air infiltration. While these considerations are valid, it is worth remembering that the objective is to achieve a building that is healthy and comfortable for the occupants. The ability to control one's immediate environment is also important, so access to ventilation is necessary.

There are different heating options, but until very recently gas central heating systems have been favoured and still account for higher (less efficient) EPC ratings. As gas is to be phased out, starting in 2025, other solutions have to be developed for tenements.

The options of **air source heat pumps** do not currently make sense for all tenement flats as such units must be installed externally and close to the flat. Installing heat pumps mounted on walls is not commonly permitted currently, so typically it is only economical and efficient to heat ground and first-floor flats with ASHPs. Maintenance of heat pumps is essential so installing them high up on a tenement wall would be problematic. Permission must be sought from those that share any common outdoor space prior to their installation, and in listed buildings and conservation areas planning permission is likely required. Heat pumps provide a lower temperature of heating than a conventional radiator system so it is advisable to carry out other energy efficiency works such as insulation at the same time or prior to installing a heat pump. Heat pumps require a thermally efficient building fabric to operate cost-effectively, which is often a challenge for traditional tenement buildings.

Ground source heat pumps using shared ground loop arrays are another possibility. These are commonly used in other countries although some have been installed in Scotland (and some using waste water found in abandoned coal mines).[14] Holes which vary in depth from 100 to 300 metres are drilled into the earth and a sleeve inserted. Flow and return pipes are inserted which allow water in the pipe to reach a slightly higher temperature. It is then fed through a heat pump to raise the water temperature further and to preheat any hot water cylinders. These ground loops can be shared amongst several flats, as drilling to any depth is quite expensive.

The advantage is that it can be suitable for flats, although, like air source heat pumps, larger radiators may be needed as the water temperature is lower than for gas fired heating, although the ambient temperature can be similar. It may be possible to avoid increased radiator sizing if the heat loss from the building is sufficiently improved. At present we are unaware of any incentives which would improve the take up of such systems although they also have the advantage of avoiding noise and being very low maintenance.

Electric storage and chemical battery heating may be considered if flats are very well insulated.

Combined heat and power (CHP) systems that are communal would seem to be the approach best suited to tenements as they are usually formed in blocks, the flats are close together, existing water based radiators can be reused and CHP systems can be located in backcourt areas and connected to existing gas grids. Heat metering would ensure that owners would pay for their use and individual hot water cylinders could be controlled by electric immersers to provide boosted hot water. Gas fired CHP systems with large thermal stores can reduce carbon by 33 per cent and can be carbon neutral if wood or wood pellets[15] are used.

107 Niddrie Road, Glasgow – EnerPHit refurbishment

Introduction by Chris Morgan at JGA

107 Niddrie Road is a stone tenement built in the early 1900s consisting of eight one-bedroom flats over four storeys and located in Strathbungo East, in the inner south side of the city. Southside Housing Association (SHA) approached John Gilbert Architects (JGA) to look at a refurbishment of the building in 2019. Unusually, SHA owned all the flats in the close and, even more unusually, all flats were empty.

14 Glenalmond Steet in Shettleston, an existing tenement housing project, heats 16 houses.

15 Use of wood pellets is falling out of favour, primarily as the source wood used is typically freshly felled timber rather than waste timber. Also getting permits for flues in high-density city-centre areas / low emission zones has become challenging.

This presented an almost unique opportunity to undertake a much more comprehensive refurbishment than is normally possible. It is also a quirk of the EnerPHit (Passivhaus standard for existing buildings) that the whole block needs to be upgraded – it is not possible to upgrade a single flat within a block. For this reason JGA proposed an EnerPHit retrofit, along with two other, less rigorous options.

Initially, it was decided to pursue a relatively conventional retrofit. However, a consortium including SHA and JGA along with CaCHE (UK Collaborative Centre for Housing Evidence) at the University of Glasgow, and Prof Tim Sharpe at the University of Strathclyde Department of Architecture were successful in bidding for a significant tranche of funding from the Scottish Funding Council to disseminate best practice in retrofit ahead of the 2021 COP Climate Summit in Glasgow.

The project was upgraded to an EnerPHit project and re-cast as an important exemplar of what to do with the 75,000 iconic pre-1919 sandstone tenements across Glasgow. The age, built form, condition and multiple forms of ownership of these tenements constitute major challenges for the achievement of net zero emission targets across Scotland. To ensure wider lessons are learnt from the project, the project has been carefully evaluated and will be monitored to assess the success of the strategies employed.

EnerPHit is not just about reducing carbon emissions. It is also about addressing fuel poverty and, in addition, JGA have developed several solutions to address wider comfort and health aspects as well as fully address heritage aspects.

Retrofit measures

In keeping with all Passivhaus and EnerPhit projects, the project features the following five key attributes:

- Insulation – very high levels of general insulation to roof, walls and floors.
- Thermal-bridge-free – thermal bridges are areas where heat can 'leak' through parts of the building which have not been insulated to the same degree. Much of the design effort in achieving an EnerPHit project is in getting these details right.
- Airtightness – very low levels of infiltration or air leakage must be achieved to be certified as an EnerPHit project. Unusually for this project, a lime plaster was used to achieve airtightness in preference to a membrane.
- Triple glazing is required for all projects in a cool temperate climate region like the UK.
- Mechanical Ventilation with Heat Recovery (MVHR) is required for all projects.

External insulation was installed on the rear and side facades of the building and extended carefully below the plinth level and into window reveals.

It was felt that even though the building is not in a conservation area, external insulation to the stone facade would not be appropriate, so an internal wall insulation solution was designed. Internal wall insulation is tricky though because it can cause increased risks associated with moisture so great care was taken and the whole element tested using WUFI hygrothermal calculations. This demonstrated that all elements were safe, except the timbers in the street side, internally insulated wall and so it was decided to remove these. Removing joists increased costs but ensured, firstly, that no future decay could come back to haunt the client and, secondly, simplified the forming of the airtightness layer. Instead of being buried in a bressumer beam, a new perimeter beam was installed with insulation behind; this also ensured that the otherwise tricky floor-to-wall detail was wholly thermal bridge-free.

It was originally proposed to retain the existing gas boilers, but the funders were keen to see a demonstration of renewable technologies. However, it was not possible to agree with the planning department a fully renewable solution, so a compromise was reached where air source heat pumps (ASHPs) were located on the adjacent ground and supply heat to the four lower flats, while gas boilers supply the upper flats.

In addition to recovering heat from the outgoing air, JGA also specified waste water heat recovery technology to recover heat from the outgoing hot water because with space heating demand low, the hot water would become the largest component of any future heating bills.

The rooms themselves were reconfigured to improve layout and meet modern space standards, old wiring was entirely stripped out and large areas of poor plaster finishes were replaced. A number of structural repairs were made and decayed timber was repaired. While a conventional chemical company quoted around £20,000 to then spray the timbers of the building with chemical fungicides and biocides, JGA opted instead to engage a building pathologist with whom suitable specifications were agreed which meant none of those chemicals were needed while ensuring the timber is entirely safe.

The whole face of the building on the street side was repointed in lime and damaged stone was replaced with matching sandstone components. This was in part a conservation effort but also ensures that the inside of the insulated walls are kept as dry as possible.

Process

From a design perspective, by far the biggest issues were encountered in achieving planning permission. This took over a year and involved a great deal of dialogue and compromise. By contrast the building control process was very simple.

Contractors CCG were engaged from the outset and provided cost feedback and practical input. Works commenced in the spring of 2021 with the intention of completing by the time of COP in November of that year. In the event, construction took much longer, completing in the autumn of 2022.

More areas of internal finishing needed to be replaced which prolonged the early process, but otherwise works progressed without undue issues. The only major issue was that, ultimately, the necessary airtightness level was not achieved meaning that the project could not be certified as an EnerPHit. A number of other options are, however, available and those are being explored as this is written. With most existing buildings using between 150 and 300 kWh/m^2/year, the EnerPHit requirement is to be no larger than 25 kWh/m^2/year. Even with the higher air leakage results, 107 Niddrie Road is modelled to achieve 25.9 kWh/m^2/year so it is agonisingly close, and this will

[left] Bressumer beam

[right] Floor to wall making it thermal bridge-free

[upper left] Individual portion might be bridge free

[upper right] The outer wall, aside from the plaster

[lower left] Filling the void under the sill

[right] The rear elevation

mean that occupants of the eight flats will be able to keep comfortable for something like a tenth of the cost for an average person in the UK.

The building will be monitored and 'real life' energy consumption will be recorded. The project has been principally funded by Glasgow City Council and by Southside Housing Association (who own the property), with additional grant funding from the Scottish Government.[16]

Other approaches

The best possible approach is for housing associations to have a look at installing insulation for passive house standard. If all housing associations were forced to do this, all owners would be forced to accept a reduced target of 80 per cent of the heat retained (grants to be made available to home owners and landlords).

Whilst insulation and draughtproofing should be the first step we should be taking to reduce our carbon load, clearly this needs to be done provided the fabric of the tenement is well maintained.

The Tenement Maintenance Working Group has made recommendations to parliament that would help to make maintenance an essential part of owning a tenement. It recommends:

1. A full tenement inspection every five years by a professional to inspect each tenement to BS 7913:2013 and to produce a report which identifies those items that need immediate attention, are urgent, or necessary or just desirable and to prepare a workplace so that owners can budget their finances.

2. The owners of each close should form a Stair Association under the Development Management Scheme to ensure the owners' full cooperation and to act as a client with a bank account (termed a 'treasurer's account') to commission any work.

3. To develop sinking funds for each tenement, albeit in a phased way. The amount paid by each owner should be either a flat rate decided by Scottish parliament or a proportional amount corresponding with the common share.

4. To make any split vote (such as can occur in a tenement with eight flats) count as favouring the decision to carry out any essential maintenance.

This proposal has been with Scottish parliament since 2019.

Newbuild housing on greenfield sites should be taxed, but housing on reclaimed urban sites should be incentivised. Much of our cities new urban housing is expensive because of the need to remediate land that is polluted by previous industries as well as having to adjust the location of existing services, all of which are additional costs which make developers prefer a greenfield site. Land in urban sites tends to be more expensive and, if in a popular area, there may be objections from neighbours not wanting more houses. This is why developers choose agricultural land – it is cheaper and there is less likely to be an objection from neighbours, although they will still need to make road and service connections to the houses.

Such suburban living brings with it increased car ownership and road use which creates more CO_2, particularly as the houses are not within walking distance of any shops or schools.

In the 1850s, the rich and the middle classes left the cities for the suburbs, but there was still a desire to be able to live in places that had facilities and access to the public world of

16 https://vimeo.com/687118855.

commerce and entertainment. Somehow, we need to remake our urban areas into places that people want to live in, and can afford to live in. In Edinburgh, although the rich moved to villas outside the centre, they also built quality tenements (for those who could afford it).

We need to be building good-quality, zero-carbon, affordable housing in urban areas fit for a new generation, as well as retrofitting our existing tenemental stock with better insulation, draughtproofing and heat recovery systems.

A final word

We should recognise the dispiriting effect of poor housing and environmental conditions. The public realm and all our shared spaces could benefit from good maintenance and much more care than we presently bestow on it. We believe that people care about the public realm and the places we all share. Tenements are no different – the roof, external walls and close, as well as life, are all shared with neighbours. Our traditional tenement housing is a key part of living in Scotland; we all benefit from housing that is well maintained, and the care we take makes our neighbourhoods places we like and want to live in.

Appendices

Appendix 1: Conservation areas

The first conservation areas were selected after 1970 and there are now over 600 designated conservation areas in Scotland; they cover public parks, battlefields, historic land and also include tenemental areas. In addition, there are 45,770 listed buildings in Scotland. The main effect of owning a property in a conservation area is that it is not possible to demolish a building or make significant changes to it without first receiving planning consent. Each local authority will have prepared a conservation plan for the area which describes the boundary and details the importance of the area's character as well as key features that need to be retained, or, if possible, brought back to the original design. The local authority, in preparing the plan, will make a character appraisal of the site listing its main elements and features, as well as any listed buildings. Article 4 directions control minor alterations that can cumulatively lead to the erosion of the character and appearance of the area. When a conservation area has an Article 4 direction, then planning permission must be obtained for works associated with the direction.

The main requirements will usually be to retain or reform any roofs in the original material; to restore and retain (or possibly replace) original sash and case windows; to only use traditional materials such as cast-iron rhones and conductors; to use lead of the correct weight in areas which were leaded before; to use only matching lime mortar when repointing and to select a matching stone when carrying out an indented stone repair. Items such as the restoration of shopfronts are usually encouraged in areas that are designated **Township Heritage Initiatives** (THIs).

Conservation Area Regeneration Scheme (CARS)

This was a Historic Scotland programme (now closed and replaced with the Heritage & Place Programme) which offered grants of up to £2 million to support heritage projects within conservation areas, mostly for public realm works but it also covered training and management of staff involved. CARS work was usually programmed over a five-year period.

The replacement programme, the **Heritage & Place Programme**, is an area-based funding programme that aims to contribute to the development of vibrant and sustainable places in Scotland through community-led regeneration of the historic environment.

Listed buildings

Listed buildings originally began in Scotland under a provision in the Town and Country Planning (Scotland) Act 1947. Presently they are made under the Planning (Listed Buildings and Conservation Areas Scotland) Act 1997. They are listed by Historic Environment Scotland.

Listed buildings in Scotland fall into three categories:

A: Buildings of national or international importance, either architectural or historic, or fine little-altered examples of some particular period, style or building type.

B: Buildings of regional or more than local importance, or major examples of some particular period, style or building type which may have been altered.

C: Buildings of local importance, lesser examples of any period, style, or building type, as originally constructed or moderately altered; and simple traditional buildings which group well with others in categories A and B.

When working on altering or extending a listed building, listed building consent is required to ensure that any changes are appropriate and sympathetic to the character of the building. If you are simply replacing old materials with new materials on a 'like for like' basis and if the repairs do not affect the appearance of the building, you may not require listed building consent. However, it is always best to check with your local planning department first.

Preservation trusts

Building preservation trusts (BPTs) are charitable organisations whose fundamental aim is the preservation and regeneration of historic buildings, particularly those that are unviable for the private sector. Since 1999, trusts have been funded by Historic Environment Scotland and the Heritage Lottery Fund; sometimes any shortfall is met by local councils.

There are considerable wider benefits realised by preservation trusts, including an ability to lever additional funding and supporting local employment, although grant funding is reducing and becoming harder to secure. There is a clear demand for organisations with the necessary skills to restore redundant buildings and bring them back into use, ideally for local communities. In many cases, private sector intervention for these buildings is not viable.

Heritage trusts

There are a number of heritage trusts in Scotland, from local towns, villages and cities. Some are city trusts, but they are all part of the heritage trust network. The Scottish Lime Centre Trust in Charleston, for example, was established in 1994 because of concern that there was a growing skills shortage in traditional mortars and stone repairs. There are grant-giving trusts and trusts which do not give grants. Seven city heritage trusts are supported by ongoing funding from HES to deliver local projects.

Appendix 2: Tenement features

Corners

17th- and 18th-century tenements with steep roofs might have corner turrets, sometimes corbelled out at the first floor. This corner feature nearly died out in Georgian classical times, but reappeared in the Victorian period.

The Georgians liked to turn corners with a curve, or create a crescent, and sometimes corner bays had enhanced upper floor windows. The corner of York Place and Queen Street in Edinburgh is just such a corner.

[left] The big corner bay at York Place

[centre] The orthogonal bay at Lauriston Place, Edinburgh

[upper right] The corner of Portobello High Street and Marlborough Street, Portobello

[right] The corner of Gallowgate in Glasgow

[far left] The corner of Manor Place in Falkirk

[left] Baron Taylor's Street in Perth

[right] Corner of 1021 Cathcart Road, Glasgow

[far right] The Granary in Kilmarnock Road in Glasgow

It has a large stone bay with three windows on the third and fourth floors and eight windows on the floor below, extending down to basement level with large vertical windows and an entrance. This was unusual; more common were corner bays and even a six-sided bay at Lauriston Place, Edinburgh.

Others, as at the corners of Ainslie Place in Edinburgh, had traditional Georgian classical layout with arched rusticated window openings on the ground floor and the main tallest windows on the first floor, reducing in height for each storey level. All windows were interspersed with stone pilasters and a deep cornice at the sills of the third-floor windows.

In the late Georgian period some tenements did develop a more Victorian boldness. The corner at Rothesay Mews in Edinburgh has a magnificent four-window circular bay corner which rises for seven storeys, and five-storey bays at the front on Rothesay Place.

A rather grand corner at Morningside Road and Church Hill Place, Edinburgh, combines a curve with an oriel and a pediment, the bow headed wallhead dormers set between stone balusters. Here the gutter is formed under the top floor window resting on its own corbel course.

Turrets

The Victorians liked to emphasise corners, so they often built slated turrets in a variety of shapes, sometimes faceted, four- and six-sided slated roofs, sometimes conical with enhanced projecting cornices and pilasters and often springing from projecting oriels with exceptional stone carved figures acting as corbels. Around 1890s the wallhead chimneys all joined the corner turrets to enhance the corners.

Curves

Stone was able to be cut to any angle and even smoothed to form curves providing a very sinuous architecture. This allowed more flexibility in laying out streets and introduced crescents and curved corners. In the 1890s, some bays that were curved also had curved windows and curved glass.

[far left] Fordell's lodging: Inverkeithing

[left] High Street in Dunblane

[right] Bull Inn, New Street, Paisley

[far right] High Street in Hawick

[above] Forbes Place in Paisley

[right] Gardner's Crescent, Edinburgh

[far right] Mill Street, Alloa

Bays

Bays developed for the middle classes, when slightly larger tenements were built. Some feu conditions required bays, which invariably meant that front gardens were needed. Slender mullions were developed and made off site. It provided more light into the principal room as well as allowing a fuller view of the street below.

[right] Odd corner bay in Gorgie, Edinburgh

[right] Bays in Portobello, Edinburgh

[far right] West End, Glasgow

[below] Exuberant Virginia creeper in West End, Edinburgh

[right] A saved one and a corner one

[far right] The bay developed from this type of feature in Leith

Oriels

Although oriels first appeared in the 1830s, they did not become common until after 1880. They were often found when the tenement was built to the pavement line. Bays were even changed to oriels if shops were added to some bay-fronted tenements.

[left] The start of oriel in Claremont Gardens in Glasgow

[centre] Glasgow bays

[right] Edinburgh bays

[far left] Depressed oriels at Church Hill Place, Edinburgh

[left] Henderson Street, Edinburgh

[right] Leith

[far right] Corner oriel, Portobello

Turnpike stairs

Because turnpike stairs used to be accessed at the rear after passing through a close or wynd, the close entrance was sometimes emphasised to give it more importance.

New Town

Stone architraves and pediments

Stone *architraves* and *pediments* around windows were often used by Victorians to bring importance to the elevation, usually on the first and second floor levels, in more upmarket tenements such as Morningside in Edinburgh. Otherwise the openings around windows were quite plain with more emphasis given to close entrances which might have cornices and console brackets, although some windows also got this treatment. Decorative elements such as stone carving might be used on corner features or *oriel corbels* often becoming features in the townscape.

[left] Straight architrave

[centre] Scroll ends supporting an entablature

[right] Lintol and jambs to receive lime coating

[left] Stops and starts in the tenement walls

[centre] Door opening with a cast iron lintel

[right] Alexander 'Greek' Thomson's work found in Alloa

Stone types, finishes and coursing

[far left] Straight scabbling

[left] Downpipe inset in splayed all round blocks

[right] Rusticated course

[far right] Reticulated block at the basements

[far left] Galetting of stones

[left] Old stonework

[right] Random rubble wall designed to take lime

[far right] Some whinstones with natural red sandstone

Appendix 3: Useful contacts

The Institution of Structural Engineers International HQ, 47–58 Bastwick Street, London, EC1V 3PS
www.istructe.org

Royal Institute of Chartered Surveyors
3rd Floor, 125 Princes Street, Edinburgh, EH2 4AD
www.rics.org/uk

Royal Incorporation of Architects in Scotland
15 Rutland Square, Edinburgh, EH1 2BE
www.rias.org.uk

Underoneroof
Impartial advice on repairs and maintenance for flat owners
www.underoneroof.scot

The Cockburn Association
Trunk's Close, 55 High Street, Edinburgh, EH1 1SR
www.cockburnassociation.org.uk

Inverness City Heritage Trust
Town House, High Street, Inverness, IV1 1JJ
www.invernesscityheritagetrust.org

Dumfries Historic Buildings Trust
24 Queen Street, Dumfries, DG1 2JF
www.dumfriestrust.org.uk

North East Scotland Preservation Trust
Viewmount, Arduthie Road, Stonehaven
Aberdeenshire, AB39 2DQ
nespt.org

Glasgow City Heritage Trust
54 Bell Street, Glasgow, G1 1LQ
www.glasgowheritage.org.uk

Scottish Historic Buildings Trust
Riddle's Court, 322 Lawnmarket, Edinburgh, EH1 2PG
www.shbt.org.uk

Perth and Kinross Heritage Trust
Lower City Mills,West Mill Street, Perth, PH1 5QP
www.pkht.org.uk

Stirling City Heritage Trust
Cameron House, Forthside Way, Stirling, FK8 1QZ
www.stirlingcityheritagetrust.org

Glasgow Building Preservation Trust
Room 16, 120 Sydney Street, Glasgow, G31 1JF
www.gbpt.org

Historic Environment Scotland
Longmore House, Salisbury Place, Edinburgh, EH9 1SH
www.historicenvironment.scot

Built Environment Forum Scotland (BEFS)
61 Dublin Street, Edinburgh, EH3 6NL
www.befs.org.uk

National Federation of Roofing Contractors
31 Worship Street, London, EC2A 2DY
www.nfrc.co.uk

Stone Federation Great Britain
Channel Business Centre, Ingles Manor, Castle Hill Avenue
Folkestone, Kent, CT20 2RD
www.stonefed.org.uk

The Lead Sheet Association
Unit 10, Archers Park, Branbridges Road, East Peckham
Tonbridge, Kent, TN12 5HP
www.thenbs.com

The Lead Sheet Training Academy
Unit 10, Archers Park, Branbridges Road, East Peckham,
Tonbridge, Kent, TN12 5HP
leadsheet.co.uk

British Geological Survey
Keyworth, Nottingham, NG12 5GG
www.bgs.ac.uk
www.webservices.bgs.ac.uk

Scottish Lime Centre
2 Rocks Road, Charlestown, Dunfermline, KY11 3EN
www.scotlime.org

Dendrochronicle is led by Dr Coralie Mills, based in Edinburgh.
Email: coralie.mills@dendrochronicle.co.uk

Repairing Scottish Slate Roofs by Moses Jenkins
www.buildingconservation.com

Scottish Stone Liaison Group
www.sslg.co.uk

Changeworks
36 Newhaven Road, Edinburgh, EH6 5PY
www.changeworks.org.uk

Charles Goad Fire Insurance Plans of Scottish Towns 1880s to 1940s
National Library of Scotland, Edinburgh
https://maps.nls.uk

Appendix 4: Model tenements

Glasgow Fire Service plans drawn up during Second World War

Showing examples of typical Glasgow tenements.

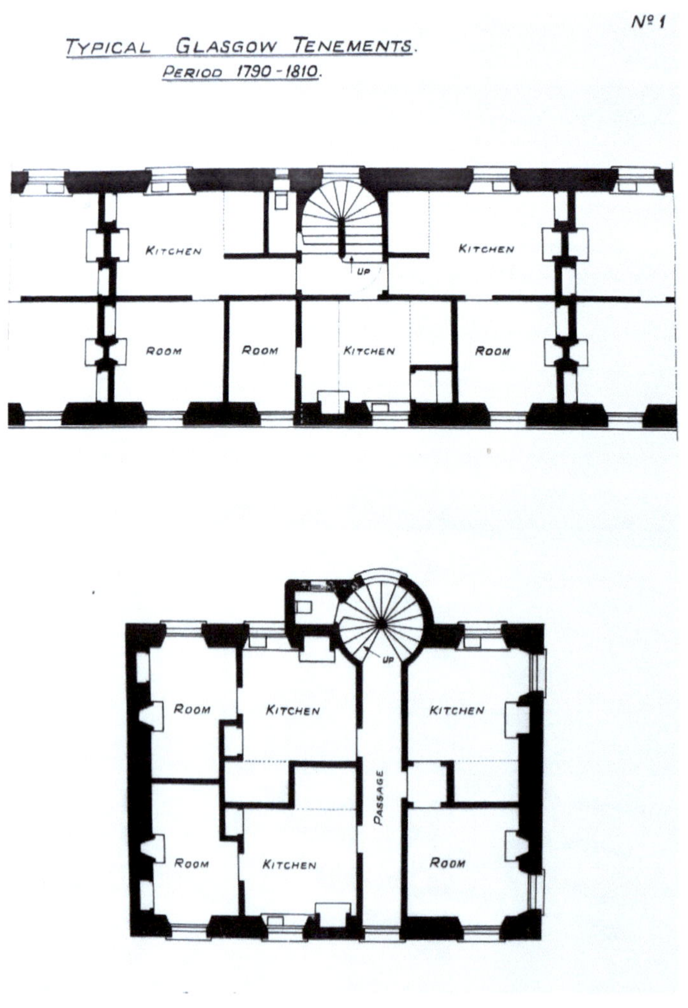

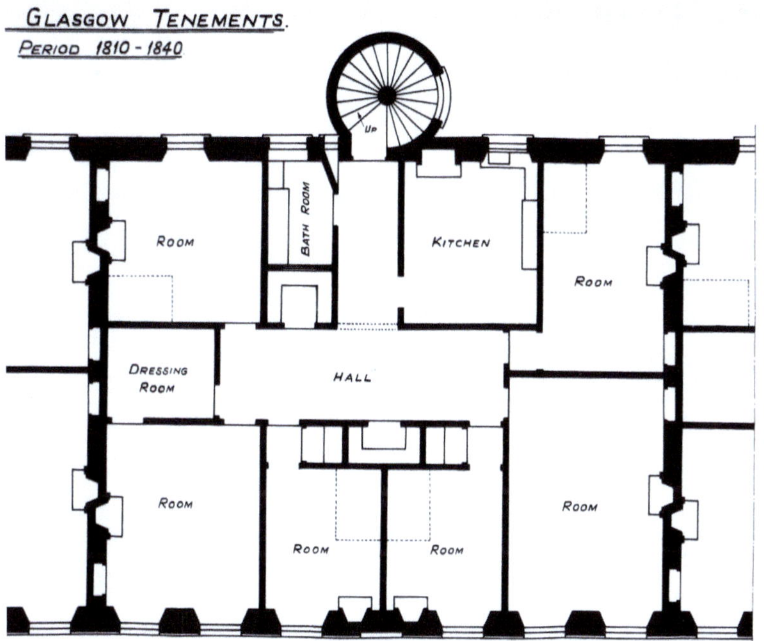

1810–1840

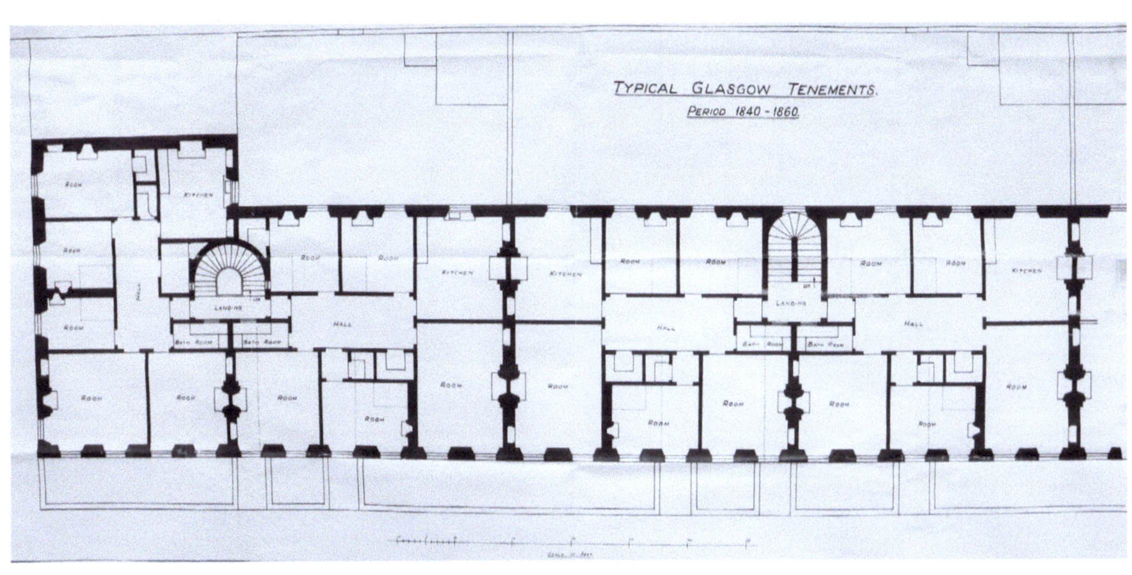

1840–1860

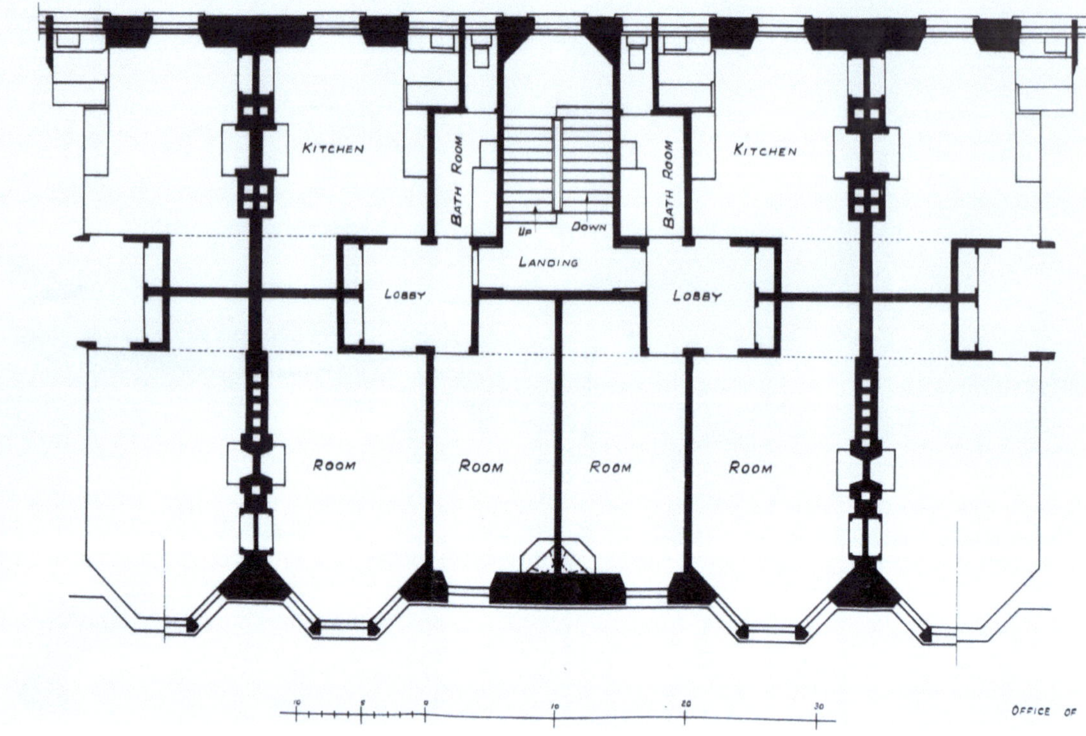

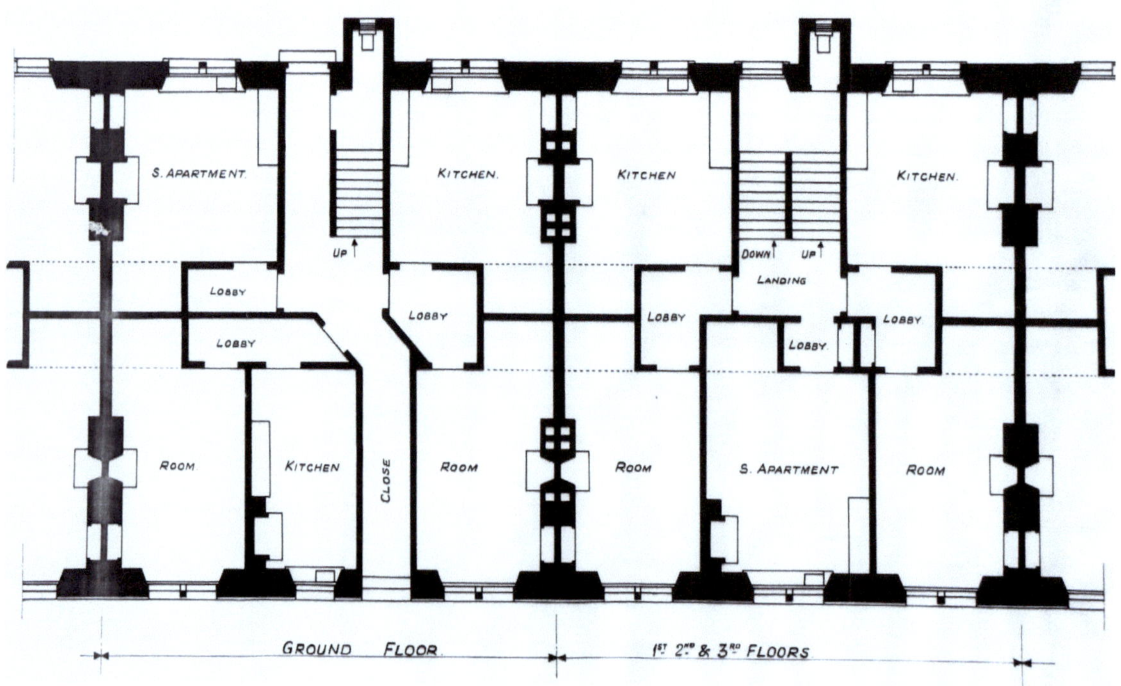

1890–1910 two versions

Charles Goad's fire insurance maps from 1880 (from NLS)

Bridgeton Cross in Glasgow

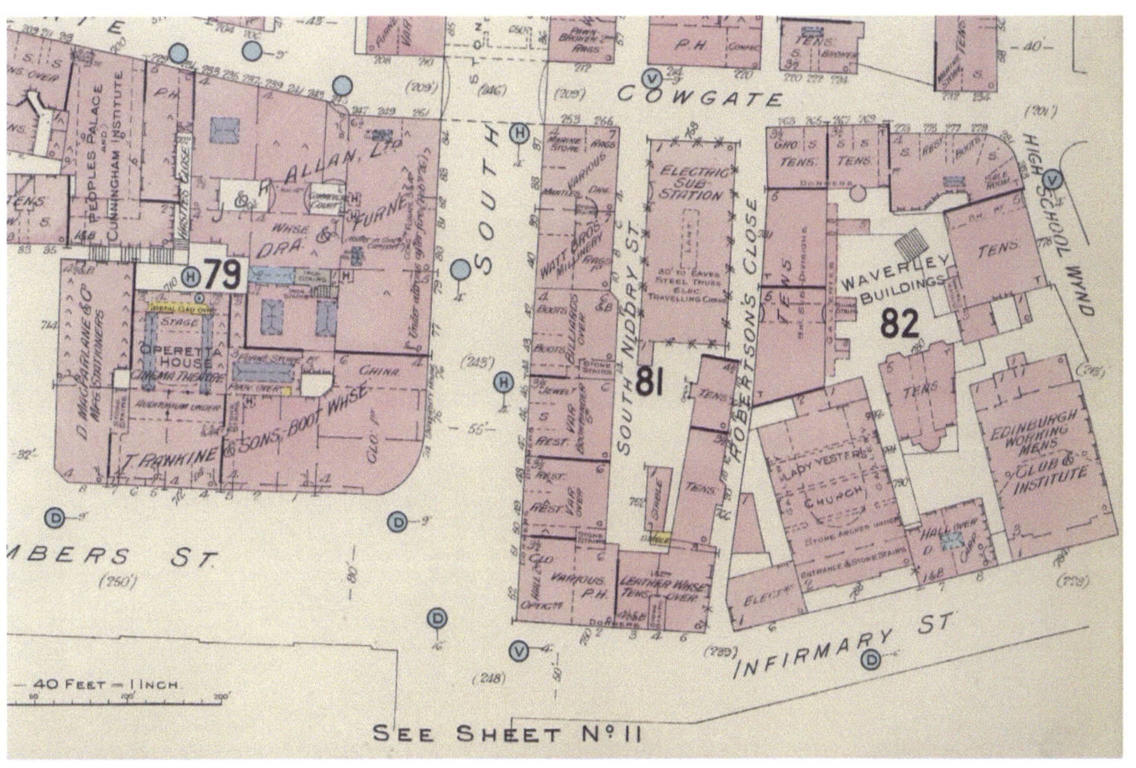

Robertson's Close in Edinburgh

Charles Gourlay's typical tenement plans and elevation for Glasgow students of building 1903

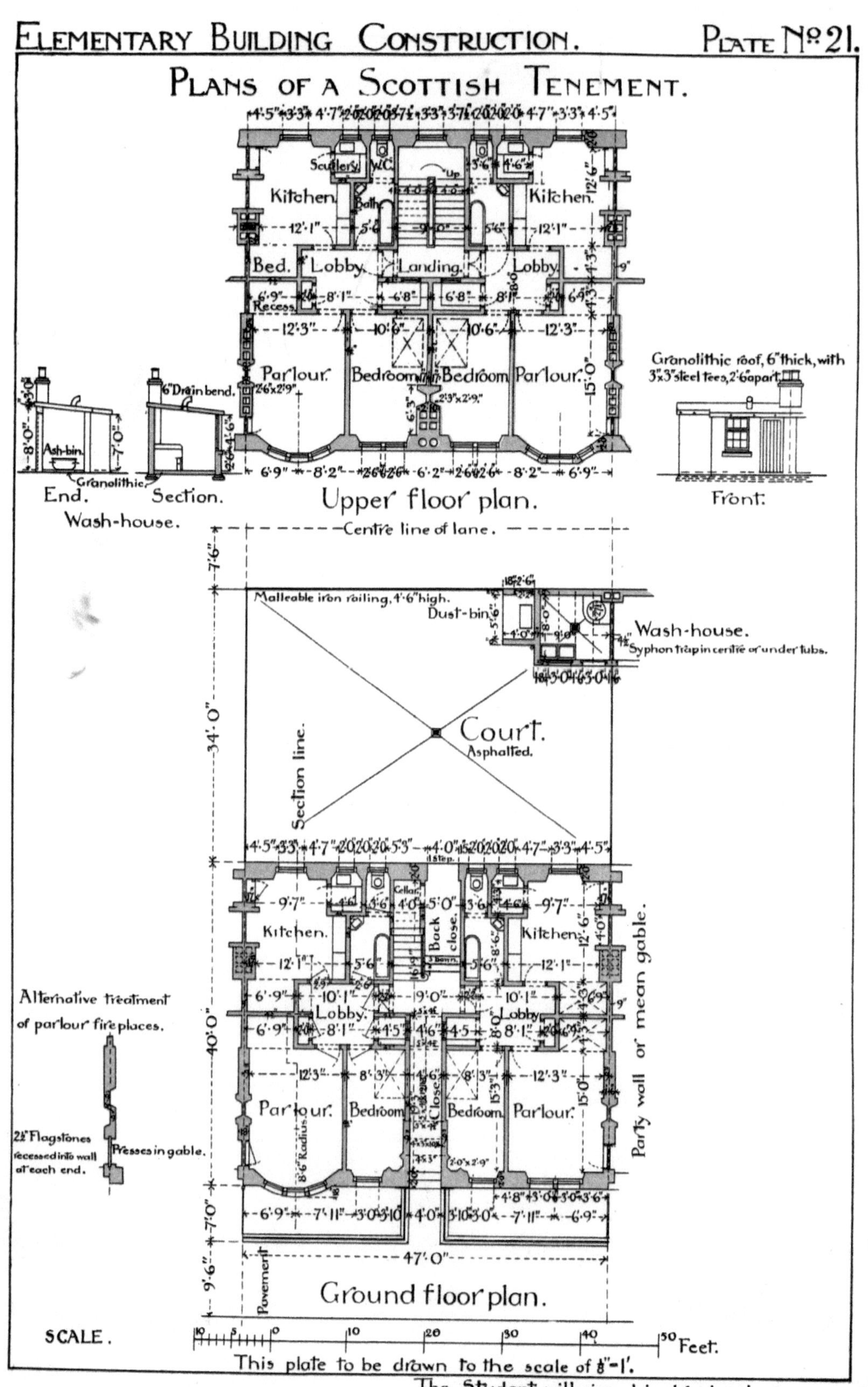

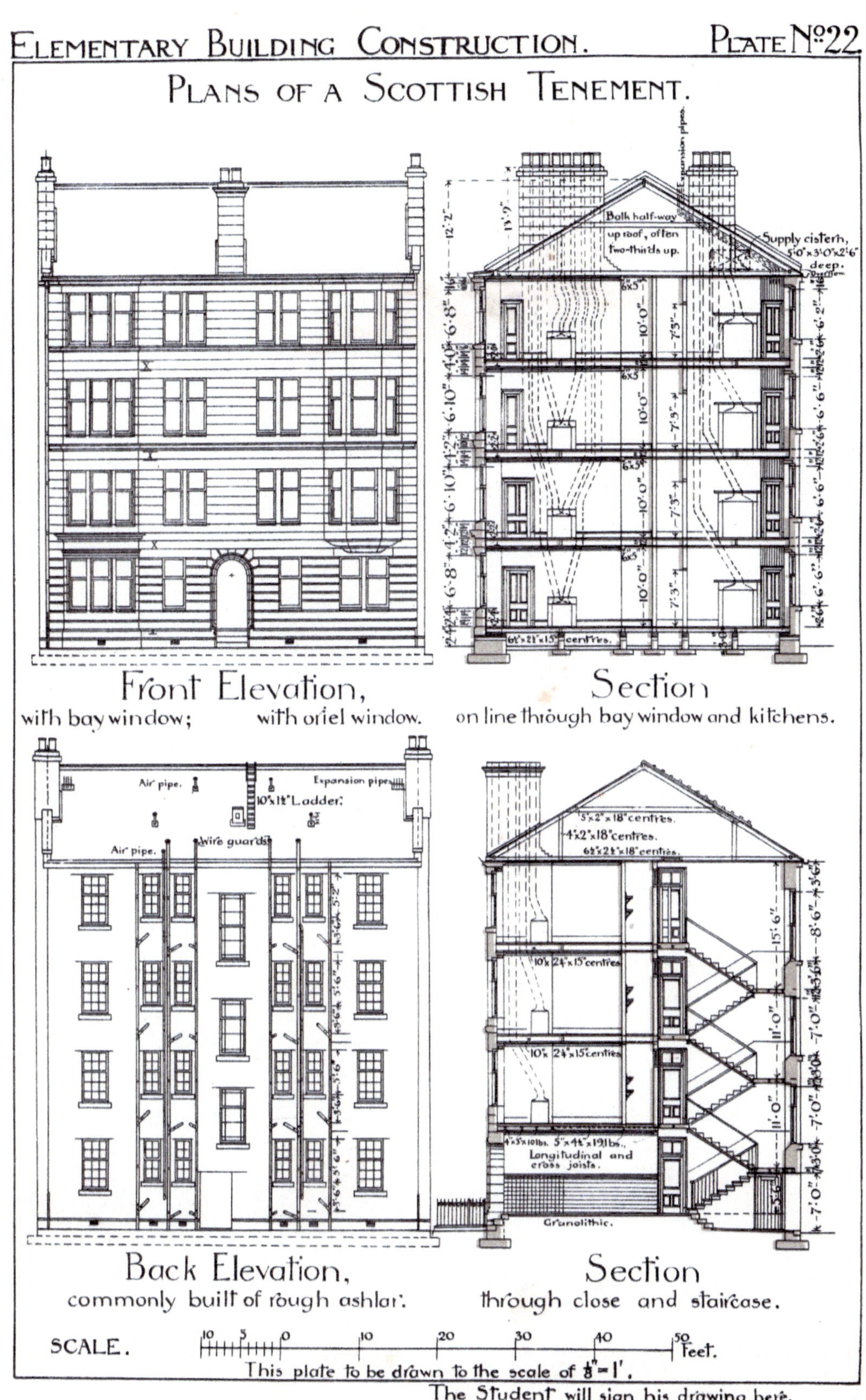
ELEMENTARY BUILDING CONSTRUCTION.
PLATE No 22.
PLANS OF A SCOTTISH TENEMENT.
Front Elevation,
with bay window;
with oriel window.
Section
on line through bay window and kitchens.
Back Elevation,
commonly built of rough ashlar.
Section
through close and staircase.
SCALE.
Feet.
This plate to be drawn to the scale of ⅛"=1'.
The Student will sign his drawing here.

Houses for the working classes of Edinburgh 1860

A model plan was put forward by the Committee of the Working Classes for Edinburgh in 1860. It showed a four-storey tenement with each floor having four flats, all single aspect, looking out to the front or to the rear. Each was equipped with a pantry and a WC. It was adopted for the next 40 years in Edinburgh, although the plan was improved to include baths and a WC in 1887 in Dalry.

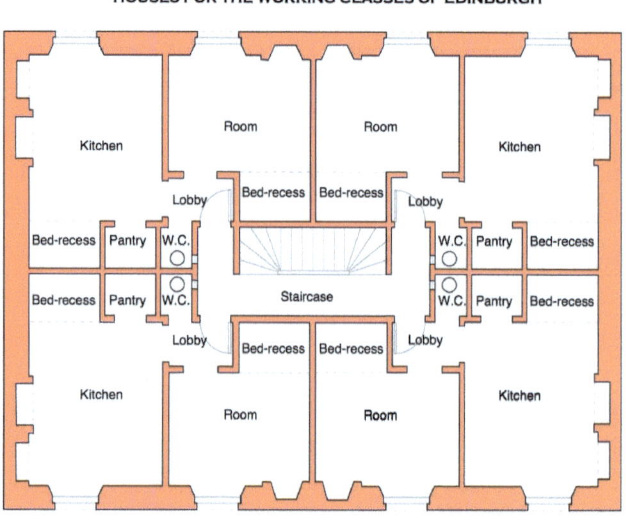

Ground floor plan shows four rooms and kitchen on each of four floors, sixteen houses in a close

Model dwelling houses, Edinburgh 1886

A two-storey tenement was built by James Gowans at the International Exhibition in Edinburgh in 1886. The plan contained four flats, each with two bedrooms, a kitchen and scullery and individual toilets. Two of the flats had a bath. None had a living room.

It showed what was possible at that date and provides a list of contractors and specialists involved, 44 different trades, some of which were as follows:

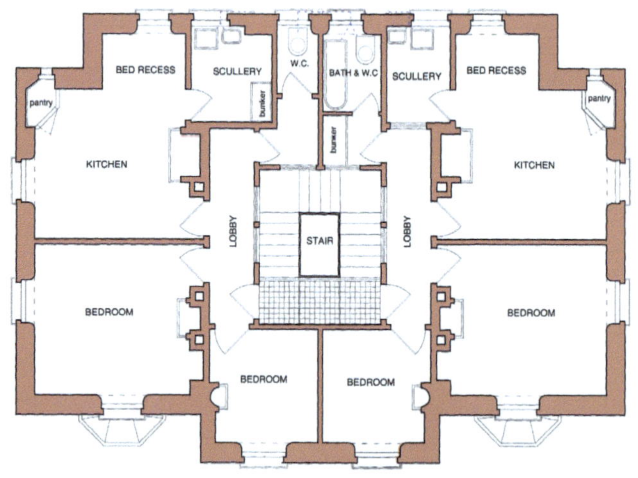

Upper floor plan, mid floor

Gas fitting and ventilation: of two houses using wrought iron tube mounted on wall surface rather than using block tin which would be buried under plaster and difficult to fix and find any gas leak: R. Laidlaw & Son, Edinburgh. They also fitted 'Bower's patent Regenerative Gas Lamps' and fresh air inlets using 'Tobin's Ventilating Tubes' and 'Sheringhams Ventilators'. Gas fitting and ventilation in other two houses by Armstrong & Hogg.

Carpentry and joinery: William Beattie.

Locks and keys: Globe patent and rim locks and fasteners for casement windows from Low and Methven, Edinburgh, for two houses, and Bell and Donaldson for two houses.

Bell hanging: William Bryden. This was a separate trade. The close door was fitted with cranks and wire pulls on two houses and, on the other houses, electric bells, pushes and dials, similar to those fitted in Edinburgh and Paris.

Disconnecting and ventilating drain traps: William P. Buchan of Glasgow.

Rhones, conductors and soil pipes: Walter MacFarlane & Co, Saracen Works, Glasgow.

Sewerage pipes and traps supplied by Wylam Walker of Hexham.

Cast-iron bath; and other water closets and cisterns from Shanks & Co, Barrhead, near Glasgow.

Ovens: Carron Company, Falkirk, also 'Simplex' close or open range oven provided by David Foulis of Edinburgh.

Downie kitchen range oven and boiler and Kinnaird grates supplied by Bailie Walcot, Edinburgh.

Mantlepiece: William Clunas (polished slate).

Glazier: Dickson & Walker. Windows glazed with sheet glass, roof lights with rolled plate glass, bathroom windows with fanlights and side ventilators.

Lathing work (to prepare a base for plasterwork): George Douglas, Edinburgh.

WC: 'Doulton's Combination Closet' which can be used as a toilet with a wooden seat or when seat raised, a 'slop-sink' or a urinal: Doulton's of London.

Cement and lime: Portland cement for the foundations, pointing and drains and hydraulic lime for building walls and partitions: Alex Shaw, Leith.

Slates and slating by Andrew Skater, Canongate, Edinburgh. Stair steps, plates and plinth course: Carmyllie quarries, Forfarshire: Duncan Falconer & Co.

Stones for plinth copings: stone from Myreton Quarry, Dundee.

Whin setts for face of building from Ravelriggs Quarry: provided by J. W. Stratton, Edinburgh.

Pavement: from Caithness with whin setts at the base: John Dobie, Leith.

Linoleum for parlour floor from Kirkcaldy Linoleum Co. Also floorcloth for passages from Shepherd & Beveridge, Kirkcaldy.

Malleable iron stair railings: (malleable iron was cast-iron annealed and given the properties of wrought iron) by William Omit, Edinburgh.

Appendix 5: Scotland's quarries and stone

Craigleith sandstone

A light grey, finely grained, close textured sandstone which was described as 'nearly imperishable'. It has stood the test of time as it was used in the Edinburgh New Town as well as being exported. Much of it was quarried as '*Liver Rock*' as the beds were not apparent, so this would be used for ashlar work. Craigleith was also quarried for rubble stone. As a replacement, Hazeldean sandstone is sometimes used, quarried from Alnwick in Northumberland.

The Craigleith Quarry was closed in 1946 but the quality of the stone was excellent and it has weathered well. In 1985, the quarry was filled in and a Sainsbury's supermarket placed on top of it.

Binny sandstone

Fine to medium grained sandstone which is grey or buff in appearance. Because of bitumen and iron carbonate minerals, it can contain darker rusty brown and black streaks. In urban environments it can discolour rapidly. Quarry: all quarries that produced Binny, Humbie and Straiton are now inactive.

High Nick sandstone

Very light buff sandstone containing irregular iron oxide banding (*Tiger Stripes*). A uniform sandstone is available provided that 'Tiger Stripes' variety is not specified. Quarry: Stancliffe Stone: Alfreton, Derbyshire.

Dukes sandstone

Dukes is an English sandstone which is medium grained with clearly defined bedding planes. It is a pinkish buff and durable although it will weather gradually. Quarry: Dukes Quarry, Whatstandwell, Derbyshire.

Dunhouse sandstone

Often used in Scotland to match buff sandstone. It is fine grained with a uniform texture and grey or buff in colour. It is relatively durable. Quarry: Darlington Quarry, co. Durham.

Craigleith Quarry, now the site of a supermarket

Darney sandstone

Darney dates back to the 1900s, a light gold to buff sandstone of fine grains. Quarry: Hexham in Northumberland.

Stanton Moor sandstone

Another popular sandstone often used in Scotland. Fine to medium grained stone with clearly defined bedding planes. Colour is buff to pink or grey and relatively durable, weathering gradually. Quarry: Dale View Quarry, Derbyshire.

Flagstones (pavements)

Rather like thick slate. Stone formed over 379 million years ago from beds of sediment separated by thin layers of volcanic dust. This makes them easy to split along the grain and so Caithness flagstones have a unique riven (non-slip) finish, which has been used in pavements and buildings throughout Scotland.

Splitting stones in Caithness Quarry. Photo taken by Joan Walsh on 16 September 1997 in a quarry owned by A&D Sutherland Ltd

Caithness flagstone

Hard-wearing flagstone, grey to black. Commonly found in close floors, damp-proof courses and pavements. Quarries: A&D Sutherland, Caithness Stone, location Watten in Thurso. Also in Castletown at Stonegunn.

Arbroath flagstone

A thicker stone commonly used at fireplace hearths. Grey in colour and smooth. Quarry: Carmyllie quarries, six miles from Arbroath.

Red sandstones

Corsehill sandstone

A red Triassic sandstone popular in the early 19th and early 20th centuries. Fine to medium grained sandstone composed of a mix of quartz felspars, mica and iron-rich minerals. It has clear bedding plane and is quite durable but if face bedded it is prone to *delamination*. Quarry: Annan Quarry: Dumfriesshire.

Gatelawbridge sandstone

A red sandstone which was used from the 17th century and much used in Glasgow. Fine to medium grained; may be affected by salts. Quarry: Newton Quarry at Gatelawbridge, Thornhill, near Edinburgh.

Closeburn sandstone

Closeburn quarry near Thornhill in Dumfriesshire. Coarse grained with a deep red colour. Quarry now closed.

Cove sandstone

A red to brown Triassic sandstone which is fine grained and has been in use since the 1890s. Cove is relatively durable although it can be susceptible to salt and frost weathering. Colour variations vary with bedding. Quarry: Cove, near Lockerbie, Dumfries.

Corncockle sandstone

Fine to medium grained red sandstone, composed of quartz with minor amounts of feldspar. Much used in Edinburgh between 1880 and 1900. Quarry: Lochmaben, in Dumfries (now closed).

Locharbriggs sandstone

A red fine-grained sandstone with quartz grains cemented by silica. Clearly defined bedding planes where clay minerals are concentrated. A durable sandstone which tends to weather uniformly although face bedded stones prone to delamination. Quarry: Locharbriggs, Dumfries.

[left] Corncockle stone being split at Lochmaben Quarry

[right] Mauchline red sandstone quarry

St Bees sandstone

A red/brown sandstone with a uniform texture, fine grained, giving a dull red colour. Although durable it can be susceptible to weathering through salts and frost. Quarry: Birkhams, St Bees, Cumbria.

Ballochmyle sandstone

Varies from bright brick-red to an orange-red. Was easy to work with and used on many Glasgow tenements. Quarry: Mauchline near Ayr.

Granites

Aberdeen granite

Extensively quarried from the Rubislaw Quarry to build most of Aberdeen. Rubislaw Quarry closed down in 1970. The stone is a durable, medium to coarse grained stone which is usually pale grey in colour with flecks of mica, but a reddish granite and variations between red and grey can occur. Commonly from Caithness and Orkney. Quarry: Kemnay, Aberdeenshire.

Peterhead granite

Similar to Balmoral granite, which comes from Finland. A coarse-grained durable granite that is pink to grey in colour. Often polished. Quarry: Stirling Hill, Peterhead, Scotland.

Creetown granite

This quarry in Dumfriesshire opened in the 1920s and is still going today. It stocks a variety of stones although many are now imported.

Slate

Although much of the west of Scotland used the West Highland slates, often from Ballachulish or Easdale, Welsh slates were used in many places. In Wales, slates were standardised by 1750 into sizes such as the 'Empress' (26x16in.); 'Princess' (24x14 in.); 'Duchess' (24x12 in.); 'Marchioness' (22x11 in.); 'Countess' (20x10 in.); 'Viscountess' (18x9 in.) and 'Ladies' (16x8 in.); also with intermediate sizes. These slates were termed '*sizeables*'. '*Undersized*' slates, also known as 'peggies'[1] (10x 6 in.) in Scotland, were often sold to the Scottish market, possibly at a reduced

1 Also 'best peggies' which were sized, typically at 24x12 in.

cost. In medieval times, slates were secured to thick marked roofs using timber pegs, hence the term 'peggie'.

Currently there are no slates being quarried or produced in Scotland. They are now sourced from England, Wales and Spain.

Ballachulish

A dark grey/blue/black slate, usually quite thick with a rough surface and evidence of *pyrite crystals*. Came in a variety of sizes so larger ones were laid at the eaves reducing in size as they rose up the roof slope.

Easdale

A blue and blue/black slate. Not quite as thick as Ballachulish and tend to have more pyrites than the Ballachulish slates so they have a rougher finish.

Banff and Aberdeen slate

A sombre grey slate with a grain.

Highland Fault slates

From Arran and Bute, Luss (grey blue and blue), Comrie, Birnam (blue slates), Aberfoyle. Mostly used locally.

Westmorland slates; Cumberland

Typically a greenish slate and blue/green from Burlington at Broughton Moor and Elterwater although also available as blue/grey from Kirkby quarry. The slates are sized or random and are generally thicker than Welsh slates. Often used to replace thicker Scottish slates.

Penrhyn: Welsh slate

Used in Glasgow and around Scotland in the late 19th century. Often rougher in texture than the usual Welsh slate. Two thicknesses were made, the 'Capital' grade which was 5–6mm thick and the 'County' grade which was thicker at 8mm and so more suitable for Scotland.

Spanish slates

A uniform black slate and not produced in random sizes. A much thicker 'Cupa' slate is available for the Scottish market.

Appendix 6: Tenements and the law

by Annie Flint

The tenement in Scotland is not just a particular type of building created by land prices, geography and the availability of stone which allowed building high. It is also a product of Scottish land ownership laws that both allowed its development and governed its subsequent management.

Land law in Scotland

Essentially, the Scottish system of land ownership developed to allow people to own slices of air but with obligations to provide support to the building above and to maintain shelter to the building below. Each owner had a share in common with their other owners of essential parts of the building.

The key feature to note is that tenement owners in Scotland own their flats outright. There are multiple owners within each block. In England and Wales, the situation is that each flat is let on a long-term tenancy – a lease – and the building remains in single ownership. While the Scots version may be more democratic, the English situation provides clarity through a single responsible body to maintain and repair the property.

Multiple ownership requires each owner to be responsible for repairs and the maintenance of the building for the benefit of other owners. Each owner also has rights – specifically in terms of decision-making.

A tenement is now defined legally as any building which is divided into two or more parts horizontally under separate ownership. This incorporates not just the traditional stone-built block of eight flats but also subdivided town houses, modern apartment blocks and 'cottage flats'. Shops and commercial properties may also be incorporated into tenements.

Tenements first developed under a common law that was based on the specific concepts of burdens, servitudes, and common interest. A servitude is a right over a neighbour's land. This could be a right of access, or it could be a specific right of support called 'oneris ferendi', the obligation to provide support. The flat which had to provide this servitude is known as the 'burdened' property.

Common law on tenements

A more detailed set of rules developed covering the 'common interest' of the tenement owners. In particular, the top-flat owner was held to be the owner of the roof above their flat and therefore responsible for its maintenance. The ground flat owner had ownership of the land – often including the outdoor space. This brought a duty to allow access and support to owners above.

The shared areas of the building, specifically the common passageway, the stairs and the roof above the stairs, belonged 'pro indiviso' (without being physically divided) to all the owners who had use of it, all of whom had rights to vote on actions and responsibilities such as to pay a stated share of costs. The dividing line between ownerships was taken to be the mid-point of floor beams or walls.

Such rules were sufficient to allow development but did not stand up to the practical issues involved in repairing and managing a property. In Glasgow in particular, a practice developed of adding a Deed of Conditions to the Title Deed that provided for more detailed rights and obligations. As with all Title Deeds, the written provisions take precedence over the common law.

The Deed of Conditions usually specified who owned which bits of the building, with parts being designated as common (the property of all owners), individual (the property of a single flat owner) and mutual (owned by a number but not all owners). A process of decision-making was also set out which called for votes to be taken of all owners, commonly at a meeting. Decisions were normally taken by a majority. Each owner's share of costs was also specified, often linked to the assessed rental value or rateable value of each flat. The use of a factor, sometimes even named, was also a common provision.

The feudal system of land ownership

The origins of land ownership lay in a hierarchical system of rights to use land in exchange for the provision of services and goods. Initially, the 'vassal' would provide military service to the 'feu superior'. By the 19th century, when tenements started to be developed, the system had become one where the 'feu' was an annual payment. The feu superior would not just receive regular payments but would set out a feuing plan which was designed to maximise the value of the land holding. The consent of the feu superior would be required (normally for a fee) for making changes to buildings and the feu superior would also act as judge if there were disputes over the various burdens and servitudes.

Land owners would sell a 'feu' to builders/ developers. The 'feu' was paid annually to the landowners (or feu superior). Where there was high housing demand, the feu superior could ask a higher feu. So, in Edinburgh, ground for large villas in 1917 could bring £40–£60 per acre per annum in feu duty but tenement buildings could generate feus of £200–£300 per acre per annum. These costs were, of course, passed on in rent.

Scottish parliament: land reform and tenement legislation

In Edinburgh and other parts of Scotland, as buildings aged and started to require major repair, the common law proved burdensome, particularly to top-flat owners. One rumour has it that the need for legislative change gained impetus due to the unfortunate experiences of a senior law lord who also happened to be a top-flat owner of a property adjacent to Edinburgh Castle and who found himself responsible for the cost of reroofing the whole historic property. No matter its origins, a process of law reform was initiated, which ultimately culminated in the Tenements (Scotland) Act 2004.

At this time, in the early days of the new Scottish parliament, the role of feudal law was also being questioned. A range of land reforms were being called for but the rights of feu superiors (often by insurance companies and other land investment companies) to charge their 'vassals' a large fee to alter their properties and to reclaim land once the original intended land use, such as land provided for school sites, became obsolete.

The Title Conditions (Scotland) Act 2003 was one of the first pieces of legislation to be passed by the newly devolved Scottish parliament, and abolished feu superiors completely. But this left a gap. Who was to enforce burdens and hear disputes? The Act shaped burdens and servitudes into a system of real burdens (real as in tied to the land rather than to a person) which allowed benefitted owners, individually or as a group, to enforce burdens. The Lands Tribunal was to be the judicial body which heard disputes. Powers to change outmoded Title Deeds were also included and these included powers to appoint and dismiss property managers. The Act also included provisions for a Development Management Scheme (DMS).

The Tenements Scotland Act 2004 set out clear responsibilities for owners to maintain their buildings, and introduced a Tenement Management Scheme (TMS), which replaced and codified the common law. Now, where Title Deeds were unclear or unworkable, owners had a clear schedule of rights and responsibilities to govern their actions. The Act also introduced some new rules which covered all owners. These included a duty to maintain (which replaced the common interest burden of support and shelter); a new provision that properly made decisions were to be binding on all owners; and the introduction of compulsory building insurance for the common parts of the building. Where shares of repair costs were not set out in Title Deeds or were not workable, costs were to be shared equally between owners. The exception was where flats were of very different sizes, in which case costs were shared according to floor area.

It was not considered possible to change who owned what in a tenement – this would be a breach of human rights – but the concept of 'scheme property' was superimposed on top of original ownerships. The key parts of the tenement structure and the roof were held to be scheme property which was the responsibility of all owners to maintain.

Repair responsibilities under the Tenements Scotland Act

Common repairs

- **The ground** (solum) on which your building stands (but not always the garden).
- **The foundations.**
- **The external walls.** If the wall in question is partly an external wall and partly an internal wall, then the individual owner should take responsibility for the internal face including the laths or timber frames that hold the plaster or

plasterboard. The stonework that forms the load-bearing structure should be paid for as a common repair.

- **The roof** (including the rafters).
- **Other structural parts** of the building such as beams, columns and load-bearing walls.
- **The close and stairs, including the close windows and close skylight** (when they are not mutual).

In a limited number of circumstances, owners may be responsible for the maintenance of parts of the building and its surroundings that they do not own, such as the garden or loft.

Mutual repairs

- Owners are responsible for the **external wall** to the midway point. But if the wall in question is a party wall between two different tenements, then the owners of both tenements will mutually share the cost of repair. Where there are **mutual chimneys** (at party walls), the repair of these chimney heads is treated as a mutual repair, although the individual flues and chimney pots are treated as individual repairs.
- **Wallhead chimneys**, and chimneys mid-roof, usually take flues from fireplaces in different flats. In this case they are treated as mutual repairs shared between the owners whose flues are connected to the chimney.
- **Close and stairs**: some buildings contain flats or shops which have their own entrance, e.g. main door flats. Owners of these parts of the building may not be responsible for maintaining the stairs and entrance ways. However, every flat in the building that has a door to the close or stair – even if it is not used – is almost always responsible for stair and close repairs.
- **Drainpipes** which serve flats on one part of the building only. Although normally considered as common property, like the gutters, they may only serve the roof on one side of a tenement, so they might be the responsibility of owners of that side of the tenement. Internal downpipes serving a valley gutter would be common.

Individual responsibility

- **Floors**. However, ground floor timbers are usually treated as part of common responsibility, as are ceiling joists.
- **Inside face of walls**: the inside face of party walls and close walls to the half way point.
- **Windows**, including small skylights associated with a flat.
- **Services** (water supply, drainage, electricity etc.) from the point where they leave the common services.
- **Front gardens**: normally the responsibility of the ground-floor flat owner.

[left] An Edinburgh tenement. In this case, the mutual chimney head, shared between two adjacent tenements, is coloured purple. The individual ownerships are coloured yellow and the common parts blue.

[upper right] A Glasgow tenement. The wall contains a mutual chimney head, which needs dividing between the two closes. All floors are individual, with the joist acting as a division line between the upper and lower flats.

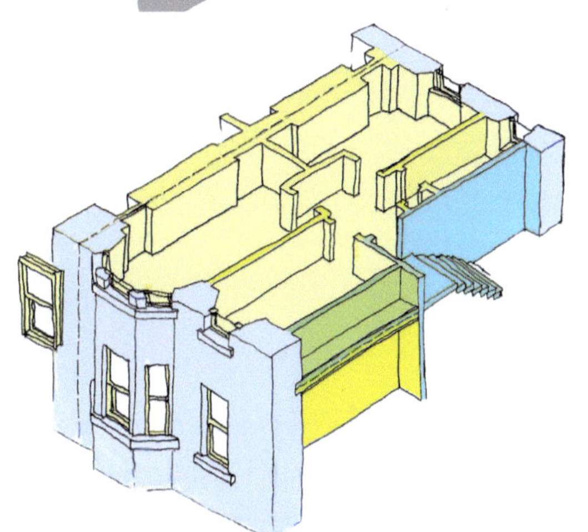

[right] Parts belonging to a home. Individual flat owners own most internal walls and surface finishes of external walls. Windows are the owner's responsibility. Party walls may be mutual or common. Floor joists shared between flats above and below.

Private landlords and the building of the tenements

During the 19th century, Scotland's population became increasingly urbanised through the push and pull factors of famine and growing industrialisation. In Glasgow, for example, the population increased tenfold in the 100 years between 1801 (77,385) and 1901 (784,496). Before the opening of the Glasgow subway in 1897, people needed to live close to their work and the rise of shipbuilding in Partick, heavy engineering and chemical works in Shettleston and Parkhead, and locomotive building in Springburn led to the development of tenement suburbs in these areas. In Dundee, it was jute-making that fuelled the increase in the population.

However, tenement housing conditions were poor and the ill health of the population was an issue of national concern. Pandemic cholera was blamed on overcrowding; not surprising, given the small, airless, even windowless rooms, and the lack of open space between tenements.

The Burgh Police Act of 1892 brought a requirement for buildings to have a certain minimum size of room, large windows to provide light to the back of deep tenement buildings, and sanitary facilities. Buildings were also 'ticketed' in an attempt to regulate occupancy levels. But in increasing housing standards, the Burgh Police Act also increased building cost.

The tenement building industry was vertically integrated. The developers of the day were landlords who expected to have a continuing involvement with the building through rental. In Glasgow, for example, many of these were small landlords. In 1900, it was estimated that Glasgow City's landlords numbered 6,661, with 89 per cent being private individuals (mostly women) and trusts. These landlords held an average of just over three flats each. (It is said that in Edinburgh, by contrast, tenements were built by larger property companies who might own several blocks.)

Landlords, who were advised and assisted in obtaining finance by their property managers, known in Scotland as 'factors', had to face fluctuations in profit levels and cyclical uncertainty. There were cycles of housing shortages and then over-building. For example, between 1891 and 1911, the number of households rose by 16 per cent but the number of available dwellings rose even more, by 31 per cent.

A further element which led to political antagonism towards landlords was the landlords' predilection to be litigious (especially when compared with their English counterparts). Between 1886 and 1890, London landlords issued one warrant (most likely for rent arrears) for every 1800 inhabitants. Glasgow landlords issued one warrant for every 54 inhabitants.

Finance could also be precarious in the days before fixed-rate mortgages. Landlords could be faced with the short notice recall of bonds if interest rates rose. Their response was to encourage the great majority of working-class tenants, especially skilled workers, to commit themselves to annual lets. This system dominated working-class housing. In 1905, 83 per cent of all tenancies were let with an annual rental below £20. Tenants had difficulties in moving if their circumstances required. In 1911, the system of annual tenancies was abolished by the Housing Letting and Rating Act 1911.

This abolition of annual rents was just a minor setback in the development of the private rented sector. The 1909–10 'People's Budget' raised tax on vacant land and ground annuals. Rates also increased and the result was a decline in rent levels and reduced profit margins. There was a steady process of concentration among the largest landlords which left 25 per cent of all houses in the hands of 3 per cent of owners.[1]

The end of the 1914–18 war brought a period of new housing shortage and scarcity rents. However, this time, the response from tenants was to strike, and the 1915 Glasgow Rent Strikes led to the 1915 Rent Restriction Act, which transformed a temporary building slump into the equivalent of a capital tax on landlords and private mortgagees.

In 1917, the Royal Commission on the Housing of the Industrial Population of Scotland (Ballantyne Report) blamed slum conditions partly on the

Rent Strike poster

1 Dr Duncan Sim, 'The Scottish House factoring Profession', *Urban History,* 1996, 23 (3), pages 351–71.

failure of the speculative builder and partly on the tenement tradition, using the criteria of overcrowding based on the number of people living in each room. This seemed to ignore the fact that Scottish tenements had fewer, larger rooms than houses in England and this gave rise to an unfavourable comparison between the two nations.

The 1919 Housing, Planning etc. (Scotland) Act (Homes for Heroes) brought more powers to allow local councils to demand that house owners repair. The Act also gave further powers to councils to build – and much of this was in 'cottage' form.

Most of the tenemental areas remained, however, but generally with declining repair condition. Landlords (often through their factors) argued that rents were too low to allow often poorly constructed properties to be properly maintained. In the 1950s, the then Glasgow Corporation declared a series of Comprehensive Development Areas, almost all directed towards the older tenemental areas built up around the heavy industries of the 19th and early 20th centuries. These declarations had the result of further blighting tenement areas.

Further rent acts, such as in 1957, brought in 'controlled rents'. The result was further sales of rented properties, often to only semi-willing purchasers who could afford little better, either to own or rent. The 1959 Housing (Scotland) Act brought in improvement grants, but the factors argued that it would be better to subsidise rents and this would allow them to do a better job of improvement. However, tenement improvement was bedevilled by a range of complications – small flats which needed to be amalgamated and whose occupants needed to be rehoused and compensated, the problems of multiple ownership as well as the sheer degree of work required to repair and update.

By 1961, levels of private rent had dropped from 90 per cent to just one third of all stock, with 42 per cent of all stock owned by local councils and a quarter being owner occupied.

Estimates for 2022 show that the tenure continues to diversify, with 58 per cent of tenement dwellings being owner-occupied, 23 per cent social rented, 4 per cent vacant or second homes, and just 15 per cent being privately rented properties. However, two-thirds of privately rented properties are flats, and within tenements, levels of private rent are much higher, with an estimated third of all older stone-built and more modern tenement flats being rented.

The managing agents

The role and even the existence of today's property factor is almost completely dependent on the role of the tenement in housing provision. The factor had initially been the rural estate owners' agent, managing their property to maximise income. As populations moved to the cities, so too did factors who became landlords' agents in a time when 90 per cent of households lived in rented property. So, the 19th and early 20th-century history of factoring is linked with the fate of private renting, particularly of the prime form of working-class housing, the tenement.

Since the 1960s and 70s, when much tenement housing had been sold into owner occupation, the factor has become the individual homeowner's agent, developing skills in the management of buildings in multiple ownership. But in the future, factors will need to adapt to manage increasingly complex buildings built using new and sometimes unproven technology to provide services to large complexes of residences with a wide range of owners who may be scattered across the globe.

What did the factor do?

The role of factor was imported and adapted from rural Scotland. Here the factor had been the landlord's agent and he was responsible for all estate management functions, including agricultural matters, and tenancy issues including eviction. The role played by factors in the Highland Clearances, particularly the brutality of Patrick Sellar, the factor to the Duke of Sutherland, led to factors being widely regarded as bogeymen. Even before the height of the Clearances, Robert Burns had written scathingly of factors:

> I've notic'd, on our laird's court-day,
> An' mony a time my heart's been wae,
> Poor tenant bodies, scant o'cash,
> How they maun thole a factor's snash;
> He'll stamp an' threaten, curse an' swear
> He'll apprehend them, poind their gear;
> While they maun stan', wi' aspect humble,
> An' hear it a', an' fear an' tremble!
> I see how folk live that hae riches;
> But surely poor-folk maun be wretches!
> Robert Burns, 1785[2]

In 19th-century urban Scotland, the role of the factor, like that of their rural counterpart, was also that of landlord's agent. But rather than being an employee of the Laird, the factor was an independent agent providing social distance between tenants and private landlords and dealing exclusively with residential management. The factor selected tenants for flats, kept properties maintained, and set and collected rents. This is similar to the role of today's letting agent. However, tenants' rights were far more limited than they are today and the factor would have had a position of considerable power, including the power to evict and even to seize possessions where rents were not paid.

In addition to tenancy management, the factor was also responsible for maintaining the property, and the payment of ground rents, feu duties, etc. The factor would have paid these fees and costs from the accumulated rental income held and forwarded a net payment to the landlord twice a year. In 1911, the Royal Commission on Housing in Scotland suggested that the going rate for collecting rents and management was 1¼ per cent of the rental income (this compared with an allowance of 10 per cent being set aside against repairs).

Factors, however, could derive their income in a number of ways apart from just collecting rent. They also collected rates for the council and were paid a commission for this work. They invested the rents and rates that they were collecting and kept the interest. Finally, they would take a five per cent commission from tradespeople who carried out repairs.[3]

This role as a rent collector is evidenced in an examination of the Post Office Directories of the day to find out about the role of those who were recorded as being house factors.

2 Rabbie Burns (1759–1796), 'The Two Dogs'.

3 Dr Duncan Sim, 'The Repair and Maintenance of Properties in Mixed Ownership: A Study of House-factoring in Glasgow', Urban Studies, 1997, 34 (2), pages 255–73.

Analysis of the Post Office Directories for the major Scottish towns of Glasgow, Edinburgh and Dundee shows that all three cities had a body of house factors.

From the records for the other occupations of factors in Glasgow and Edinburgh, a fifth of factors also described themselves as insurance agents. This was an activity that sat well with rent collecting, as insurance premiums in pre-NHS days were often collected on a weekly or monthly cash basis.

The role of the factor was in fact somewhat wider than that of a letting agent. Some factors did build and own properties themselves, but with only a five per cent holding, were more minor players as landlords than many tenants gave them credit for. Other factors became more involved in the property development process, bringing a range of small investors and builders together to facilitate tenement building.[4] This perhaps was a more common practice in the west of Scotland where there were a greater number of small investors than in the east of the country where landlords were better able to afford the building of a whole tenement property.

In Edinburgh, 14 per cent of factors also listed themselves as being joiners, surveyors, and architects, implying a close involvement in the development process. Edinburgh house factors were also likely to describe themselves as valuators or appraisers – presumably setting rental values for properties. Only one per cent of factors in Glasgow were likely to describe themselves in this way, despite the fact of their clear involvement in the development process, particularly the introduction of the Deed of Conditions which they pioneered.

The purpose of the Deed of Conditions was to facilitate better management of properties with multiple owners. This was effectively a secondary Title Deed, which would set out which parts of the property were held in common and which each individual investor was responsible for maintaining. It also covered how votes should be taken and the share of repair costs each owner should pay. These Deeds of Conditions would sometimes even name the factor, ensuring a stream of work going into the future. Effectively, the factor was involved in the vertical integration of the rental industry – from procurement and funding, building development and setting out details of ownership through to the continuing management of the tenants and property.

Who were the factors?

These factors, while being a mass of individuals and small firms, linked themselves together through factors' associations which were set up in Edinburgh and Leith, Glasgow, Airdrie, Rothesay, Paisley, Dundee, Partick and Govan (1907). The Glasgow association had 100 members while Partick and Govan had some twenty to thirty members factoring 90 per cent of property in the burgh.

Some of these factors had relatively large holdings of between 2,000 and 5,000 properties with portfolios being built up through mergers or seeking larger institutional landlords as clients. Such institutional landlords included, for instance, John Brown Shipbuilders housing in Clydebank, the National Coal Board, various cooperative societies and trusts such as James Gardner's Trust and the Robert Lawson Trust in Partick, Glasgow.

4 N. J. Morgan and M. J. Daunton, 'Landlords in Glasgow: A Study of 1900', quoted in Duncan Sim 'The Scottish House Factoring Profession', Urban History, 23 (3), 1996, pages 351–71.

Other factoring firms sought to expand their role. Duncan Sim[5] recounts the story of Winning & Fulton (Paisley); this firm was involved in factoring and house letting but were also agents for Phoenix Insurance. Their expansion was into the estate agency business. They factored housing for JP Coats, the local textile manufacturers. Subsequently, they purchased the Coats portfolio and sold off the individual houses. They also managed local council housing in Paisley and then became sole surveyors for the Paisley Building Society, subsequently the Dunfermline Building Society.

The role of the factor – for good and bad

With 90 per cent of properties being privately rented prior to the First World War, the vast majority of the population would have encountered a factor. As the public interface between the tenant and the landlord, the factor was often held to be the culprit in the developing situation of unaffordable rents and poor house conditions. The factors had thus little chance of escaping their bogeyman role. But was this totally deserved?

In 1971, the Francis Committee looking at the working of the Rent Acts concluded that the level of controlled rents in Glasgow was particularly low. However, they also noted a high proportion of well-maintained dwellings in Scotland compared with London and attributed this to the professional factoring system in Scotland and the advice factors gave to landlords and owners.

Some factors did try to improve housing themselves directly and there is a good case study in 388 Tollcross Road in Glasgow. This was a property purchased by the Glasgow Property Factors Association in 1935. In 1962, the factors' association drew up a plan to renovate the block and replan it so that twelve flats became eight flats with bathrooms and kitchens.

The plan was dependent on the Glasgow Corporation providing decant facilities and rehousing of the tenants. But this was before anything that the Francis Committee report might do to salvage the factors' reputation and it was their reputation as bogeymen, aligned with the equally disliked private landlords, that seemed to have held sway with the council. The council refused to cooperate in respect of decant and rehousing. As a result, the whole replanning scheme fell. In 1972, following the area being declared a Housing Treatment Area under the 1969 Housing (Scotland) Act, the factors sold the property to the new Tollcross Housing Association.[6]

The advent of community-based housing associations also led to a diminution of the role of factors as small flats were taken under the housing associations' ownership. Nonetheless, David Watson, the local Govan factor, provided factoring services to the local housing association and was patient over being paid their expenses. One of their employees was also the secretary of the housing association.

The Law of Agency misunderstood

There is a widely held view that factors are more able to carry out work than the owners they act as agents for. However, factors need to operate within the Title Deeds, the Deed of Conditions and the law generally, and in almost all respects they can do no more than owners can. They

5 Dr Duncan Sim, 'The Scottish House Factoring Profession', *Urban History*, 23 (3), 1996, pages 351–71.
6 Miles Horsey (1990), *Tenements and Towers – Glasgow Working Class Housing 1890*, RCAHMS.

cannot act without the agreement of owners and, given that the factor no longer holds a pool of rental income, the factors' business model needs them to secure payment in advance for all commissioned works and payments to external bodies (insurance payments for example).

The one key advantage that factors have is that they can legally enter into contracts on behalf of all owners. Owners' associations, as currently constituted, do not have the legal persona that allows them to do this. If owners seek to commission common repairs themselves, then this has to be done by one individual who will need to take all contractual risk including being responsible for recovering costs from other owners.

The decline of factors

The study of the Post Office Directories carried out for this book shows that, over the period 1875–1940, 970 single individuals or firms of factors were found in Glasgow. Over the same period, in Dundee, 184 factors firms were identified, and there were 63 in Edinburgh. In 1911, there was one factor for every 3,600 of population in Glasgow, one for every 3,400 in Dundee, and just one for every 23,000 in Edinburgh. It is interesting to note how the differing geographical patterns of investment by landlords in the past has had an impact on today's distribution of factors where factoring firms continue to predominate in the west and have a much lower profile in Edinburgh and the east of Scotland.

Analysis of the number of factoring firms over time shows that the peaks of activity were in 1885 and then 1905 – dates that coincide with tenement building booms.

In all three cities, the majority of factors (43 per cent to 59 per cent) were seen in the records for less than five years. In Glasgow and Dundee, over a fifth of factors lasted for over twenty years. In Edinburgh this figure was just seven per cent. Most of these firms were lost by merger. By 1940, there remained 208 house factors in Dundee, Edinburgh, and Glasgow combined.

Analysis of the current Property Factors Register, established under the Property Factors (Scotland) Act 2011, shows 77 established, commercial factoring firms managing over 400,000 properties between them. (This total excludes the sole traders, community-based companies, registered social landlords (RSL) and those factors with under 100 properties in management.)

Registered social landlords and other community-based organisations though counted 125, managing approximately 140,000 properties. The number of firms and RSLs across the whole of Scotland now equates roughly with the number found in the three cities of Glasgow, Edinburgh and Dundee alone in 1940.

The role of the factor has also changed. Today, a factor's prime role will be to maintain the repair condition of properties. To do this, they will need, as a core competence, 'a solid knowledge and understanding of property law, titles and burdens, tenement law, building technology, embracing the various elements that constitute these, at times, complex structures, while paying particular attention to the common parts and their on-going management, maintenance, and repair requirements. Health and safety is also always a critical concern here, involving as it does residents, contractors, and property professionals alike. This issue recently gained added prominence, given the range of consequences that are likely to emerge from the Grenfell Tower tragedy.'[7].

7 See Peter Apps (2022), *Show me the Bodies*, Oneworld Publications.

What happened without factors?

In cities such as Edinburgh, the development of tenements by larger landlords led to buildings being owned in their entirety by one owner. There was no need to develop a Deed of Conditions to manage relations between multiple owners; factors had a lesser role in development finance, and larger landlords seemed to have employed their own rent collectors or used law agents.

Local council acts in main towns

With the break-up of flats into homeownership, a process that took place throughout the later part of the 20th century, problems started to arise. While the Deed of Conditions and factors stayed in place in the west of Scotland, ensuring that the tenement management system could still function, the lack of standardised break-off Title Deeds in Edinburgh, combined with the absence of property factors, resulted in greater complexity for these flat owners and reliance upon common law remedies and common law arrangements over ownership of parts of the tenement.

Edinburgh Council promoted its own Acts of Parliament, such as the Corporation (later City) of Edinburgh District Council Confirmation Order Acts of 1933, 1967 and 1991 which allowed the council to serve repair notices following visits by its team of building inspectors. The council also became more active in promoting repairs through agency agreements with owners that allowed the council to undertake major repairs on the owner's behalf and reclaim the VAT – which effectively paid for the council's administrative services.

The council also established its own tenement management section – The 'Stair Partnership'. This was to be more proactive – not just serving repair notices but also prompting owners to get together with their stair neighbours to form associations and to undertake regular preventive maintenance. The council was also active in providing detailed tenement advice on its website and sending out leaflets to a huge number of tenement owners to explain the 2004 Tenements Act. However, the work of this section was negated by a corruption scandal with allegations of bribery to accept and extend contracts in the Building Control Department, which managed many of the major repair schemes undertaken on agency schemes.

The City of Edinburgh Council took a few years to set up another dedicated tenement support service – the 'Shared Repairs Service'. Like many other councils, Edinburgh uses its 'missing shares' powers to assist owners by covering the repair costs of owners who can't or won't contribute to common repair schemes. In addition to this though, the council has pioneered a relaxation of data protection legislation to allow them to put owners in touch with their co-owners. They have also promoted a smartphone-based app which is now being rolled out Scotland-wide which allows owners to communicate and organise repairs more easily.

In other areas, councils also took active steps to work with tenement owners. In Argyll, for example, the council had areas of rundown tenements, built very similarly to the Glasgow tenements, in areas such as Campbeltown and Rothesay. The Council here promoted a number of Townscape Heritage Schemes in the central areas of these towns. This brought together several conservation and housing grant schemes under the aegis of a local officer to try and bring up conditions in the worst areas of their towns. The council also offered small grants to owners who set up owners' associations, and also published and circulated tenement management and repair leaflets in appropriate areas.

Self-factoring

Where there is no factor managing a tenement, owners have to 'self-factor'. Evidence gathered by the Office of Fair Trading (OFT) (2009) investigation into the factoring market estimated that some 35 per cent of all tenemental property, at that time, was in fact 'self-managed' or 'self-factored'. However, there can be many degrees of organisation amongst those who self-factor. Some are well organised having written owners' association constitutions, bank accounts to which all owners make regular contributions, and commission regular inspections of their building and act in accordance with the inspection recommendations. But equally, a tenement may effectively have no management with no common insurance policy and no one making any attempt to carry out preventive repairs or sometimes even to deal with critical repairs. This can have serious implications allowing neglect to lead, in some cases, to the collapse of building frontages following damp and rot. And where there are major house fires, rebuilding can take years where some owners have no or inadequate insurance.

A number of bodies sought to rectify this situation by educating owners. The Scottish Consumer Council produced several versions of *Common Repairs, Common Sense* and *The Tenement Handbook* was published in 1980. *The Tenement Handbook* was republished in 1993. More recently, the Under One Roof website is also available to assist owners. But these publications did not just educate the self-factors – many factors and architects have reported using such publications to educate themselves and their staff.

Factoring today

Today, factors are seen by some as being essential tools in good management of tenements, and by others as little different to the bogeymen versions of the past that require regulation and control in order to ensure they always serve the best interests of their clients.

Many factors themselves are seeking to professionalise their industry, instituting qualifications in property management that are specific to the role and legislation in Scotland. Factors are also highly involved in the development of new tenement legislation that would make it easier for owners (and factors) to maintain tenement buildings.

One significant change in recent times was the passage of the Property Factors Act in 2011. This act had three major parts to the regulation of factors. Firstly, it set up a compulsory register of all property factors operating in Scotland. Factors who fail to meet their obligations can be removed from the register. Secondly, the act set out a code of conduct which sets out minimum standards of practice and to which all registered property factors must comply.

Finally, a route for redress was created through the Homeowner Housing Panel (now incorporated into the First Tier Tribunal, Housing and Property Chamber). Homeowners can apply to the Tribunal if they believe that their factor has failed to comply with the statutory code of conduct, or otherwise failed to carry out their factoring duties as set out in the Written Statement of Services.

Training the factors

Another modern change to property management in Scotland has been the drive to professionalise. The factors' representative body – the Property Managers Association of Scotland (PMAS) – has spent the past few years investigating how they can better arm their members for the future. The development of a professional qualification for factors in Scotland has progressed to the point that a curriculum and course has been designed which factors can take either online or in person to prepare them for a Professional Diploma in Property Factoring qualification provided by the Institute of Residential Property Managers. The curriculum covers expected topics such as legislation, technical and commercial, finance and management issues. Importantly, taking the bogeyman reputation head on, the course also covers issues of ethics, consumer focus, best practice in communications and transparency, confidence building, liability and trust and an understanding of the varied and often conflicting perspectives in tenement management.

Future tenement legislation

Despite the legislation set out above, and the advancements in factoring practice, tenements were approaching what the Royal Institute of Chartered Surveyors (RICS) dubbed a 'Condition Cliff Edge'. This highlighted that poor repair conditions continued in tenements and that it was almost impossible to deal with energy efficiency improvements in older types of tenements. Newer 'apartment blocks' also brought more problems through the use of complex structures and innovative materials and it became apparent that the existing legislation could not deal with the overcladding crisis that was the fallout of the Grenfell Fire tragedy.

In 2020/2021, a Scottish Parliamentary Working Group set up to look at legislative change to improve tenement maintenance. This group, made of organisations throughout the housing sector including factors and other professions as well as politicians, ultimately proposed a number of changes that would affect all tenement owners in Scotland. The three specific headline proposals were:

Making it a legal requirement to set up an owners' association

This new style of owners' association would be a corporate body, registered as a special type of limited company under UK Company Law. This would give owners the power to enter into contracts which so far they had been unable to do as a group. The association would be able to make decisions about the management and maintenance of the tenement in the best interests of all the owners. The idea is that this owners' association should be based on the concept of the Development Management Scheme (DMS), which was set out in the Title Conditions Act. Other provisions of the DMS allowed for the creation of sinking (building reserve) funds and a streamlined decision-making process that revolved around the approval of an annual maintenance plan with implementation delegated to a manager. It is very likely that factors will take on the position of manager but in this they will face competition from other professions such as business managers and surveyors.

Owners would be legally required to undertake a regular five-yearly property condition survey

This would allow for a professional report to be prepared for all current owners and prospective purchasers which would describe the actual condition of the building with recommendations on what work was needed along with a programme of maintenance works.

Introduction of mandatory building reserve funds

Often referred to as a 'sinking fund', this would ensure that owners make regular monthly contributions so that when a major building repair issue arises, there is a ready pot of cash available. The building reserve fund would remain with the building and form part of its value.

The above proposals were accepted by the Scottish government and were being examined in further detail by the Scottish Law Commission as of 2022, with potential legislation coming in 2026–7.

While these are the three headline proposals of the working group, a number of other changes have also been identified that could potentially help tenement owners maintain buildings.

Appendix 7: Police Acts

Police Acts and bye-laws led to severely stringent building codes. After the Burgh Police (Scotland) Act of 1862, successive improvement acts enable burghs to frame their own building regulations which spread to other burghs: 'Partick, Dundee (1873), Govan (1874), Irvine (1876), Leith (1877) Port Glasgow, Kirkcaldy, Perth (1878), Edinburgh, Aberdeen, Kilmarnock (1880).'[1] The Dean of Guild Court was used to effectively control developments and further Police Acts were required after the Public Health (Scotland) Act of 1897. Guidance was given on the size of chimney flues, the quality of materials for mortar, wall thickness, foundations, window size, damp courses, roof joists, chimney construction, guttering and the quality of materials. This all meant an increase of 25–45 per cent in material costs compared to London housing. Views expressed by architect T. L. Wilson that the Glasgow Acts of 1892 and 1900, although they had raised building standards and improved sanitation, had led to an increase in rents and that: 'as the requirements of the London Act are sufficient for all purposes of safety it is relevant that the Glasgow Act involves an absolute waste of money and to that extent interferes with the erection of cheap dwellings.'

The other controls set by the Police Acts were as follows:

- Minimum floor to ceiling heights (nine feet).
- The need for ventilation (no more single ends).
- The requirement for a piped supply of drinkable water to each house.
- A toilet connected to the main drainage system.
- Communal stairwells to be ventilated.
- Backcourts to be no less than one and a half times the height of the tenement.

1 Richard Rodger, *Scottish Housing: Crisis and Confrontation in Scottish Housing 1880–1914*, page 36.

Appendix 8: The English tenement around 1730–1930

Back-to-back housing

'By building tight-packed parallel streets devoid of green areas, developers could achieve a density of nearly seventy-seven houses per hectare. The combination of a minimum standard with private development for profit, however, effectively determined the design of every working-class street: the developer would build to the highest density within the by-laws.' [1]

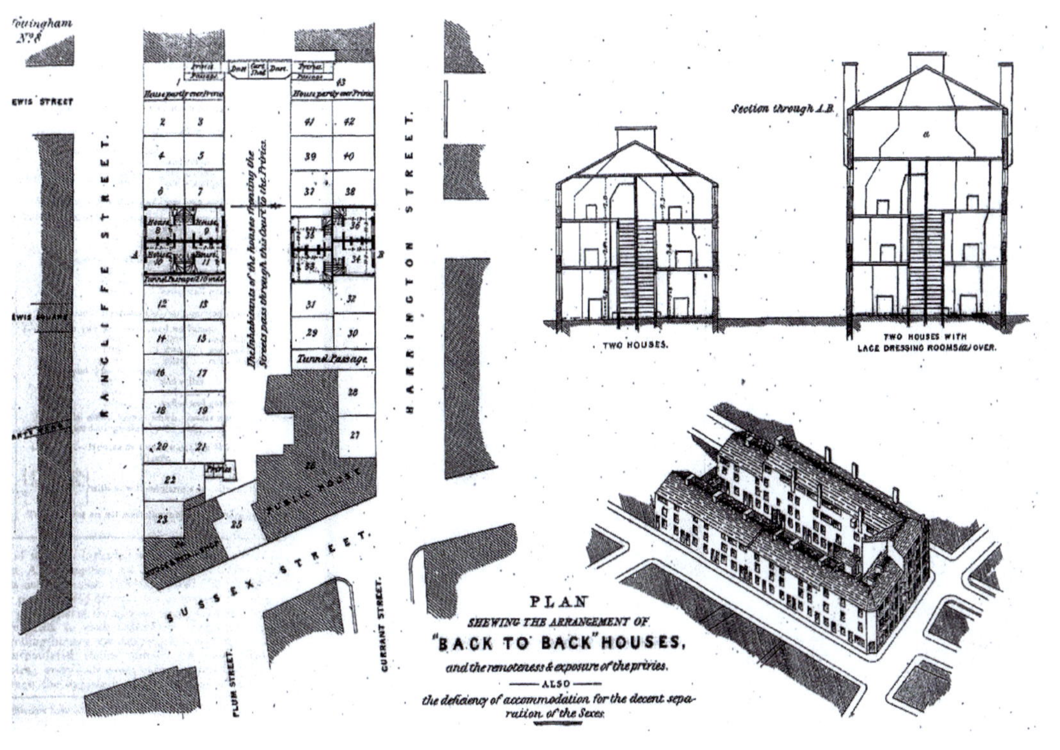

Back-to-back housing

1 *Architecture; From Prehistory to Climate Emergency*, Calder Barnabas.

[left] Back-to-back housing, Crosbie Street and Blundell Street 1848, Liverpool

[right] Huntly Street, Liverpool

In 1706 back-to-back housing started in Bermondsey, followed by Birmingham, Sheffield, Liverpool, Manchester and Liverpool in the 1780s. There were two house types: the court and cellar building. They had no proper sanitation and drainage, or adequate ventilation. There was a standpipe water supply in the street and the streets were rarely cleaned.

In Liverpool it wasn't until 1864 that this type of housing was finally banned but then the city consisted of 17,825 court houses.

In Leeds, the first purpose built back-to-back housing was constructed in 1787. The Leeds Improvement Act in 1842 legislated that each new house must have a proper privy, and it must discharge to a public sewer. In Leeds in 1890 back-to-backs were being built. I stayed in one of these houses in 1973.

They worked by charging the tenants an entrance fee and weekly subscriptions. In the 1840s permanent societies were being formed which were able to offer a more flexible approach.

In 1844 the Metropolitan Building Bill aimed to ban back-to-back houses in London although the bill was abandoned. So back-to-back housing persisted until, in the 1880s, the Metropolitan Association for Improving the Dwellings of the Industrious Classes built Gibson Gardens. (Established in 1841, this association was a forerunner of the modern housing association. Their first project consisted of 21 two-roomed and 90 three-roomed flats in four-storey blocks; although built in four storeys, it was one of the few survivors that made it into the 20th century.)

Modern flats

The first block, designed byHenry Darbishire in a red-brick Jacobian style, was for Peabody's Trust. It opened inCommercial Street, Spitalfields, on 29 February 1864.

Water closets were grouped in pairs, with one WC shared between each of the two flats.

In its early days, the Trust imposed strict rules to ensure that its tenants were of good moral character. Rents were to be paid weekly and punctually; there was a night-time curfew and a set of moral standards to be adhered to.

The Peabody Group are still building homes in London and the south-east after 150 years and the company has an active research programme on housing issues.

The Housing of the Working Classes Act of 1890 gave local authorities the power to build housing although few did much building until the 1920s.

In poorer areas, larger houses were simply subdivided into flats. The grand and terraced houses that were deserted by their owners who moved to the country, were 'made down' in much the same way as occurred to large houses in Scotland. Once abandoned by the middle classes, the houses were divided into small rooms for individual families. This was happening in the 1830s and, in 1859, George Baldwin, editor of *The Builder*, noted that 'vast numbers of houses are rising up every month in this metropolis built with a view to their occupation by single families which, even from the time they are finished are let in tenements of two or three'.[2]

Peabody Trust Estate, Spitalfields, Commercial Street, London 1863, *Illustrated London News*

2 George Godwin (1859), 'Town Swamp and Social Bridges', *The Builder*, 1859.

City of Liverpool, Labourers Dwellings, Hornby Street area, 1903

So the English referred to a tenement as a subdivided house at this time, usually without even an outside privy, or with a single privy serving an area which drained into a cesspool which served hundreds of families and was rarely cleaned out. To quote the French physician Louis Simond: 'these narrow houses, three or four storeys high, – one for eating, one for sleeping, the fifth at the top for servants, – the agility, the ease, the quickness with which the individuals of the family run up and down, and perch on the different storeys, give the idea of a cage with its sticks and birds … a suite of apartments all on one floor, even of a few rooms only, looks much better, and is more convenient.'[3]

Tenement housing was built in Liverpool in the 1920s to house families who were living in single rooms.

3 David Olusoga and Melanie Becke-Hansen (2020), *A House Through Time*, Picador, page 81.

[above] Hughes Field Estate, Greenwich, 1926

[right] Nos 10 and 11 Park Place, St James', London 1904

[below] Harcourt House elevation, London 1907

Painting of Harewood House

It was the beginning of balcony flats, built in Liverpool as labourer dwellings in 1905 to replace the back-to-back flats.

Hughes Field estate built in 1926 was a forerunner of the block housing that had some features of the block housing that was built in Ferguslie Park.

Investors did start to build tenements around 1890, although mostly for the better off in urban areas. They tended to be large apartment flats, six or seven storeys high, and were called 'mansions', rather than tenements or flats.

Eventually inner-city flats were built, mostly in London and for the better off. Shaw Sparrow in 1907 noted that speculators in flats 'would do well to note that the growing exodus from London into the Garden Cities proves beyond doubt that flats have popular rivals to contend against; and these rivals will gain the day if flat-owners show less tact than the railway companies do in their business dealings with persons of small income'.[4]

4 A flat dweller's point of view. Walter Shaw Sparrow (1907), Flats, *Urban Houses and Cottage Homes*, Hodder and Stoughton, page 18.

Appendix 9: Timeline Of Events

Timeline of events from 1400 to 2019

1400	Early mention of 'tenement', which refers to land ownership, not the structure. Name derives from the then standard unit of landholding, the 'burgage plot', or 'tenement' (Adam, 1978).
1450	GB runs out of oak.
1544	Henry VIII declares war on Edinburgh and many timber houses are burnt down, after which the city council encouraged the enclosure of open galleries and stairs, preferably in stone.
1552	Parliament lays down guidelines for sharing the burden of costs for rebuilding the burnt lands and tenements caused by the English invasion.
1559	John Knox's sermon in St Giles, Edinburgh was part of the start of the Reformation in Scotland.
1561	Mary, Queen of Scots, arrives in Scotland at Leith.
1570	28 Saltmarket Glasgow: timber fronted tenement built; cleared in 1870s.
1585	Edinburgh. North and south of High Street ridge, tenements built: 8, 10 and 12 stories high.
1596	Glasgow. Dove House had walls 2.5 feet thick and stunted oak trees used for joists as shortage of timber.
1600	Ireland well wooded; part of reason for Plantation was to secure oak for English ships.
1603	Union of the Crowns of Scotland and England under James VI of Scotland, who became James I of England and Ireland on 24 March on the death of Elizabeth I.
1621	An Act of Parliament required roofs of slate or tiles on new and replacement roofs.
1624	Scottish Parliament bans thatch from roofs.
1652	Fire destroys a third of Glasgow. Boxed in wooden balcony projections created the problem in High Street, Gallowgate and Saltmarket.
1690	Time of famine in Scotland.
1660–1807	Atlantic slave trade.
1674	Edinburgh. After fire in High Street, wood banned from facades.
1677	Glasgow. Fire in Candleriggs: Council banned wood front and rear and gable to be stone.
1681	Viscount Stair's *Institutions of the Law of Scotland* published with tenement law provisions.
1683	Edinburgh. Exchange Building constructed.
1690	Edinburgh. Mylnes Court built.
1697	Edinburgh. Width of Parliament Close widened to 4 feet as it was being used as a toilet.
1698	Edinburgh fire: five storeys max by Act of Scottish Parliament, also external walls at least 3 feet thick. Scottish Darien Colony in central America in difficulty.
1700	Edinburgh. Fire in Exchange building and 15-storey tenement burnt down. Darien Scheme to establish a Scottish colony in present day Panama fails, bankrupting many investors.
1707	Union formed with England.
1715–1745	Jacobite risings.

1727 Glasgow. Street lamps installed in High Street, Saltmarket, Bridgegate; use rapeseed oil and hemp oil.

1734 Perth bans forestairs.

1736 Witchcraft Act repealed.

1742 Dumfries bans forestairs.

1746 Dundee bans forestairs.

1748-1851 Window Tax: affected homes with more than seven windows.

1750 Edinburgh. Major campaign to remove forestairs. Tenements in the Old Town became dilapidated and the better off moved to the developing New Town and more prosperous areas.

1750 Glasgow. Tenement development continued despite overcrowding. Ironwork such as grilles and gates made industrially from 1750 to 1840.
1750 toxic glues and arsenic were used in wallpaper. There was a wallpaper tax up until 1830. Paints also contained lead, arsenic and cinnabar (mercury sulphide). White paint wasn't available until the 1940s. Black paint was not available either.

1750 Cheaper coal available so more chimneys.

1751 Edinburgh. Wood projections outlawed; tenements must be at least three stories.

1755 Glasgow was the biggest tobacco importing city in UK.

1756 Paisley agrees to buy lamps for street lighting.

1759 Carron Iron Works founded.

1775 American War of Independence ended. Paisley bans encroaching onto street. Tobacco Lords diversify to transporting African slaves to Caribbean to pick cotton.

1784 Tax on bricks. The tax increased to 1839 so bricks mainly used for internal walls. Abolished in 1850. Speculative building in Edinburgh becomes commonplace due to rise of a middle-class.

1785 Shop Tax introduced; bakeries excluded.

1787 Glasgow. Calton weavers strike and riot.

1790 Cast iron columns from 1690 to 1910.

1790 The Forth and Clyde Canal opened, crossing central Scotland. Joined the Union Canal at Falkirk; however, by 1842 much of the freight was using the railways.

1793 Glasgow bank failures.

1796 Inverness removes any forestairs.

1800 First Glasgow Police Act. Required laying down of foot pavements and owners to remove all forestairs outshot that obstructed free passage in the said streets. Did away with piazzas.

1803 Wrought iron from 1803 until 1850 when steel was introduced. Wrought iron little used after 1890 although still being made up to 1910.

1817 Building of the Union Canal authorised; ran from Edinburgh to Falkirk initially and opened in 1822.

1820 The Scottish Insurrection of 1820

1823 Glasgow Police Act: open arches and colonnades blocked up. Development of terraces in Blytheswood; by 1820s had become commercialised.

1824 A great fire in Edinburgh.

1824 Joseph Aspdin, a bricklayer, invented portland cement, although concrete had been used since ancient times (Salisbury Cathedral's foundations are concrete, as are those of the Pantheon in Rome).

1827 City Improvement Act Edinburgh.

1831 Glasgow's population rose to 202,426.

1832 Cholera epidemic in which about 10,000 die in Scotland.

1832 Reform Bill; all streets numbered and paved and most lit by gaslight.

1833 Slavery Abolition Act 1833. This made the purchase or ownership of slaves illegal within the British Empire, with the exception of 'the Territories in the Possession of the East India Company'.

1832 Cylinder glass invented by Chance brothers, but little used because glass tax was charged by weight of glass.

1837-1901 Queen Victoria's reign: Victorian Age. Up to 300 people shared a common stair, with no sanitation.

1832 Glasgow. Cholera outbreak, and typhus was common.

1833	Police Act for water supply and drainage. Sewage still being discharged into water courses and in Edinburgh, into the Firth of Forth.
1833	Thomas Neigh, a Dundee merchant, invented the mechanical process of spinning jute by first soaking it in whale oil.
1837	£452,900 invested in start of Glasgow, Paisley, Kilmarnock and Ayr Railway Co. 10% of total attributed to slave owners.
1837	Slave Compensation Act. The Act compensated slave owners in the British colonies of the Caribbean, Mauritius and the Cape of Good Hope in the amount of approximately £20 million for freed slaves. Compensation payments finally ceased in 2015.
1839	Chance Bros patented plate glass for more refined jobs.
1839	Parliamentary Report on Housing in Britain: in Glasgow 'so large an amount of filth, crime, misery and disease existed in one spot'.
1840	Thatched roofs die out in Glasgow and Edinburgh, but not in the other parts of Scotland.
1840	Edinburgh. Doors got larger with added fanlights or transom windows, although fanlights at close doors in Edinburgh New Town and some Georgian tenements were used by 1780s. Harling also fell out of favour to be replaced by stone.
1840	Eviction of Highlanders reached its peak, many emigrating to Glasgow and the Central Belt.
1842	'The Sanitary Conditions of the labouring population': 70% of Glasgow's population lived in degradation yet even when cleared, more tenements were built in the backlands. A single close ('the Rookery') had 500 people in it.
1842	6,000 people in Paisley sign a petition in effect to get rid of the corn laws. They have no money.
1843	Indian Slavery Act. The British Empire abolished slavery legislatively through this Anti-Slavery Act although a large numbers of former slaves traded their status for that of perpetually bonded servitude. This was in part due to the fact that the British did not abolish debt-bondage; instead, they regulated it.
1845	Glass Tax repealed.
1845	Irish Potato Famine began and lasted seven years. The population fell by 25%, many dying or emigrating during this period. Between 1841 and 1851 the Irish population of Scotland increased by 90%.
1846	Glasgow Police Act. Burghs like Gorbals, Anderston and Calton absorbed into Glasgow.
1847	*Edinburgh. Towns Improvement Clauses Act*: party wall construction, walls and roof finishes to be incombustible.
1850	*Edinburgh Burgh Police Act*. There were 597 people living in Blackfriars Wynd; 1025 people living in 142 buildings. Tax on bricks abolished.
1852	First use of plate glass in Glasgow at 2–28 Bothwell Street.
1856	Henry Bessemer accidentally invented a way of making steel effectively. However, Eiffel used wrought iron for the armature to create the structure of the Statue of Liberty. By 1889 steel was replacing iron everywhere.
1857	Loch Katrine supplies water to Glasgow (following an Act of Parliament in 1855), inaugurated by Queen Victoria in 1859.
1861	Edinburgh. Tenement collapse in High Street; 35 die.
1862	*General Police & Improvement Act (Scotland)*: section 210 dealt with water supply and sanitation. Henry Duncan Littlejohn appointed the first Medical Officer of Health in Edinburgh (to 1908).
1864	Edinburgh. City Council enforces installation of a water supply in houses of the poor.
1865	*Edinburgh.Littlejohn's Report on Sanitary Conditions. Slavery abolished in America.*
1866	*City of Glasgow Improvement Act*. Glasgow undertakes slum clearance. Ticketing of closes introduced. WCs required for male and female at least on landings if not inside.
1867	*City Improvement Act Edinburgh*. Public Health (Scotland) Act reinforced clauses in 1862 Act requiring improvement in lighting, ventilation, drainage and repair, failing which occupation would be denied.
1868-1871	Thomas Annan photographs the streets and closes of Glasgow.
1870	Greek Thomson builds Great Western Terrace, Glasgow.
1873-75	Edinburgh. Plasterers strike resulting in wood boarding being used instead.

1875	A Glasgow plumber, Wallace by name, employed by Mr Lockhart, hit upon an idea to vent a trap to a high point, to get rid of smells in 1855. William Paton Buchan, employed by Wallace, developed the trap and patented it in 1875. It was used in many backcourts from 1870 onwards.
1878	Scotland building slump. City of Glasgow bank crash. Stonemasons Union lose their money and Union suffers as wages cut.
1879	Flogging sailors was outlawed on British ships in 1879 although American sailors were spared the lash after 1860.
1879	**Edinburgh Municipal and Police Act: Public Health Act?** Penalty for throwing soil, filth, etc. from window 40 shillings. Owners must ensure close was swept daily and washed twice a week. Close walls to be whitewashed once a year (clause 98). Tenement height limited to no more than 1.5 times width of road. No more than one habitable storey in the roof. No encroachment or projections into the street. Plans sections required at 1.5 inches to 10 feet to Dean of Guild. Party wall to be carried 2 feet above any timbers on roof. Drains to be cleansed yearly.
1880- 1890	Red sandstone introduced to Glasgow, transported by rail from Ayrshire (Glasgow, Paisley, Kilmarnock and Ayr Railway Co. was built partly from slave owning investment in 1837).
1880	Cologne Cathedral, started in 1248 and completed in 1880. Arbroath stone pavement used to floor the cathedral.
1890	Tiled closes begin to appear in Glasgow. Locally quarried freestone becomes scarcer so replaced with red sandstone from Dumfriesshire which could now be transported by rail.
1892	Burgh Police Act applied to ALL burghs: but excluded Edinburgh, Glasgow, Aberdeen, Greenock and Dundee'
1892	Glasgow embarks on sewage processing plant in Dalmarnock.
1897	Edinburgh Lodge of Masons strike for an eight-hour day, which they eventually win, but as other trades did not strike as well, workers continue working nine-hour days and masons suffer.
1898	*Burgh Police (Scotland) Act.* Tenements to have no more than 12 houses, unless access is from an outside stair with balconies, then twenty-four flats may be built. Floors to be deafened, all apartments given three coats of plaster. In Dundee this resulted in the construction of so-called 'platties', tenements with external stairs and deck access to flats. Joists at every hearth to be bridled and where practicable supported by a brick arch or concrete … No timber, joist, beam to be inserted into a wall near to fireplace … closer than 12 inches. Gas streetlights in Glasgow. Every habitable room on ground floor to be at least 9 feet 6 inches. Floors above at least 9 feet and attic rooms at least 8 feet high. Glazed area to each room at least a tenth of room area. Walls to have a damp proof course, including party walls, and a layer of asphalt in solum. All external walls, party walls, passage and partition walls constructed with incombustible materials. Party wall to be carried through roof by one foot and to be coped. 'Mortar used on new or altered buildings shall be "fresh burnt lime and clean sharp pit sand, grit or ground bricks, or freestone shivers".'
1890	The Housing of the Working Classes Act of 1890, granted under part one, the right of the clearance of unhealthy areas and the provision of dwellings for those displaced as part of an improvement scheme.
1892	*The Burgh Police Scotland, Act of 1892*: limited the number of dwellings to each common stair and made compulsory both the provision of running water inside dwellings and the introduction of internal or shared water closets. In addition to the mandatory provision in new buildings erected after this point it also altered the rear elevation of many existing working class tenements, with the building of brick towers containing water closets, accessed from the landings on the stairs.
1892	*Glasgow Building Regulations Act and bye-laws.* In 1892 the first Building Regulation Act was passed and allowed the requirements for new house construction to be specified; this also moved to preventing the replication of the back land closes of earlier decades. Under this Act also, the heights of ceilings were specified and the number of houses in a block leading from one common stair was restricted to 16. The use of built-in beds was also prohibited.
1900	*Glasgow Building Regulations Act.* Set down minimum areas of 1, 2 and 3 apartment houses and prohibited enclosed bed recess spaces. Glasgow Corporation then drew up some additional bye-laws which eventually were contained in the 1953 Building Standards (Scotland) Regulations.
1905	Concrete becomes more commonly used in construction.
1910	Lloyd George Finance Act: increased tax on land values.
1910	Tenements with no bathroom still being erected in Denniston.
1914	Use of cast iron girders ceases.

1914-1918	First World War.
1914	Glasgow Women's Housing Association formed in 1914.
1915	Glasgow Rent Strikes 1915–1919, resulting in the Rent and Mortgage Restrictions Act, which brought in a long period of rent control, in different forms, which were not abolished until 1988. Affected landlord profitability thus improvement and repair.
1917	Royal Commission on the Housing of the Industrial Population of Scotland, 1917: Industrial and Rural, concluded that the private provision of housing for the working class had failed. Known as the Ballantyne Report after the commission chairman.
1918	Representation of the People Act passed giving vote to men over 21, and women over 30. Neither required to be property owners to secure the right to vote. It was another decade before women secured the same voting rights as men.
1919	Housing Town Planning (Scotland) Act 1919, the so-called Addison Act, made provision for state subsidies to support the construction of council housing, was the beginning of statutory intervention on improving existing privately owned housing. The case began to be made during 1920s and 1930s for financial assistance. This was established for rural areas only during this period.
1935	Housing (Scotland) Act made subsidies available for slum clearance resulting in the construction of tenemental so-called 'slum clearance' estates, built as and long known as 'stigmatised places'.
1937	Scottish Special Housing Association established to develop good quality Scottish housing. It was transferred to Scottish Homes; abolished in 1989, and some functions transferred to Scottish Homes or other housing associations. Scottish Homes was disbanded in 2001 and taken over by Communities Scotland, which itself was disbanded in 2008 to be taken over by Housing and Regeneration directorate of the Scottish Government.
1940	During Second World War: cast and wrought iron railings removed to support the war effort. Most of it was then dumped in the sea.
1947	Scottish Housing Advisory Committee report *Modernising our Homes* made the case to subsidise grants for property improvement to offset future slum housing crisis.
1947	The Town and Country Planning Act: allowed Comprehensive Development Areas (and Redevelopment Areas), a statutory designation to tackle the issue of urban blight, by directing future development according to a coherent plan. The Hutchesontown-Gorbals Comprehensive Development Area (CDA) was the first to be formally approved in 1957 and there was an ambitious plan to redevelop 26 Outline Comprehensive Development Areas where there were old tenements.
1953	Moss Heights built in Glasgow; had district heating.
1958	Basil Spence's Hutchesontown E 20 storey Brutalist high rise built; demolished in 1993.
1964	Red Road Flats built; 28 floor point blocks, steel framed, demolished in 2015.
1967	The Scottish Housing Advisory Committee published *Scotland's Older Houses*
1968	Glasgow storm. This blew down chimneys and roofs on a large scale. Led to the creation of a new Glasgow Council department to co-ordinate an emergency repairs programme, headed by Theo Crombie, later to be a key player in tenement improvement programme.
1969	Housing (Scotland) Act marked first time public subsidy, in the form of grant, was made available to tenements. Also made provision for Housing Treatment Areas to undertake improvement on an area basis. Was the first Act to specify 'The Tolerable Standard', now replaced by the Housing (Scotland Act) 1987.
1969	Ronan Point collapse.
1971–1978	Founding of Glasgow's community-based housing associations who sought to improve and later renovate entire tenemental neighbourhoods.
1971	Edinburgh New Town Conservation Committee was established in 1971 to tackle the economic and physical demise that threatened the Georgian New Town. Scottish Office provided a separate stream of funding to support this work.
1972	First tenement improvement project starts in Govan; Govan Housing Association formed.
1974	Housing (Scotland) Act refined the grant-making provisions and established the Housing Action Area that was critical to the preservation of tenements. There were two Acts in 1974 of major importance for tenements. The Housing Act (UK wide) which created the Housing Corporation and Housing Association Grant; and the Housing (Scotland) Act which replaced Treatment Areas with Housing Action Areas with enhanced powers for councils and levels of grants for owners.
1976	Laurieston Tower Blocks built.

1992	Lead finally banned in paint in the UK; despite knowing that lead poisoned water since 1933, lead water pipes in tenements (and lead lined cold water storage tanks) were not banned until 1969.
2004	Law of the Tenement Act passed as part of the legislative programme to abolish feudal title in Scotland, one of the very last countries to do so in Europe.
2006	Housing (Scotland) Act abolished the HAA and also effectively withdrew grant funding for the improvement and repair of property.
2009	The Climate Change (Scotland) Act 2009 requires the preparation of strategic programmes for climate change adaptation, as soon as reasonably practicable.
2015	Compensation pay-outs for slave owners ceased in 2015 and amounted to £17 billion.
2017	Grenfell Tower Block fire killed 223 people. Materials approved by BRE (Building Research Establishment) after a fire test which was rigged: BRE had been a government body but was privatised in 1978.
2019	Climate Change (Emissions Reduction Targets) (Scotland) Act 2019. Scotland sets a target date for net zero emissions of all greenhouse gases by 2045.
2023	Rishi Sunak has a rethink: government extends life of petrol and diesel vehicles from 2030 to 2035 and yet he says he still plans to meet the 2050 Net Zero carbon target. Also licences new oil fields off Shetland. Also, the government drops the requirement for landlords to insulate the property they rent, so tenants will continue to have cold damp houses in the UK.

Glossary

Aberdeen bond a pattern of stone bedding communion in North East Scotland. Large rectangular blocks of granite alternate with three or four smaller square stones, like granite or black gabbro or a dark whinstone

Air admittance valves are 'negative-pressure-activated' one-way mechanical valves

Airborne noise noise which travels through the air

Andesite fine grained volcanic rock similar to basalt

Apron flashing leadwork to be dressed over slates; found at skylights, the base of chimneys, and junctions between steep mansard slating and a slated roof above

Aqualite a bitumen based flexible flashing used in the 1980s as an alternative to lead

Arbroath pavement about 70mm thick stone used as front and back hearths

Architrave 1. Wooden architraves are moulded or lain facings that are formed around the door frame, and windows, to cover the joint with the plaster. 2. Stone architraves are usually one vertical stone upright around windows and doors, usually formed into a decorative section.

Arrises the sharp edge of any cut stone

Artex is a surface coating used for interior decorating, most often found on ceilings, which allows the decorator to add a texture to it. Until 1984 the Artex coating was made with white asbestos to strengthen it.

Article 4 direction is made by the local planning authority. It restricts the scope of permitted development rights either in relation to a particular area or site, or a particular type of development anywhere in the authority's area.

Ashlar stone stone which has been hand dressed to a rectangular format and tightly bedded with thin mortar joints. Usually about 300mm high by 400–600mm broad

Astragals slender wooden glazing bars rebated to hold glass which would be putted into place

Balance flue balanced flue appliances are the only appliances that do not make use of the room air to feed the fire. A balanced flue pipe is double skinned with the exhaust from the fire travelling through the core and fresh air being drawn from outside the building through the outer wall of the pipe.

Balks, or baulks a large squared beam of wood

Balusters Usually cast iron, but also wrought iron. Uprights secured into treads to which a banister rail is secured. Also stone balusters at parapets

Basalt black, fine grained, igneous rock, may include iron oxide, feldspar and pyroscene

Batten rod a removeable timber batten secured to the inside of a sash and case window, to contain the lower sash

Bed recess beams	timber beams that temporarily supported the floor joists, were supported on studs. Size approx. 7in.x2in.
Bilgates	Usually about 6in.x3in.x1in. thick blocks of wood to fit into the brickwork coursing and to which the door frame is secured. Usually three blocks on each side of a door
Bitumen felt	William Briggs developed the tarred paper in Dundee in 1865
Black gabbros	course grained igneous rock containing feldspar and magnesium
Blind windows	in Georgian times, elevations needed to show repetition and order, so sometimes blind windows, surrounds and entablatures in stone as well as the backing, were installed, even though there was no opening. Also installed if there were flues or staircases behind
Blocking courses	the course of stones set above the cornice stone and which forms space for a secret gutter behind it
Boasted	roughly dressed or scabbled
Bolection moulding	a rebated panel housing, the face of which stands proud of the framing
Bonding plaster	a granular first coat of plaster that adheres to the backing and is scratched as a base for the following one or two coats of finishing plaster
Borescope	device which allows a closer inspection of hidden spaces by inserting the borescope into a small hole. Contains a light and optical fibres
Borrowed lights	small windows in a close to obtain borrowed light from the close to light an old toilet
Bottom rail	would be 9in. deep. All rails tongued into the stiles
Branders	battens set counter to the floor joists, to support the finished ceiling. Also found in renovation to form a level ceiling
Bressumer beams	large timber baulks acting as structural beams, often carrying floor joists and sometimes found at shop windows. From the French 'sommier', a pack horse, so it meant to carry a great load
Bridling	a joist which is set perpendicular between two joists to carry the other 'trimmed' joists
Brighton patter sash fastener	classic method of fastening and locking sliding sash windows with a screw and a turn knob which slots into a receiver and is then tightened
Buchan trap	crudely described as a 'buckin' trap, allowed a fresh air inlet before a trap.
Bullnose edge	a half round edge to a stair nosing, formed in concrete, stone or timber
Bunce	material extracted from a building site which is often sold by the builders
Cant	laid at an angle, or referring to the bedding of stone not laid face bedded
Cast-iron lintels	found externally in Edinburgh tenements at ground floor openings, often shaped to mimic the stone. Thought to be used to stiffen the structure when the ground was suspect
Cast-iron plate	used in the 1900s to tie an oriel back into the wall
Cat-slide roof	term given to a small dormer set into a steeply pitched roof, with a slightly less steep slated roof over the dormer
Caulked	to fill, or stop an open joint very tightly. Railings would be caulked using molten lead
CCTV survey	a cable driven video camera using a long optical fibre, allows a recorded view of the drainage pipe showing any blockages
Ceiling roses	a decorative ceiling centrepiece formed in plaster typically round in shape, displaying a variety of ornamental designs. derived from 'sub rosa'
Cementitious grout	a fluid mix of cement compound which is used to fill voids underground, pumped in under pressure
Chain timbers	timber wall plates buried in half brick walls to stiffen them
Chase	recess in a wall to house a downpipe
Check nails	old 'T' profiled slate nails used to be available to ensure side nailing gripped the edges of each slate
Checks	set back in stone openings to allow space for the window frame. Also 'checks' in other materials referring to recesses
Chippings, also called 'shivers'	small pieces of stone
Clerestory lights	high level glazing to light a space

Clipping of lead	lead cover flashings which are dressed into rages, may be clipped with folded lead or secured with stainless steel screws before pointing the gap
Close	narrow alley leading to a courtyard originally, but now refers to the common close and stairs to access all the apartments
Close stairs	one end of each tread is bedded into the close wall, the other end is open and rests on the stone tread below. The force is passed from each stone to the landings.
Club skews	the lowest stone at the eaves course in a skew wall, built into the gable wall to prevent the skew stones from sliding down. Early ones were often decorated
Codes of lead	the weights of lead and thicknesses are: code 3-1.32mm; code 4-1.8mm; code 5-2.24mm; code 6-2.65mm; code 7-3.15mm; code 8-3.55mm
Coffin tanks	large lead lined cold water tanks
Cold bridges	in the layers of a wall, a cold bridge can develop where the insulation is discontinuous
Collars, or collar ties	horizontal ties that couple the two rafters together but secured to rafters mid-point
Compound wall	external stone walls in tenements were not solid stone but made from a smooth ashlar stone (rubble at the rear) at one face and a rubble stone inside, then backfilled with lime mortar and smaller stones
Comprehensive redevelopment areas	a statutory designation under the Town and Country Planning Act (1947) permitted the local authority to tackle the issue of urban blight, preventing piecemeal uncoordinated development, and directing future development according to a coherent plan
Concrete reinforcement	came into use when steel bars were used rather than wrought iron after 1890
Conductors	Scottish term for a downpipe
Console brackets	a large ornamental bracket forming a frame for a shop sign, or either side of an entablature
Consolidation	the process of stabilising poor ground condition
Copper bowl	in the backcourt, a 'copper' would be used for washing clothes, supplied with water and a drain, and heated be a small stove under the copper. Smoke would be directed up a small chimney in the wash-house
Corbel, corbelling	a corbel course is a projecting stone course on which the gutter normally rests. Corbelling is the way an oriel projects from the faces of a building
Cord clutch	a traditional device set below the inner sash pulley to temporarily hold the cord to allow the window to be hinged in when secured to simplex hinges
Cord grip	a small metal device which hooks onto a screw set into the sash fame
Cornices	at eaves level, projecting stone course, with a 'backing course' on top to act as a counterweight; depth of cornice would often go full width of wall
Counter battens	battens secured to sarking but counter to horizontal tile battens laid on top of them
Counterweights	usually lead counterweights to balance the weight of each sash
Coursed stone bedding	stones laid in roughly equal horizontal courses
Cramps	also known as 'dogs'. Steel or wrought iron formed into a staple shape and bedded into slots in the stone
Crane or Cran	in Scottish dialect can mean a heron, swift or crane; simply refers to a bent tap
Crown glass	glass blown and spun in discs up to 50 diameter
Crowsteps	found at gable ends, stones that form steps, up to the chimney
Cupulas	glazed skylights which may be rectangular, elliptical or circular, letting light into a central close stairwell
Cylinder glass	1700–1860 glass drawn, blown and made into a cylinder, then flattened
Cylinder sheet glass	machine drawn cylinder sheet glass was invented in 1903
Dado	the lower part of a wall up to about 1200mm in a close. In a room, the dado would be above the skating and below the dado rail
Damp-proof course	a course of slates, bitumen or asphalt, designed to prevent moisture rising up the wall from the ground
Deafening	soundproofing material, commonly a mix of ash and plaster
Deals	regularised sawn sections of timber, smaller than planks

Dean of Guild Courts	under Scots law, one of a group of burgh magistrates had the responsibility of enforcing the Burgh's Building Regulations and bye-laws
Deep flow gutters	UPVC gutters made by Marley, often used on rear gutters
Delamination	the breakdown of layers of stone due to wetting and drainage, and incorrect bedding of stone
Dendrochronology	the science of dating events, environmental change and provenance from timber tree rings
Differential settlement	when ground conditions are poor or foundations are inadequate, differential settlement can occur
Diminishing courses	Scottish slating is best laid in diminishing courses with the larger slates at the eaves, diminishing to smaller slates at the ridge
Disconnecting trap	often a Buchan trap, formed below ground, it disconnects sewer gases from the house drain
Doff system	a hot steam system for cleaning stone, also used to remove paint, guano and Linostone
Dog cramps	wrought iron cramps formed to a staple shape to tie stone sections together. Also used in timber construction
Dogs	steel or wrought iron formed into a staple shape and bedded into slots in the stone
Dolerite	black, medium grained igneous rock, like basalt
Dooks or douks	timber wedges driven into mortar joints in stone or brick walls to allow fixing of battens. Also brick sized doors to restrain door frames. Also 'docks'
Double banked	mutual chimneys where the flues were on either side of the division between the two tenements. The wall thickness here would be about 800mm (2ft6in.)
Double tanks	twinned and breached hot water cylinders that were located in narrow loft spaces formed above bathrooms
Dovetailed pockets	pockets set in stone stairs were wider when deeper, to ensure the caulking was restrained
DPC	damp-proof course
Drip	a throating on the underside of a coping to ensure water is thrown from the face of the wall surface under the coping
Droving	also 'tooled', stone finished with a sharp chisel forming a series of parallel lines, usually vertical but sometimes horizontal (broached), sometimes angled
Dry grit blasting	dry grit was sometimes used to blast stone walls, but mostly dry grit was used to clean rust from metals
Dry rot	decay in timber caused by *Serpula lacrymans*, a wood-destroying fungi
Dry sand blasting	early sand blasting simply blasted dry sharp sand onto stone walls, sometimes eating into the stone and damaging arrises. It produced a lot of dust and became a health risk.
Dunny	a narrow passage at basement level
Dwangs	small pieces of framing timber set between timber stud uprights to stiffen the structure; small sections of joists that were fixed between joists to stiffen or retain fixings
Ear wings	Cast-iron downpipes can have cast-in lugs to provide a fixing directly into the stone; these 'lugs' may be called 'ear wings'
Ecocide	destruction of the natural environment, especially when wilfully done
EnerPHit	a Quality-Approved Energy Retrofit standard using Passive House Components
Entablatures	the horizontal members spanning the two solums or pilasters, often with moulded cornices
Face bedded	stonework laid with the natural bed upwards and outwards, prone to delamination
Feldspar	incorporated in rocks, a silicate of sodium, potassium and calcium
Feu duty	the sum tenants would pay every year to the feu superior to reside in a house
Feu system	the tenure of land in perpetuity for an annual payment of a fixed sum of money to the owner of the land. Abolished by the abolition of Feudal Tenure (Scotland) Act 2000
Fire break	typically the party wall between tenements and the close walls in a stairwell are fire breaks

Fire suppression systems	there are a number of fire suppression systems such as the water mist system which are going into use for special types of housing
Firebricks	bricks that are fired to resist a high temperature
Fitch pattern sash fastener	a fitch sash lock operated in the same way as a regular sash fastener but was scallop shaped to ensure the sash windows were tightened together. In use from 1930s
Flanking noise	noise which can travel up a wall and result in noise in theft above (or below)
Flashband	a proprietary product consisting of a thin aluminium tape with a bitumen backing that can be applied over cracks in flashing to fix leaks (for a time)
Flaunched	a sloping mortar fillet, usually between slates and wallhead or skew, to throw off moisture
Floating floor	resilient cushions to support an added floor finish, designed to reduce impact noise
Flooring brads	annular ring shank nails provide greater holding power over smooth shank nails
Flue bridges	also called 'feathers'; stones which are bonded into the chimney to create flues; if stone it would be bonded in to external stone, if brick, brick wall ties may secure the bricks
Flue liners	Stainless-steel flue liners, usually crimped to allow them to be flexible, installed into existing flues to ensure the flue is intact for a gas boiler or fire
Foamed slag	a lightweight cellular concrete aggregate produced by special treatment of molten blast-furnace slag, crushed and graded into commercial sizes. Produced in 1881 at Motherwell iron works
Footing	offsets in the external load-bearing walls that spread the load onto the foundations
Fore-stair	in medieval times outside stair to street side of house, giving access from street level to the principal apartments.
Fauceted joints	Cast-iron gutters were widened at one end to accept the bedding of the adjacent gutter. Also found in clay drainage pipes
Freestone	the best stone as it is homogeneous and without any clearly defined strata. Also referred to as 'Liver rock'
Galetting	a 'galet' was French for a pebble. Small stones would be formed into the mortar beds during construction
Galley kitchens	long narrow kitchens that were formed to open into living rooms, access by folding louvre doors
Gauge	in brickwork, the distance between four courses of brickwork
Georgian wired glass	glass that has a core made from steel mesh which holds the glass together in a fire
Glazing bars	slender wooden sections, rebated to hold glass, known as astragals, come in various shapes and sizes
Gooseneck	the curved shape of the lead vented pipes that came from hot water cylinders and exhausted onto the roof (also onto cold water storage tanks)
Granolithic	a floor surface consisting of cement, coarse sand and fine stone clippings such as granite
Grey areas	in the 1960s, certain areas were treated as grey areas by mortgage lenders because their future was unclear; that is, they didn't see a future for the tenement in that area. People had to pay cash, rather than obtain a mortgage.
Greywacke	fine to course grained hard sandstone with mainly angular quartz and rock fragments
Gunmetal	a type of bronze, alloy of tin, copper and zinc, which resisted rust
Gypsum plaster	not the same as lime plaster as it contains gypsum, which is a modern plaster and much faster at setting
H beams	the most common section of a rolled steel joist, formed as a steel beam with two flanges and a web, usually about 8in.x4in. but larger sizes at shops and bigger openings
H frames	H frames were timber door frames that were floor to ceiling height and supported the floor joists. Sometimes a fanlight above and sometimes brick infill with a relieving arch
Hammer dressed	dressing of stones to remove sharp edges or corners; usually contains square or rectangular shaped marks
Hangers	the vertical timbers that stiffen the rafters to the ceiling joists
Hanging stair	a central close stair with an open well, the interlocking treads transfer the load to each landing

Harling	the act of 'throwing' lime mortar mix onto a wall surface to ensure it adheres
Heat pumps	a device used to heat a building by transferring thermal energy from a cooler, external space, to a warmer space using the reverse refrigeration cycle
Heat Recovery Ventilation Unit	works by extracting moist and stale air from wet rooms in your home; it recovers the lost heat from the extracted air and heats the incoming ventilation
Holderbats	a metal fixing for cast-iron downpipes. Can be driven into a mortar joint, or some plugged into the stone and then a strap bolted around the downpipe. Also 'haudfast'
Honeycombed wall	small stub walls which are built as perforated walls to allow through ventilation and support the ground floor joists
Hopper	a receptacle for collecting water from gutters and channelling it to downpipes
Hot applied asphalt	Trinidad asphalt consists of 40% bitumen and calcium carbonate as well as asphalt. Combined with a proportion of crushed stone it is laid hot in two coats
Hot work permit	required by contractors doing any hot work on a building site because of the risk of heating timber and later causing a fire
Housing action areas	areas designated as having houses below the 'tolerable standard'
Housing Corporation	set up under the Housing (Scotland) 1987 Act to encourage and provide funding for the growing Housing Association movement
Impact noise	noise which travels through the structure, usually caused by hard floors
Inbands	at a window opening, a smaller stone that with 'outbands' forms the window surround
Indented	defective stone can be cut out and new matching stone cut to size and indented, placed into the space of the decayed stone
Ingoes	the reveal at a door or window. Can refer to inside or outside reveals
Intercepting trap	a trap fixed on a drain to prevent the passage of sewer gas into the drain
Intermediate rail	also known as middle or lock rail, 9in. deep. All rails tongued into the stiles
Intumescent collar	a fire-resisting collar around a pipe, made from intumescent material which expands to seal any gaps when there is a fire
Jawbox	derives from jaw or jawp, a splash or dash of water. Jawboxes at kitchen sinks were common around 1870 and eventually replaced by fireclay sinks
Jettied	deriving from the old French word 'jet', a medieval technique for allowing timber framed buildings to project over the floor below. Commonly found in medieval French towns
Jumper	hand tool for making deep holes in stone to break up the stone
Keene's cement	a Gypsum plaster, like Plaster of Paris which is then steeped in a solution of alum and subjected to intense heat, ground to a powder, and sifted. Invented around 1840. Used on close dados and arrises
Keyed	keying a backing may mean roughing it with a pick or scratching a coat of mortar to provide a 'key' for the next coat
Keys	the plaster that holds the wall or ceiling plaster to the laths, see 'rivets'
Knot holder	a metal ring which the sash cord was secured to and which slotted into the sash frame to allow easy removal
Krames	small wooden buildings occupied by goldsmiths, watchmakers and jewellers
Laths	thin strips of wood to form a base for plasterwork. Laths would be riven and likely Baltic fir
Laylight	secondary glass fixed light which lets light into a top floor hallway and stops dust falling into the hallway as well as preventing heat loss
Lead mastic	is designed for pointing joints between lead and masonry or brickwork. it is a silicone sealant which can be applied using a mastic gun
Lead substitutes	different makes such as Lediax', 'Ubiflex' and 'Masterform' are lead substitute material, not always approved by conservation officers, and rarely approved for listed buildings
Ledged and braced doors	found on store doors and back doors. The ledges were horizontal battens carrying the vertical boards and the bracing were the diagonal timbers which strengthened the door. Also framed, ledged and braced doors

Lime putty	the soft lime produced as a result of the slaking process. Together it forms the main ingredient of lime mortar
Limestone chippings	can be sorted pre-bagged, known as 'Quietex'
Limewash	an extra protective coat (or two) of lime wash, helps to protect the lime harling; can be coloured
Liners	bailies and lesser officials monitored all aspects of town life. The 'Liners' were clerks for the burgh
Linkhorns	once the link boy had escorted someone to the house, they were used to extinguish the link boy's torch
Linotol	modern composition flooring laid over a stone or concrete surface, consisting of a binder and a base fibrous and resin material. Magnesite flooring is similar, made from magnesium oxychloride and a fibrous binder
Liver rock	best grade of sandstone which can be worked in all directions, also known as 'freestone'
Lock rail	also known as middle or intermediate rail, 9in deep. All rails tongued into the stiles
Luckenbooths	small timber lock-up shops
M-Type roofs	Georgian roofs originally consisted of two roof trusses spanning onto a central beam which would contain a central valley and possibly a cupola
Mansard roofs	French style to make use of attic spaces, utilising the top floor with an initial steep slated pitch, then another pitched roof, often with wallhead dormers and found in Aberdeen and throughout Scotland
Margins	stone margins around doors and windows, often with a small check to receive harling, eventually lime washed over
Mid-feathers	the divisions between flues; if stone it would be bonded in to external stone, if brick, wall ties may secure the bricks
Midden rakers	people who would go around the bins to check if there as anything of value
Modified felts	modified bitumen felt is made from three layers: boards, tar and mineral. It is laid over a naked flame
Mosaic	small fragments of stone formed in various designs and laid on a cement bed, then polished smooth
Mullions	the vertical divisions between windows, usually stone, but not always, usually pre-cut and shaped to size
Muntin	a vertical division in timber framing, usually in a panelled door
Mutual chimneys	adjoining tenements sharing the same chimney stack
Mutual downpipe	a downpipe shared between two tenements, taking water from both tenement gutters
Nepus gable	small gable usually facing the front which carries above eaves level; often acts as a chimney and contains a dormer window
Night soil	tenants would deposit their faeces in a midden in the back court, where it would be collected and sold on as compost
No-fines construction	in situ concrete with large aggregate but no fine aggregate and no reinforcement
Nogging	brick infilling between timber framing
Nosing	the exposed edge of a stair tread. Brass nosings are usually formed if the stair is being overlaid with Lintel, as well as non-slip strips
Nuralite	originally a flexible rubberised flashing used in the 1980s, now a New Zealand company selling a range of waterproof products such as tanking membranes
Oncome	at the fireplace where the smoke enters the flue, the cross section begins to diminish at the 'gathering' of the flue
One-hour fire resistance	modern building regulations and Housing Acts require the main door into flats to be one-hour fire resistant. This requires doors to be solid core, to have intumescent seals at perimeters and to have a self-closer
Oriels	a projecting bay cantilevered from first floor level
Outbands	at a window opening, a larger stone that is bonded into the wall to strengthen the wall, together with 'inbands', to form the window surround.

Oxter piece	short triangular section near eaves formed by a vertical strut and the rafter and ceiling joist or main collar
Pantiles	Scotch pantiles were s-shaped in profile with a projecting nib, hung on battens
Paraffin wax	popular in the 1920s and used as a stone sealer. However, although it didn't appear to damage the sandstone, soot adhered to the surface making it appear black
Parapets	take different forms; a raised parapet will contain a hidden gutter behind. Some parapets were formed from stone balusters
Parge coat	or 'parging': refers to a base coat of lime plaster applied directly to a stone or brick wall
Parlour	from the French parlor meaning a sitting room in a private house
Parting bead	a slender timber bead which is set into a chase in the case and separates the operating parts of the two moving sashes
Passive house standards	EnerPHit sets standards for passive houses which requires thorough post inspections and analysis to meet the approved targets
Patent glazing bars	usually, but not always, aluminium. Glazing bars with integral flashings and fixings
Patination oil	is a white spirit-based surface treatment that prevents lead carbonate from forming on the surface of lead sheet
Pavement lights	Cast-iron frames set into the pavement with glass prisms to distribute light to basement level
Peggies	smaller slate about 12 to 14in. long by 6in,
Penchecked	a splayed rebate used in stone treads to interlock them together
Pend	covered way or passage
Penny gap	a small gap between sarking boards or boards on parapets or platform roofs, that allowed ventilation
Piazzas	the ground floors of tenements were arched with an open area under and the space sometime incorporated a small shop stall or booth
Pilasters	a thin rectangular pier projecting from the face of a wall
Pitch pine beams	in America, known as the southern pine, imported from the Southern states, is resinous and hard to cut
Pitched faced rubble	hammer dressed, or rock faced, usually coursed rubble
Plastic repair	not actually plastic, but refers to an applied mortar, which may be cement, lime mortar or a special mix like 'Lithomex'
Plate girder	built-up girders formed by riveting together steel plates and angle irons to form beams and standards
Plate glass	plate glass was originally made by grinding large plates of glass until clear. It was made in the 1950s using the float glass method
Platform construction	where the floor construction bears the loadbearing walls
Platform roofs	flat roofs, usually leaded, formed above slated roofs
Platties or pletties	external concrete decks accessed by a common stair giving external access to each house; tenements built under the 1892 Police Act were allowed 24 houses of a common stair provided access was from an external deck
Plinth block	also referred to as a base block. Timber which is fixed to the base of a timber architrave
Plug and feathers	steel wedges (plugs) with half round steel strip (feathers) on either side of plug, used to split stone by driving in plugs between feathers
Pocket hole	a rectangular hole in the case to allow access for recording the weights; also a pocket cover which would be checked to fit flush into the case
Pocketed	holes cut into the stone to receive balusters or railings
Pole plate	when the rafters and ceiling joists meet at the eaves, the timbers may be joined by a square pole plate notched into the timbers. Sometimes set at an angle
Press	an internal cupboard inset into a party wall. Lined and with a door
Projecting shops	shop units built onto existing tenements that extended over a garden or frontage area to form a new shop

Pulley wheel	a steel or brass pulley wheel set into a recess in the window sash to accept the sash cord for the counterweights
Purlins	larger timber beams that provide a mid support to rafters. Rarely found in tenements
Pyran glass	a fire-resistant plain glass used in fire doors
Pyrite crystals	are iron sulfides with the chemical formula FeS_2 also known as fool's gold. Pyrite crystals are usually found in quartz veins in sedimentary rock, and metamorphic rock
Quoins	cornerstone at external angle of wall
Racebond joint	see 'risband' joint
Raggle	a chase or slot formed in stonework into which lead flashing or asphalt upstaged is inserted
RCDs	Residual Circuit Breakers
Red lined	an area that had been 'red lined' possibly for demolition, so no mortgages would be approved in that area
Register grate	open grates for heating living rooms consisting of a cast-iron frame
Relieving arches	brickwork shaped into a low arch to spread the load above to the sides and 'relieve' the lintel
Reticulated	a pattern formed in ashlar stone formed in a wormlike fashion with irregular sinkings
Rhones	from the Rhine. Scottish word for gutters although sometimes applied to downpipes as well
Ridge board	the ridge board to which the rafters are secured and a ridge pole secured to the head of it
Risband joint	a vertical joint between different buildings having no bonding. Also referred to as a 'racebond' joint
Rising main	also 'riser', a term for the main water supply that rose up a tenement to supply cold water tanks in the loft. Originally in lead, now replaced in polyurethane pipes
Rivets in plaster	the protrusion of plaster when applied to laths that forms through the gaps
Rock asphalt	bitumen mixed with pulverised rock and run into moulds. Then reheated and mixed with grit to form a tough and waterproof surface
Rodding canes	bamboo or polyurethane rodding canes may be used to clear blockages, although water jet cleaning is now more common
Rolled steel joists (RSJs)	Rolled steel joists (RSJs) come in a range of standard sections such as the H beam and the I beam, also T beams. Patented in 1849 in France, in common use by the 1890s in Scotland
Roman cement	contains pozzolanic ash and is a hydraulic mortar
Rothounds	labradors trained to sniff out dry rot
RSJs	Rolled Steel Joists (see above)
Run	plaster cornice that is formed in situ using a preformed template
Rusticated	large blocks, usually at basement level, which have a roughened surface
Rybats	hewn stone at the side of door and window openings, usually formed as inbands and outbands. Also called scuntions
Saddles	Small slated pitched roofs found at the rear of wallhead chimneys, mostly, but not always, in Edinburgh
Safe lintels	usually pitch pine, formed at window and external door openings, usually two with the outer lintel being stone. Often with relieving arches above
Sarking	rough boarding about 170mm wide laid on joists with a 'penny gap' between boards onto which the slates were fixed. Derived from 'sark' or a 'sail'
Sash and case windows	sash windows where in common use in England around 1670 but only came into common use in Scotland after 1680 when counterweights were designed to ease the lifting of each sash
Sash fastener	a fitting to secure lower and upper sash windows
Sash hook	a hook on a pole to lower the upper sash window
Sash lifts	come in a variety of shapes, from simple pulls, ring pulls, handles and bars

Sash screw	a removable screw that was set into a brass ferrule and allowed the more permanent locking of two sashes
Scabbled	roughly faced stone, usually with a pick
Scale burners	LED Luminaire which burns during the hours of darkness, controlled by a solar time switch. Originally would have been gas scaled burners. Located at each landing level
Scalpings	small stones, similar to gravel or shivers
Scarcement	projection from the foundation forming a ledge which carried the ground floor joists
Scottish bond	a brick bond used in Scotland to build brick thick wall, typically at least 9in. thick. Three rows of stretcher bond bricks and a fourth row of header bricks
Scratch coat	initial base coat of plaster or cement which is scratched to provide a base for the next coat
Scuntion	the side of a door or window opening consisting of inbands and outbands
Secret valleys	lead valleys found between M-type roofs and behind parapets that could not be seen from the street. Central lead valleys often stepped with outlets and overflows
Shivers	small cuttings of stone
Shoring	a temporary timber structure to support a fracture or structural element
Shunt duct	duct system which made connection to the extract duct a floor above to limit noise transfer
Shutters	internal timber window shutters, made in pairs, that fold into a pocket
Shutters to cast in-situ concrete	shutters made in timber to provide a formwork for cast in-situ concrete, as in forming copes on chimneys
Sill	at a sash and case, a timber window sill which would be checked to receive the case and bedded, originally in plaster, and pointed with burnt sand mastic. Also stone sills
Simplex fittings	invention circa 1890: two brass slotted hinges which allowed the lower sash to engage into the hinges and once a cord was released, allowed the lower sash to hinge in for cleaning
Single banked	where the flues of adjoining tenements are built in line, often at the same time. Thickness here would be about 400mm (15.75in.)
Single end	small single aspect flats, usually facing the front
Sinking fund	a bank account owned by a Stair Association which saves funds provided by each of the tenement owners, so that the funds can be used for tenement maintenance
Site welded	also called lead burning – a welding process used to join lead sheet. It is a manual process carried out by gas welding, usually oxyacetylene
Sizeables	slates sorted into specific sizes
Skews	external gable walls and party walls between tenements had to rise about a foot above the making to provide a firebreak between tenements. These walls were then capped with stone and are called skews (from Danish 'skiaev' meaning oblique)
Slimlite glazing	modern double glazing which has two thin sheets of glass and a filling of xenon gas. The width of the glass enables it to be puttied into place, so is often used in conservation areas
Smokeless fuel	anthracite, coke, and charcoal, they contain little impurities and don't produce smoke when burned
Snecks	small stones that level up the riser in snecked rubble
Soakers	a small piece of lead flashing, about the size of a slate, used to protect the joint between slate and a wallhead, or between changes in slating
Soffit lining	when timber shutters were not installed, the ingots might be firmed with soffit linings to mimic shutters
Solum	space within the walls of a tenement which is below the finished ground floor. Usually ventilated and treated with bitumen
Spine wall	the central spine wall, usually stone, in a tenement stair
Spring, springing	the point from which an arch springs, or where a projection extends out from the wall

Stagger bonded simply the laying of stones on top of one another but with the vertical joints staggered. Also applies to other materials

Stall risers the small upright panel under a shop window, sometimes allowing vents to the basement

Standards vertical columns or stanchions at shopfronts which support structural beams

Starts found in tenements possibly 1890, may be particular to Edinburgh – a tall thin rybat, with a larger and deeper stone (a stop) set between two 'starts'

Stone cornice gutter Georgians often made stone cornice gutters, the stone sections originally asphalted together, but often leaked

Stone jambs stone uprights that form a flexible opening for a fireplace. would also have a stone lintel and possibly a brick relieving arch

Stonehard a type of epoxy concrete used to renovate worn stone stairs, as the epoxy allows the material to be laid thinly

Stooled sills/stooling the seating formed at the end of a stone sill which is set into the wall. The seating forms a joint with the architrave and is slightly raised to prevent moisture penetrating the joint

Stoop and room old mineworkings where voids were cut to remove the coal (rooms) and pillars of stone were left (stoops) to support the ground above

Stop cock an externally operated valve regulating the flow of the water main supplying a house

Stops stones between 'starts' which are bigger and bedded into the depth of the stone wall

Storm doors inside a close, a set of twin inner opening doors that protected a small lobby with a glazed door behind

Storm roll where there is a discharge of water from a dormer roof into a flashing, there may need to be a storm roll, which is a raised ledge under a lead flashing that prevents a deluge of water from discharging under the slates

Stormont windows small skylights set near the ridge to provide access for chimney sweeps

String course horizontal line of projecting stone course carried along the face of a building

Stugged to roughen the surface of stone or brick, usually to provide better adhesion for lime mortar, but some rubble was roughly 'staged' as a texture

Sumps at the end of parapet gutters, a sunken box-shaped outlet which connects to the main downpipe, and possibly another overflow outlet at a higher point in case of blockage

Sunspaces glazed in balconies which are south facing, so gain heat from solar radiation

Sweeper upper employed by the council, to sweep up the bin stores after the bin men had been

System building a method of building in which prefabricated components are used to speed the construction of buildings

T beams standard rolled steel section which is T shaped

T-Pren expansion joint proprietary expansion joint with two lead sheets joined with a rubber joint. Leadwork is welded to the T-Pren leadwork to form an expansion joint. Avoids the need for stepped joints but has a limited life expectancy

Tanked a waterproof membrane dressed up a wall and onto the floor area in a basement or area below ground level

Tessellated tiles small differently coloured tiles laid in a geometric pattern and laid on a cement bed

Thackstanes projecting stone on a chimney head that protected the joint when the roof was thatched

Thatch thatching varies depending on location. Typically it might be a reed thatch on a turf base secured over wattles

Thermal mass heavy mass walls may be poor insulators but can store energy, so they are best placed inside with the insulation outside. The thermal mass retains the heat and releases it slowly

Throating found underneath copings, commonly called a 'drip'

Tilting fillet a small angular section of timber used to tilt up slates at skews or at lead skews to form watergates

Toby the main underground stop cock, turned off with a 'toby key' by a quarter turn. Thought to derive from Irish Shelta 'tobar,' meaning a well

Tolerable standard	a basic level of repair your property must meet to make it fit for a person to live in
Top rail	topmost rail of a door which would typically be 1.75in. to 2in. thick (approx 50mm) and 4.5in. deep (90mm); all rails tongued into the stiles
Torch-on technique	using a blowtorch to melt modified asphalt, along with viscous organic liquids, onto the area
Transom	a horizontal division between windows
Turnpike stairs	narrow spiral staircase to access upper floor flats, possibly from the French, tourniquet
Turret	small conical tower, usually on a corner and often slated. Also facetted turrets
Twillath	a galvanised mesh used as a base to place over poor stonework to allow a tender system to be applied
Twinfan	used with communal ventilation systems that required more extraction
Uncoursed stone	different sizes of stone, ashlar or rubble, may be best laid uncorked, so the horizontal beds are discontinuous
Undercloak	a flashing formed over a stone corbel and under a gutter to ensure if the gutter overflows, water will not soak the wallhead timbers
Undersized	a range of slates from small to large, unsorted
Value engineering	a term that some people like to use instead of 'cost cutting'
Vapour barrier	certain types of internal insulation require a vapour barrier on the inside to prevent water vapour condensing on the wall surface, behind the insulation
Vapour permeable	the ability of stone to allow liquids, and water vapour, to travel through it
Vapour permeable roofing felt	also referred as a vapour permeable membrane. It allows water vapour to pass through it and so avoids interstitial condensation
Vermiculite	mineral which undergoes significant expansion when heated
Vitrified fireclay	pipe made from a blend of clay and shale that has been subjected to high temperature to achieve vitrification, which results in a hard, inert ceramic
Volute	a scroll in the form of a spiral, usually found at the bottom of close stair handrails
Wallhead chimneys	chimneys springing from the front or rear walls. Often quite tall as the pots had to exhaust above the ridge line to prevent downdrafts
Wallhead dormers	dormers which rise up in stone from the main elevation and contain a window and have their own roof and slated sides
Wallplate	narrow section of timber about 20mm × 140mm which is set into or onto a warhead to provide a seating for a joist
Wally close	a tiled close where the dado is fully tiled to the top. Mostly in better class tenements, but sometimes just the ground floor close is tiled. Derived from 'wallies' meaning a set of false teeth
Warning pipes	Also overflows. Outlets to secret valleys and gutters which allow an overflow of water when a pipe is blocked or in exceptionally heavy rainfall. The pipe should be located to ensure any discharge is noticed by the residents
Water caddies	In Renaissance and medieval times, 'caddies' took water from street pumps and delivered them in buckets to people in tenements
Watergate	an angled fillet or upstand, used under lead to from a waterproof valley
Wattle and daub/wattling	Wattles were rods or twigs, usually hazel interleaved and used to form a base for daub, which was a clay mixed with straw
Whale oil	oil obtained from the blubber of bowhead whales
Whinstone	generic term to describe a hard dark rock
Wynd	a small street or lane

Bibliography

1886–1966 Glasgow's Housing Centenary, Glasgow Planning Department, Corporation of Glasgow, 1966, Glasgow.
A Capital View: The Art of Edinburgh, Popiel, Alyssa, Birlinn, 1988, Edinburgh.
A Concise Building Encyclopaedia, Corkhill, T, Pitman, 1951, London.
A Heritage in Stone, Davidson, Ian Mitchell, National Trust for Scotland/Sandstone Press, 2017, Dingwall.
A Hillhead Album, Morton, Henry, University Press, 1973, Glasgow.
A History of Council Housing in 100 Estates, Boughton, John, RIBA Publishing, 2023, London.
A History of Scottish Architecture, Glendinning, Miles, Ranald McInnes and Angus MacKechnie, Edinburgh University Press, 1996, Edinburgh.
A House Through Time, Olusoga, David and Melanie Becke-Hansen, Picador,London.
A Vision of Scotland 1671–1717, Cavers, Keith, HMSO, 1993, Edinburgh.
Antwerp; The Glory Years, Bye, Michael, Allen Lane, 2021, London.
Aberdeen on Record: Images of the Past, Glendinning, Miles, 1997, Mercat Press,The Stationery Office, Edinburgh.
Annie's Loo, Young, Raymond, Argyll Publishing, 2013, Glasgow.
Architectural Iron and Steel and its Application, Birkemire, William, John Wiley & Sons, 1894, New York.
Architecture of Glasgow, Gomme, Andor Harvey and David Walker (eds.), Lund Humphries, 1987, London.
British Building Industry 1815–1979, Powell, C.G., The Architectural Press, 1980, London.
Building Construction in Scotland, Various, Scottish Vernacular Buildings Working Group, 1976, Edinburgh & Dundee.
Building Construction and Architectural Drawings, Reid, John, Blackie and Son, 1911, Glasgow.
Building Early Modern Edinburgh, Allan, Aaron, Edinburg University Press, 2018, Edinburgh.
Building Stones of Edinburgh, McMillan, Andrew, Richard Gillanders and John Fairhurst, Edinburgh Geological Sociey, 1999, Edinburgh.
Building the Industrial City (Themes in Urban History), Doughty, Michael, Leicester University Press, 1986, Leicester.
Buildings of the Scottish Countryside, Naismith, Robert, Gollanz, 1985, London.
Burt's Letters from the North of Scotland, Burt, Edmund, Birlinn, 1974, Edinburgh.
Bygone Perth, Hutton, Guthrie, Stenlake Publishing, 2005, Catrine, Scotland.
Carpentry & Joinery, Hasluck, Paul N, Cassell, 1912, London.
Climate Change Adaptation for Traditional Building, Curtis , Roger and Jessica H Snow, Historic Scotland, 2016, Edinburgh.
Common Repairs, Common Sense, Scottish Consumer Council, Scottish Consumer Council, 1990, Edinburgh.
Edinburgh New Town, A Model City, Carley, Michael, Robert Dalziel, Pat Dargan and Simon Laird, Amberley Publishing, 2015, Stroud.
Edinburgh: The Making of a Capital City, Edwards, Brian and Paul Jenkins, Edinburgh University Press, 2014, Edinburgh.
Elementary Building Construction & Drawing, Gourlay, Charles, Blackie & Son, 1903, Glasgow.
Empireworld: How British Imperialism Has Shaped the Globe, Sanghera, Sathnam, Viking an imprint of Peguin Books, 2024, London.
Enlightenment in a Smart City 1660–1750, Pittock, Murray, Edinburgh University Press, 2019, Edinburgh.
Enlightenment Edinburgh: A Guide, Szatkowskii, Sheila, Birlinn, 2017, Edinburgh.
Farewell to the Single-End 1865–1975, Glasgow Corporation, Glasgow Corporation, 1975, Glasgow.

Flats, Urban Houses & Cottage Homes, Sparrow, W Shaw (ed.), Hodder and Stoughton, 1907, London.
Getting your Hands Dirty: A Reappraisal of Scottish Building Materials, Construction and Conservation Techniques, Walker, Bruce, *Architectural Heritage Volume 17*, Edinburgh University Press, 2006, Edinburgh.
Glasgow, Maver, Irene, Edinburgh University Press, 2000, Edinburgh.
Glasgow: A History, Meighan, Michael, Amberley Publishing, 2015, Stroud.
Glasgow: The Forming of a City, Reed, Peter, Edinburgh University Press, 1992, Edinburgh.
Glasgow's East End through Time, Adams, Gordon, Amberley Publishing, 2014, Stroud.
Glossary of Scottish Building, Pride, Glen L, Scottish Civic Trust, 1975, Glasgow.
Going Underground: Edinburgh, Henderson, Jan-Andrew, Amberley Publishing, 2022, Stroud.
Guided by a Stonemason: Exploring the Cathedrals, Abbeys and Churches of Britain, Maude, Thomas, I B Tauris Publishers, 1997, London & New York.
How to Read Scottish Buildings, McCannell, Daniel, Birlinn, 2015, Edinburgh.
Hyndland: Edwardian Glasgow Tenement Suburb, Laird, Anne, Anne Laird, 1997, Glasgow.
Ironwork and other Sections, Glasgow West Conservation Manual, Various, Glasgow West Conservation Trust, 1993, Glasgow.
Leith through Time, Gillon, Jack and Fraser Parkinson, Amberley Publishing, 2014, Stroud.
Limeworks, McAfee, Patrick, Associated Editions, 2009, Dublin
Looking After your Sash and Case Windows, Historic Scotland, Historic Scotland, 2002, Edinburgh.
Lost Aberdeen: The Outskirts, Morgan, Diane, Birlinn, 2007, Edinburgh.
Maintaining Your Home. Short Guide No 9, Curtis, Roger, Moses Jenkins and Jessica Snow, Historic Environment Scotland, 2014, Edinburgh.
Medieval Scoland, Yeoman, Peter, Historic Scotland/ Batsford, 1995, London.
Miles Better, Miles to Go. The Story of Glasgow's Housing Associations, Various, Housetalk Group, 1977, Glasgow.
Model Dwelling-Houses, Gowans, James, First Edition published 1886, Cambridge University Press, 2011, Cambridge.
Modern Practical Joinery, Ellis, George, Batsford, 1902, London.
No Wood, No Kingdom, Pluymers, Keith, University of Pennsylvania Press, 2021, Philadelphia.
Old Closes and Streets of Glasgow 1868/1877, Annan, Thomas, Dover Publications Inc. 1977, New York.
Old Greenock, Monteith, Joy, Stenlake Publishing, 2004, Catrine, Scotland.
Renewing Old Edinburgh, Johnson , Jim and Lou Rosenburg, Argyll Publishing, 2010, Argyll.
Repairing Scottish slate roofs, Maxwell, Ingval, Historic Scotland, 2006, Edinburgh.
Report on the Royal Commission on the Housing of the Industrial Population of Scotland Rural and Urban, Royal Commission on Housing in Scotland, Ballantyne, Henry (chair), HMSO, 1917, London.
Revealing the History Behind the Facade: A Timber-framed Building at No. 302 Lawnmarket, Edinburgh, Crone , Anne and Diana Sproat, *Architectural Heritage Volume 22*, Edinburgh University Press, 2012, Edinburgh.
Roofing Felt, Lindhe, Jens, The Danish Roofing Advisory Board, 1998, Virum.
Roofs: A Guide to the Repair of Historic Roofs, Government of Ireland, Stationery Office, 2010, Dublin.
Scotland's Lost Industries, Meighan, Michael, Amberley Publishing, 2012, Stroud.
Scotland's Shops, Lennie, Lindsay, Historic Scotland, 2010, Edinburgh.
Scotland's Bow-fronted Shops, Lennie, Lindsay, *Architectural Heritage Volume 20*, Edinburgh University Press, 2009, Edinburgh.
Scottish Housing in the Twentieth Century, Rodger, Richard,Leicester University Press, 1986, Leicester.
Scottish Townscape, McWilliam, Colin, Collins, 1975, London.
Scottish Traditional Brickwork: Short Guide 7, Jenkins, Moses, Historic Scotland, 2014, Edinburgh.
Six Scottish Burghs, MacMillan, Andy, Canongate Press, 1992, Edinburgh.
Sixty Years in an Aberdeen Granite Yard, McLaren, John, University of Aberdeen Centre for Scottish Studies, 1987, Aberdeen.
Slater's Royal National Commercial Directory and Topography of Scotland, Slater, I, 1860, London.
Slaterwork, Urquhart, Gorden and Neil Grieve, Glasgow West Conservation Trust, 1999, Glasgow.
Stone by Stone, Curran, Joanne, Patricia Warke, Dawson Stellfox, Bernard Smith and John Savage, Appletree Press, 2010, Belfast (a book which is expansive).
Supporting the City, Wright, Howard (ed.), Whittles Publishing, 1997, Caithness.
Tenement: An Architectural History, Burns, John Joseph, Glasgow City Heritage Press, 2019, Glasgow.
Tenements and Towers 1890–1990, Horsey, Miles, RCAHMS, 1990, Edinburgh.
The Scottish house factoring profession, Sim, Duncan, *Urban History* Vol. 23, No. 3 (December 1996), pp. 351–371 , Cambridge University Press, 1996, Cambridge.
The Architects and the Carpenter, Yeomans, David, RIBA Heinz gallery, 1992, London.
The Architecture of Scottish Cities, Mays, Deborah (ed.), Tuckwell Press, 1997, Edinburgh.
The Builders of Edinburgh New Town 1767–1795, Lewis, Anthony, Spire Books, 2014, Reading.
The Care & Conservation of Georgian Houses, Davey, Andy, Bob Heath, Desmond Hodges, Mandy Ketchin and Roy Milne, Architectural Press, 1986, Edinburgh.

The Glasgow Sugar Aristocracy: Scotland and the Caribbean 1775–1838, Mullen, Stephen, University of London, 2022, University of London Press

The Edinburgh Municipal and Police Act (1879), Skinner, William and Alexander Harris, William Blackwood and Sons, 1879, Edinburgh.

The Evolution of Scotland's Towns, Dennison, Patricia, Edinburgh University Press, 2018, Edinburgh.

The Historic Shopfronts of Perth, Lennie, Lindsay, Perth & Kinross Heritage Trust, 2010, Edinburgh.

The History of Working Class Housing: A symposium, Chapman, Stanley D (ed.), David and Charles, 1971, Devon.

The History of Edinburgh, Maitland, William, 1753.

The Light of Truth and Beauty, The Lectures of Alexander Thompson,Edited with an introduction by Gavin Stamp, 1999, The Alexander Thomson Society, Glasgow.

The Industrial Archaeology of Glasgow, Hume, John R, Blackie & Son, 1974, Glasgow.

The Making of Classical Edinburgh 1750–1840, Youngson, A J, Edinburgh University Press, 1966, Edinburgh.

The Making of Urban Scotland, Adams, Ian, McGill-Queen's University Press, 1978, Edinburgh.

The Oxford Handbook of Modern Scottish History, Devine, Tom and Jenny Wormald (eds.), Oxford, 2014, Oxford.

The Plumber's Companion, Hastings, James, David & Charles, 1972, Montreal.

The Repair and Maintenance of Properties in Mixed Ownership: A Study of House-factoring in Glasgow, Sim, Duncan, *Urban Studies* Vol. 34, No. 2 (February 1997), Sage Publications, 1997, London.

The Rural Architecture of Scotland, Fenton, Alexander and Bruce Walker, John Donald Publishers, 1981, Edinburgh.

The Report of a Committee on the Working Classes of Edinburgh, Littlejohn, Dr Henry, 1865, Edinburgh.

The Scottish Nation 1700–2000, Devine, T M, Penguin Press, 2000, London.

The Scottish Town in the Age of Enlightenment 1740–1820, Harris, Bob and Charles McKean, Edinburgh University Press, 2014, Edinburgh.

The South Clyde Estuary, Walker, Frank, RIAS, 1986, Edinburgh.

The Story of Scotland's Towns, Naismith, Robert, John Donald Publishers, 1989, Edinburgh.

The Technical Development of Brickwork in Scotland 1700–1900, Jenkins, Moses, PhD thesis, University of Dundee, 2016, Dundee.

The Tenant's Handbook, Gilbert, John & Chris Orr and Stuart Hashagen, Scottish Consumer Council/ASSIST, 1983, Dundee.

The Tenement Handbook: a practical guide to living in a tenement, Gilbert, John and Ann Flint, RIAS/ASSIST Architects, 1992, Edinburgh.

The Tenement: A Way of Life, Worsdall, Frank, Chambers, 1979, Edinburgh.

The UPAS Tree: Glasgow 1875–1975, Checkland, S G, University of Glasgow Press, 1981, Glasgow.

The Weathering of Natural Building Stones, Schaffer, R J, Building Research Establishment/HMSO , 1932, London.

Thomas Annan: Photographer of Glasgow, Maddox, Amanda and Sarah Stevenson, Getty Publications, 2017, Los Angeles.

Timber! How Wood can Save the World from Climate Breakdown, Brannen, Paul, Agenda Publishing, 2024, Newcastle upon Tyne.

Tradeston to Govan in the 70s, Mortimer, Peter and Duncan McCallum, Stenlake Publishing, 2016, Catrine, Scotland.

Tradition of Edinburgh, Chambers, Robert, W & C Tait, 1825, Edinburgh.

Up Our Close: Memories of Domestic Life in Glasgow Tenements, 1910–1945, Faley, Jean, White Cockade Publishing , 1990, Glasgow.

Vernacular Building 42, The Urban Vernacular, Harrison, John, Scottish Vernacular Buildings Working Group, 2019, Edinburgh.

Vernacular Building 43; Early Flatting in Scotland, Stell, Geoffrey, Scottish Vernacular Buildings Working Group, 2020, Edinburgh.

Wrought Iron: Its Manufacture, Characteristics and Applications, Aston, James and Edwards Story, A M Byers Co., 1957, Pittsburgh.

Index